FLAMMABILITY OF FABRICS

VOLUME NINE
FIRE AND FLAMMABILITY SERIES

Edited by

CARLOS J. HILADO

FLAMMABILITY OF FABRICS

VOLUME NINE
FIRE AND FLAMMABILITY SERIES

265 W. State St., Westport, Conn. 06880

Printed in U.S.A.
Library of Congress Card No. 73-82115
Standard Book No. 0-87762-133-0

To Connie

INTRODUCTION

One of the goals of the Journal of Fire and Flammability and its supplements is to so guide their content that within five years a complete collection of issues would be the basic literature source in this field of technology. Compilations of Journal articles in areas of specific interest thus become a valuable literature resource, and the Fire and Flammability series of books embodies this concept.

Fabrics were the first materials in our society to be the subject of recent fire safety concern, in large part because of the amounts and extensiveness with which it is present in everyday life. The Flammable Fabrics Act and its amendments express the government's concern about the fire safety of fabrics.

It is a measure of the importance of fabric flammability, that twenty articles contain useful information on the flammability of fabrics, and fourteen of these articles express this focus in their titles. Much of this information is novel and unique in coverage and viewpoint, and all of it is presented in the context of our mission of helping to make this world safer for its people.

These twenty articles are presented in this volume, with the hope that it will indeed prove to be a valuable literature resource in this field of knowledge.

CONTENTS

PUBLISHER'S NOTE

Flame Resistant Cotton—A Status Report

G. L. Drake, Jr., R. M. Perkins, and W. A. Reeves

Southern Regional Research Laboratory*
New Orleans, Louisiana

(Received December 17, 1969)

ABSTRACT

Cotton—normally considered flammable—can be rendered flame resistant. Several durable flame retardant formulations and various techniques are described. A phosphorus containing chemical is usually the major ingredient, such as, THPC, THPOH, or APO. Properties of fabrics treated by some of the different processes are reported, and the desirability of a particular treatment for a given end use recommended. A special formulation for polyester/cotton sheets has been developed; techniques for treating cotton rugs, and cotton batting are now available. In part due to present legislation, test methods are being reviewed, studied, and revised, and new tests are finding specialized applications.

INTRODUCTION

Cotton in its most natural state, that is, in the boll, is highly flammable. By contrast, a very tightly woven, heavy cotton fabric, such as an 18 oz/sq yd duck, would resist flaming when a match is applied to it; however, it would glow, smoke, and slowly be completely consumed. Early legislation in English law, and in some of our initial state laws, was to the effect that no textile material should burn faster than cotton in its natural state. Actually, what was meant was that anything that burned faster than similar items of cotton clothing was considered hazardous. The realization of a need to be more specific in the way of a flammability standard on which to base legislation prompted the state of California to assign public officials to make a study of test methods. Some of the original equipment for measuring flammability of apparel textiles resulted from the efforts of the California Fire Marshal.

In 1943, the AATCC established a committee on the Flammability of Clothing Textiles and began work on test methods. By 1947, the method was almost in its final form and became the basis for Commercial Standard CS 191-53. In 1954, legislation was enacted making it a criminal offense to sell or offer for sale any textile product intended for use in apparel which did not meet certain requirements under the Flammability Test. This legislation, Public Law 88, effectively eliminated from the market those highly flammable fabrics which had caused multiple injuries. However, statistics from the Department of Health, Education, and Welfare still estimated about 150,000 persons injured as a result of the ignition of clothing and a death toll of 2,000-3,000 from fire each year. In the interest of the public, a safety-minded Congress, urged by President Johnson, amended the Flammable Fabrics Act of 1953. The new law

* One of the laboratories of the Southern Utilization Research and Development Division, Agricultural Research Service, U. S. Department of Agriculture.

strengthens the old and covers interior furnishings as well as apparel fabrics; it also now includes hats, gloves, footwear, and interlining fabrics; it encompasses paper, plastics, rubber, synthetic fibers, and synthetic foam; and it broadly covers any item of wearing apparel or interior furnishings. The new legislation offers some formidable challenges for researchers, technologists, and merchandisers. Two problems immediately present themselves: (1) whereas the Flammability Test method was available for use in the enforcement of earlier legislation on apparel items and could be tied into the law, there is no such test or standard with which to evaluate home furnishings and other materials covered by the amendment; and (2) current finishes are costly and usually adversely affect the esthetics and/or physical properties of the flame resistant materials.

The test standard, CS-191-53, on which the 1953 Act is based, is being revised in several directions to clarify the test procedure for easier execution and accurate interpretation of results: tolerances are included in the revision; the apparatus is being automated; machine washing is being substituted for the hand washing; and preparation of specimens is being designed to include narrow fabrics, small decorative items, and loose fibrous materials, such as hula skirts, and other costume items as wigs and false beards.

Another common test method that is being studied by the National Advisory Committee to the Secretary of Commerce is the Standard Vertical Flame Test. The equipment used is described in a number of variations in documents of: Federal Test Methods CCC-T-191, American Society for Testing and Materials, American Association of Textile Chemists and Colorists and the National Fire Protection Association. Although the test appears to be simple, it is not. Numerous variations exist in draft conditions (open or closed front, open or vented top, open or vented bottom, plus differences in how to vent), type of gas, size of burner, size of flame (with or without air supply), means of application of flame and time of application, and presentation of the sample to the flame.

Even though the Flammability Test and the Standard Vertical Test are widely used, other methods, both foreign and domestic, exist, and are being studied by the Secretary. One simple test that has been found extremely useful as a research tool is the Angular Match Test. In this test, a match is applied to a 1 cm strip and the angle at which it fails to support combustion is recorded. Fabrics passing the 180° angle have very good flame resistance, fabrics passing the 90° angle have modest flame resistance, and those passing the 0° angle have very little flame resistance. Fabrics passing the 135° angle or more usually pass the Standard Vertical Flame Test.

While test methods are being reviewed, studied, and revised, and while much needed information on burn accidents is being collected, the textile industry has accelerated its research and development efforts to supply flame retardant materials for varied and numerous end-uses. And it appears safe to predict that the end-use will determine the test method; for example, nightwear, blankets, other napped materials will be tested by an improved 45° Flammability Test; rugs, carpets by the pill test; industrial and military fabrics by the Standard Vertical Test; and blends and synthetics by the Limited Oxygen Index, etc.

Cotton Flame Retardant Treatments: Heat Cure Processes

The biggest market for flame retardant finishes is for the durable type. Actually one such finish is about a quarter of a century old, the FWWMR finish, which made outdoor fabrics weather, water, mildew and not resistant as well as flame resistant [1]. This finish is well adapted to military and other industrial uses and still accounts for considerable yardage; however, it is unsuitable for lighter weight apparel and household fabrics.

The most widely used durable flame retardants for apparel and household goods, and for certain military items, such as tent liner fabrics, are based upon tetrakis (hydroxymethyl)phosphonium chloride (THPC). This is a water-soluble, crystailine compound produced in high yield from formaldehyde, phosphine, and hydrochloric acid. In the U. S., THPC is sold by Hooker Chemical Co., and in Europe by Albright and Wilson Ltd., England. The methylol groups of THPC react readily with amines and amides [2]. It was during a study of the physical and chemical properties of aminized cotton that the value of THPC as a flame retardant was first recognized [3]. A flame retardant based upon THPC and an amide was then developed [4]. The formulation consists of THPC, urea, and trimethylolmelamine (2:4:1 mole ratio) plus various auxiliaries [5]. Normally about a 35% solution is applied (70% wet pick-up) to an 8-oz cotton fabric so that it passes the Standard Vertical Flame Test (pad-dry-heat cure method). The finish is durable to both laundering and drycleaning. Its primary limitations are: (1) stiffness imparted to some types of fabrics, and (2) reduction in tear and tensile strength of some fabrics. Both the reagents and concentrations of reagents can vary widely; however, optimum phosphorus fixation for laundry durability appears to be about 5% urea. Durability is inversely related to the urea concentration in the treating solution. Urea is needed to improve tensile and tear strength. Thiourea can be substituted for urea, in about equal weights, and is actually more durable to laundering and alkaline boiling than when urea is used. Trimethylolmelamine is preferred over other N-methylol amides, such as dimethylolethyleneurea, dimethylolethyltriazone, dimethylol carbamate, etc. [6].

The THPC-amide flame retardant has been modified in several ways in an effort to improve the properties of the treated fabric and/or to reduce the cost. The bromoform adduct of the allyl ester of phosphonitrilic chlorides can be produced in an aqueous emulsion using a peroxide catalyst and then added to the THPC-amide formulations [7]. The adduct in itself can be used as a flame retardant, and since it is very efficient, less of the combined formulation is required than of the THPC-amide alone. The major objection to this modification is that the overall cost is increased. Another adduct that has been added to the THPC-amide formulation is the bromoform adduct of triallyl phosphate [8]. Again cost is one of the major deterrents.

When polyvinyl chloride is used alone with the THPC-amide flame retardant, the primary benefit seems to be an improvement in tensile strength of treated fabrics. Since polyvinyl chloride in combination with antimony oxide is a flame retardant, this combination has also been substituted for some of the THPC-amide. However, greater total add-ons are required to impart a specific degree of flame retardancy, because the antimony oxide-polyvinyl chloride is less efficient than the THPC-amide. The merit in using this combination formulation is reduced cost [9].

A recent modification was developed by the Piedmont Section of the AATCC particularly for use on polyester/cotton sheets [10]. Tris(2,3-dibromopropyl)-phosphate, which is water-insoluble, can be emulsified and added directly to the THPC-amide formulation. The brominated phosphate penetrates both the polyester and the cotton and produces a very durable finish for the blend fabric. The presence of the phosphate with the THPC-amide also prevents loss in lightfastness of certain dyes [11].

THPC has been used in a formulation with tris(1-aziridinyl)phosphine oxide (APO) [12, 13]. The aziridine rings in APO are extremely reactive, especially in acid medium. Both TPHC and APO react to form polymers inside the cotton, producing a very durable flame-resistant fabric with relatively low resin add-ons. A typical formulation for treating 8-oz cotton contains about 13% APO and 14% THPC, plus about 1.5% NaOH and wetting and softening agents. The finish is durable to repeated launderings and drycleanings. When exposed to flames or intense heat, the treated fabric forms tough, pliable, and protective chars. In addition, the treated fabrics exhibit little or no adverse change in hand or appearance, and are shrink, rot, and mildew resistant, besides being glow and flame resistant. Breaking strengths are reduced by about 20% and tearing strengths about 40%. Wrinkle recovery is 250-260° (W+F). A weak feature of this finish is the yellowing caused by bleaching; however, methods for the temporary removal of the discoloration have been developed. It should be emphasized that the amount of resin add-on needed in a fabric so that it will pass the standard vertical flame test after repeated launderings is less than that of most other durable flame retardants.

The APO thioanalog, APS, has been used with THPC to produce flame resistant fabrics [12]. Properties of the treated materials are similar to APO-THPC-treated fabrics. The APS-THPC solutions are more stable than the APO-THPC, however, at present, APS is not made commercially.

APO can be used without THPC, but with a catalyst (preferably $Zn(BF_4)_2$) to produce flame resistant fabrics having high wrinkle recovery (over 300°, W+F) [14, 15]. Major disadvantages are strength losses, yellowing with sodium hypochlorite bleaching, and the lack of a commercial source of APO at this time.

When one mole of THPC is reacted with 0.8 to 1 mole of NaOH, a mixture of products results, and the major product is referred to as "THPOH." THPOH can be substituted for THPC in the THPC-amide formulation on a molar basis [16]. The THPOH flame retardant imparts significantly different properties to fabrics: fabrics retain about 80-90% of their original breaking strength; because of reduced stiffness, this retardant is applicable to a wider variety of fabrics. Wrinkle recovery is fairly high (280-290° W+F); rot resistance is good; and little or no yellowing of the treated fabrics is observed after a sodium hypochlorite bleach.

Recently, Ciba introduced Pyrovatex CP, which is a fiber-reactive phosphone alkyl amide for treatment of cotton [17]. The hand of the flame resistant fabric is good, and the finish will withstand 15 launderings at the boil with no change in retardancy. Treated fabrics suffer 20-30% loss in tensile strength and about 30% loss in tear strength. Effective treatment requires an add-on of 25-35%. Some problems with bath stability have been observed by mills. Another disadvantage is the high cost of the Pyrovatex.

AMMONIA CURE PROCESSES

The flame retardants described above cause the cellulose to be insoluble in cuene and impart some degree of wrinkle recovery. It was theorized, therefore, that these processes caused crosslinking of cellulose molecules, thus causing strength reductions. A method of polymerizing the phosphorus compounds inside the fiber, with little of no reaction with the fiber, should produce flame-resistant fabrics without causing degradation of strength. The most successful technique was to chemically "fix" the polymer, and the preferred chemical has been ammonia.

The original process utilizing ammonia to insolubilize methylol phosphorus polymers inside cotton to impart flame resistance was done in 1953 [18]. In this process, THPC, trimethylolmelamine, and urea were the major ingredients. After padding and drying, the fabric was exposed to ammonia and/or ammonium hydroxide. An immediate reaction took place and an insoluble polymer containing phosphorus and nitrogen formed inside the fiber. Strength of the treated fabric was essentially unchanged. Slight modifications of this process have been made to increase the stability of the pad bath, and to make it more attractive for application using commercial laundry equipment [19].

THPC and urea have been made into a water-soluble precondensate [20, 21]. After impregnation, the fabric is dried, exposed to ammonia vapor, and then passed through ammonium hydroxide. The initial ammonia reaction immobilized the polymer in the fiber; the subsequent aqueous ammonium hydroxide treatment insured complete penetration of the ammonia and the production of a highly insoluble polymer. The flame resistant finish is very resistant to repeated laundering and drycleaning and to boiling soap and sodium carbonate solutions. Tearing strength is slightly reduced; breaking strength remains essentially unchanged. This process has been used in Europe for about 10 years. The ammonia gas step can be eliminated by incorporating diammonium sulfite in the treating solution, which releases ammonia during the drying step and initiates the polymerization reaction.

A new, simple, and extremely attractive process is the THPOH-NH_3 [22]. The THPOH is made, as mentioned previously, by reacting 0.8 to 1.0 mole of NaOH to THPC. Fabrics are impregnated with the THPOH solution, dried to 10-20% moisture, and next exposed to NH_3 gas. An 8.5-oz twill or sateen will pass the standard vertical flame test with about a 13% add-on. Of particular interest is the fact that this process can be used on very lightweight fabrics—a 2-oz fabric needs about 20% add-on. Stiffness is essentially unaffected; breaking strength remains unchanged, sometimes increases; and, with a softener, tearing strength reduction is about 10%.

The ammonia gas fixation step can be eliminated using a slightly different approach. After the NaOH has been reacted with the THPC to produce the THPOH solution, copper nitrate is added for solution stability and subsequently ammonium hydroxide or other ammonium salts [23]. The impregnated fabric is dried at about 150°C and essentially the same flame resistant product results; however, because of the presence of the copper salts, the treated fabrics have a light bluish-green color.

Since the THPOH is the most expensive agent used in the THPOH-NH_3 processes, a means of reducing the total cost of the flame retardant was sought [24]. Trimethylolmelamine was incorporated into the THPOH treating solution; fabrics were padded and dried to 10-20% moisture as before. The impregnated fabric was then exposed to NH_3 gas for a few seconds and finally heated to 150°C for three minutes. The trimethylolmelamine can be effectively substituted for about 1/3 of the required THPOH. Properties of these treated fabrics are about the same as the THPOH-NH_3 treated fabrics, except: (1) breaking strength remains unchanged, but usually is not greater than the original, and (2) the total weight add-on is about 1/3 greater for the THPOH-

TMM-NH_3 process, that is about 18% add-on is needed for an 8.5-oz cotton sateen to pass the standard vertical flame test instead of 13% for the THPOH-NH_3. Also this process does require a heat cure in addition to the chemical fixation to provide adequate polymer formation and maximum durability to laundering.

SPECIALIZED TECHNIQUES

Although the processes described thus far have been developed primarily for cotton fabrics, some of the formulations have been used in various special techniques to impart flame resistance to products with specific end-uses. For example, a technique was developed for imparting durable flame resistance to the surface of low density cotton pile and napped fabrics [25]. The usual method for applying flame retardants to cotton fabrics consists in impregnating by padding, drying, then curing either with heat or chemical reaction. If the fabric has raised surface fibers, as with a rug or blanket, the padding operation compresses these fibers and they may become fixed in this compressed state. It is then difficult or impossible to renew the nap to its original state. One successful technique developed consisted of passing the cotton pile or nap fabrics through a fine spray produced from an aqueous flameproofing solution. The amount of solution added to the fabric was just sufficient to wet the surface fibers without wetting the major portion of the base fabric. Drying the sprayed fabric was accomplished preferably by passing it through a forced draft oven, or alternately by tumbling in a tumble dryer. Curing was done at high temperature in a forced draft oven, or by a second spraying with a chemical, such as ammonium hydroxide solution, which caused the flame retardant to immediately form an insoluble polymer inside the fiber. Flameproofing formulations tried included THPC-trimethylolmelamine-urea, APO-THPC, and APO (pad-dry-heat cure), and THPC-urea precondensate (NH_3 cure). This technique for napped fabrics requires much less added chemicals than when the entire textile product is treated, thus reducing the chemical cost. Another advantage is that the treated material retains its tensile and tear strength since the flame retardant is restricted to the surface fibers which contribute little or nothing to the fabric strength. When the flame retardant used is of the type to cause some crosslinking with the cotton, then the nap or pile is made more resilient by the treatment, and has less tendency to become packed down during laundering and subsequent use.

A related technique has been used to render Cotton Flote* flame resistant [26, 27]. The Cotton Flote process involves the spraying of cotton fibers openly arrayed in web form with a chemical system consisting of both thermoplastic and thermosetting resins. The spray damp webs are lapped conventionally to the desired thickness/density, then dried and cured. The Cotton Flote process is a versatile process whereby dimensionally stable, highly resilient products can be made in density ranges from 1.5 to more than 6 pounds per cubic foot. These products resist compaction in use and have tensile strengths far in excess of conventional cotton batting. An extremely attractive feature of Cotton Flote is the fact that it can be molded into any desired shape to conform with high-style upholstered furniture while retaining its enhanced

* Trade name registered by the National Cotton Batting Institute, Memphis, Tennessee.

resilience, dimensional stability, tensile strength, and resistance to compaction in use. It will far outperform conventional batting and at a much lower cost than polyurethane foam.

The passage of the amendment to the Flammable Textiles Act and the enactment of the Automobile Safety Act by the United States Congress prompted researchers at the SURDD laboratory to develop a flame resistant Cotton Flote. A urea complex, antimony trioxide, or diammonium phosphate with borax have been successfully incorporated into the resin-latex system. The treated products retain their ability to resist burning under a wide variety of conditions: (1) mild leaching, (2) 100% relative humidity at 100°F for 3 days, (3) 158°F for 16 hours, and (4) storage for 7 months. At least one company is now marketing a flame retardant Cotton Flote innerspring mattress and box spring combination. Flame resistant Cotton Flote samples were also treated under a series of different environmental conditions in accordance with General Motors Corp. (Fisher Body Division) Engineering Section Test Method 30-27 as revised November 13, 1967, and were found to pass the test.

Current estimates show that the annual production of Cotton Flote is about 100 million pounds per year, and that the greatest portion is going into automobile seating. Only about 2% of the total production is being made flame resistant, but the industry knows that the formulations and techniques are available for use when the demand is created by the consumer or through legislation.

CONCLUSION

Cotton can be made durably flame resistant. Several excellent formulations have been developed, especially those containing phosphorus compounds, which impart glow resistance as well as flame resistance. Two distinct means of causing polymer formation and/or crosslinking with the cellulose are discussed—heat cure and chemical cure. Depending upon the process used, the treated fabric may also have other desirable properties, such as rot resistance, wrinkle recovery, dimensional stability, etc. At this state of the art, no one finish is applicable to all types and weights of fabric, nor for all end-uses. In selecting the best treatment for a specific application, one must consider fabric construction, properties affected by the finish, end-use, chemical and processing costs, and availability of materials. Cotton has some advantages over other fibers in being easier to make flame resistant. For the same weight and fabric structure, less flame retardant is required in cotton than in rayon, and even larger quantities of known flame retardants are required in nylon and polyester fibers. But flame resistant cotton has competition on the way in man-made fibers which are being "tailor-made" during the manufacturing process. Many advances in flame-resistant materials can be expected shortly since advances are generally related to research effort, and today's effort is estimated to be at least 100 times what it was in 1965.

A critical evaluation of test methods is underway, and necessary improvements are being implemented.

REFERENCES

1. R. Bruce LeBlanc, "Flammability and Fire Resistance of Textiles," *Amer. Dyestuff Reptr. 57,* No. 27, 35-8 (1968).

2. W. A. Reeves, and J. D. Guthrie, "Intermediate for Flame-Resistant Polymers," *Indus. and Eng. Chem., 48,* No. 1, 64-7 (1956).
3. W. A. Reeves, O. J. McMillan, Jr., and J. D. Guthrie, "Chemical and Physical Properties of Aminized Cotton." *Textile Res. J. 23,* No. 8, 527-32 (1953).
4. W. A. Reeves, and J. D. Guthrie, "Imparting Flame-Resistance to Cotton by the Use of THPC-Resins," U.S. Dept. Agric., Bur. Agr. and Ind., Mimeo, Cir. Ser., AIC 364, Rev. (1954). "THPC, New Flame-Resistant Treatment, Is Permanent and Effective," *Textile World 104,* No. 2, 101, 176-82 (1954).
5. J. D. Guthrie, G. L. Drake, Jr., and W. A. Reeves, "Application of the THPC Flame-Retardant Process to Cotton Fabrics," *Amer. Dyestuff Reptr. 44,* No. 10, 328-32 (1955).
6. G. L. Drake, Jr. and W. A. Reeves, "Flame Resistant Cellulosic Textiles." In manuscript.
7. Carl Hamalainen, W. A. Reeves, J. D. Guthrie, "Cotton Made Flame-Resistant with Bromine-Containing Phosphonitrilates in Combination with THPC Resins," *Textile Res. J. 26,* 145-49 (1956).
8. J. D. Reid, J. G. Frick, Jr., and R. L. Arceneaux, "A Compounded Flame Retardant for Cotton Fabrics," *Textile Res. J. 26,* 137-40 (1956).
9. G. M. Wagner, (to Hooker Chemical Corp.) "Flameproofing of Cellulosic Material." U.S. Pat. 3,219,478 (1965).
10. Piedmont Sect., Amer. Assoc. Textile Chem. and Colorists, "A Study of Fire Retardancy of Polyester/Cotton Sheeting," *Amer. Dyestuff Reptr. 57,* No. 10, 373-77 (1968).
11. Piedmont Sect., Am. Assoc. Textile Chem. and Colorists," Effects of THPC on the Lightfastness of Dyes and Pigments," *Textile Chem. and Colorists 1,* No. 5, 124-9 (1969).
12. W. A. Reeves, G. L. Drake, Jr., L. H. Chance, and J. D. Guthrie, "Flame Retardants for Cotton Using APO- and APS-THPC Resins," *Textile Res. J. 27,* No. 3,260-6 (1957).
13. G. L. Drake, Jr., J. V. Beninate, and J. D. Guthrie, "Application of the APO-THPC Flame-Retardant to Cotton Fabric," *Amer. Dyestuff Reptr. 50,* No. 4, 129-34 (1961).
14. G. L. Drake, Jr., and J. D. Guthrie "Imparting Crease Resistance and Crease Retention to Cotton with APO," *Textile Res. J. 29,* No .2, 155-64 (1969).
15. G. L. Drake, L. H. Chance, J. V. Beninate, and J. D. Guthrie, "APO, A Durable Flame Retardant for Cotton," *Amer. Dyestuff Reptr. 51,* No. 8, 40-2 (1962).
16. J. V. Beninate, E. K. Boylston, G. L. Drake, Jr., and W. A. Reeves, "Conventional Pad-Dry-Cure Process for Durable-Flame and Wrinkle Resistance with Tetrakis (Hydroxymethyl)Phosphonium Hydroxide (THPOH)," *Textile Res. J. 38,* No. 3, 267-72 (1968).
17. Anon., "Ciba Show New Flame Retardant for Cotton," *America's Textile Reptr. 82,* No. 16, 40 (1968).
18. W. A. Reeves, and J. D. Guthrie, (to U. S. Secretary of Agriculture), "Ammonia Insolubilized Methylol-Phosphorus Polymers and Cellulosic Materials Impregnated Therewith," U. S. Pat. 2,772,188 (1956).
19. G. M. Wagner, (to Hooker Chemical Corp.) "Process for Treating Cellulosic Material with Flameproofing Composition," U. S. Pat. 3,310,419 (1967).
20. Harold Coates, (to Albright and Wilson, Ltd.) "Insolubilization of Further-polymerisable Methylol-Phosphorus Polymeric Materials," U.S. Pat. 3,096,201 (1963).
21. G. L. Drake, Jr., W. A. Reeves, and R. M. Perkins, "Imparting Flame Resistance to Cotton by Chemical Fixation," *Amer. Dyestuff Reptr. 52,* No. 16, 41-4 (1963).
22. J. V. Beninate, E. K. Boylston, G. L. Drake, Jr., and W. A. Reeves, "Application of a New Phosphonium Flame Retardant," *Amer. Dyestuff Reptr. 57,* No. 25, 47-78 (1968).
23. D. J. Donaldson, and D. J. Daigle, "Phosphorus-Nitrogen Flame Retardant Via Copper Complex," *Textile Res. J. 39,* No. 4, 363-7 (1968).
24. J. V. Beninate, R. M. Perkins, G. L. Drake, Jr., and W. A. Reeves, "An Economical Durable Flame-Retardant Finish for Cotton," *Textile Res. J. 39,* No. 4, 368-74 (1969).
25. R. M. Perkins, G. L. Drake, Jr., and W. A. Reeves, "A Technique for Imparting

Durable Flame Resistance to the Surface of Low Density Cotton Textiles," *Amer. Dyestuff Reptr. 54,* No. 15, 17-18 (1965).

26. N. B. Knoepfler, P. A. Koenig, and H. L. E. Vix, "Flame Retardant Cotton Flote Products for Automobile Seats, Mattresses, and Furniture," *Oil Mill Gazetteer 73* No. 8, 9-13 (1969).
27. P. A. Koenig, and N. B. Knoepfler, "Flame Retardant Cotton Batting," *Amer. Dyestuff Reptr. 58,* No. 17, 30-6 (1969).

APPENDIX

SELECTED FABRIC FLAMMABILITY TESTS

1. NFPA No. 702 (May 1962). National Fire Protection Association International, 60 Batterymarch Street, Boston 10, Mass.
2. Flammability of Clothing Textiles, Standard Test Method 33-1962. (American Society for Testing and Materials Designation: D1230061), AATCC Technical Manual 1963 Part II, B-114.
3. Flammability of Clothing Textiles (32). Commercial Standard 191-53, U. S. Department of Commerce, Bureau of Foreign and Domestic Commerce. (For sale by Supt. of Documents, U.S. Gov. Printing Office, Washington, D.C. 20250).
4. AATCC Fire Resistance of Industrial Fabrics (33). Standard Test Method 34-1952. (American Standards Association: ASA No. 114.107-1961). AATCC Technical Manual 1963 Part II, B-113.
5. Standard for Flameproofed Textiles, NFPA No. 701, National Fire Protection Association International.
6. Burning Characteristics of Fabric Clothing. Joe R. Yockers California State Fire Marshal, National Fire Protection Association, 107 S. Broadway, Los Angeles, Calif.
7. Federal Trade Commission. Rules and Regulations under the Flammable Fabrics Act, Public Law 88-83rd Congress, Chapter 164—1st Session, H. R. 5069.
8. Testing Apparatus for the Evaluation of Fire-resistant Textiles, B. C. Rowley and M. Sonnino, American Cyanamid Co., Stamford Labs., *American Dyestuff Reporter,* March 7, 1949.
9. Federal standard stock catalog, CCC-T-191b method 5902 Flame Resistance of Cloth: Vertical, CCC-T-191b method 5903-T "Flame Resistance of Cloth; Modified Vertical."
10. ASTM Standards on textile materials, ASTM D626-55T, "Fire-Retardant Properties of Treated Textile Fabrics," published by American Society for Testing Materials, 1916 Race Street, Philadelphia, Pa.
11. British Standard Specification, "Performance Requirements of Fabrics Described as of Low Flammability," B. S. 3121: 1959, British Standards Institution, 2 Park St., London W 1, England.
12. "An Economical Flame-Retardant Finish for Cotton," J. V. Beninate, R. M. Perkins, G. L. Drake, Jr., and W. A. Reeves, *Textile Res. J. 39,* No. 4, 368-74 (1969).
13. "Candle-type Test for Flammability of Polymers," C. P. Fenimore and F. J. Martin, *Modern Plastics 44,* No. 3 ,141-2, 146, 148, 192 (1966).
14. "Some Effects of Chemical Composition on the Flammability Behavior of Textiles," G. C. Tesoro and C. H. Meiser, Jr., 39th Annual Meeting, Textile Res. Inst., Apr. 9-11 (1969), N. Y.

G. L. Drake

George L. Drake, Jr. is Head of Special Finishes Investigations at the Southern Regional Research Laboratory of the U.S. Department of Agriculture, at New Orleans, Louisiana. He received his B.S. degree in Chemistry from the University of Georgia in 1944 and his M.S. degree in Chemistry from the University of Georgia in 1949. He has been with the U.S. Department of Agriculture since 1949, and has published extensively in the field of cotton research.

Mrs. R. M. Perkins

Mrs. Rita M. Perkins is a member of the Cotton Finishes Laboratory of the U.S. Department of Agriculture at New Orleans, Louisiana. She received her B.S. degree from Tulane University and studied chemical engineering at Louisiana State University. She has been with the U.S. Department of Agriculture since 1943, and has experience in cotton research.

W. A. Reeves

Wilson A. Reeves is Chief of the Cotton Finishes Laboratory of the U.S. Department of Agriculture at New Orleans, Louisiana. He received his B.S. degree in Chemistry from the University of Southwestern Louisiana, his M.S. degree in Chemistry from Tulane University, and his Sc.D. degree from Clemson University. He has been with the U.S. Department of Agriculture for nearly twenty years, and has experience in textile research, currently in the areas of durable-press, water-repellent, soil-repellent, strain-repellent, mildew-resistant, rot-resistant, light-resistant, and flame-resistant finishes.

M. A. KASEM* AND H. K. ROUETTE

Swiss Federal Institute of Technology (ETH)
Department of Industrial and Engineering Chemistry
8006 Zurich, Switzerland

(Received June 17, 1972)

FLAMMABILITY AND FLAME RETARDANCY OF FABRICS: A REVIEW

ABSTRACT: Flammability and flame retardancy of the commonly used fabrics are critically reviewed with reference to the importance of research on fabric flame retardation, mechanism of flame propagation, measurement of fabric flammability, progress in the science and technology of flame retardation, theory of flame retardation and the mechanism of flame and glow retardation by the currently known flame and glow retardants. Several concluding remarks have been made. There are 66 references.

INTRODUCTION

ALMOST ALL TEXTILE fibers are organic polymers, most of which decompose below 300°C, producing gaseous products; some of these are flammable and ignite under favorable conditions. Consequently, fibers constitute fire hazards. Ignition of textiles causes perhaps about 20% of the total loss of life and property due to fire of all kinds [1]. Relevant statistical surveys have recently been reviewed [1-3]. It appears that, as estimated in the United States, 150,000-300,000 persons per year suffer burn injuries as a result of clothing ignition. A high percentage of the victims are young children. Because loss of life and property and severe suffering are often associated with burn injuries in which clothing materials play some role [4], the issue of fabric flammability has become an international concern as is evidenced by a meeting [4] having representatives from many countries of the world. The consensus was that the consumer should be protected and such protective measures can be seen in the development of legislation against the use of "highly or dangerously flammable fabrics" in countries such as Canada, France, Germany, Great Britain, Japan, Switzerland and the United States. Some of this legislation have been summarized [1, 4, 5]. The concern of the consumer and the impact of legislation have shown the pressing need for flame retarded fabrics and fibers for such applications as apparel, interior furnishings, automobiles, carpets, air craft to name a few. The textile trade and allied industries are becoming more involved in research and development to reduce or retard fabric flammability, as is evidenced in the increased production and sale of flame retardants during the last

*Postdoctoral Research Associate. To whom correspondence should be addressed.

few years [4]. The estimated market for flame retardant chemicals is expected to increase from 200 million pounds in 1971 to 850 million pounds in 1975 [6].

FLAMMABILITY CHARACTERISTICS OF FABRICS

Factors Affecting Fabric Flammability

The flammability of fabrics is a complex phenomenon and is influenced by the characteristics of both the fiber and the fabric. The chemical composition of the fiber is the important fiber characteristic affecting flammability. In addition, chemical finishing operations such as mordanting during chrome dyeing of wool i.e. cross-linking [7] decrease flammability because the fragments resulting during polymer decomposition are too large to be volatile. However, increasing molecular weight of the fibrous polymer increases the decomposition temperature [8] and the exothermic nature of the decomposition reaction becomes more pronounced with increasing molecular weight. Fabric properties such as construction, weight per unit area and surface smoothness affect flammability. Nielson and Richards [9] attempted to relate fabric flammability with the above properties, and found that heavier fabrics burn more slowly than the lighter ones. For blends, the flammability is influenced by the nature and proportion of the constituting fibers; in particular by the so-called "scaffold" effect [4]. Tesoro et al. [10] found that the flammability of polyester increases with increase in the amount of cotton in the blend from 0 to 15% and remains virtually independent of the cotton content (up to 85%) in the blend. Moreover, the presence of "nonflammable fibers" in blends may reduce flammability of the flammable component of the blends, even to the self-extinguishing point. Tesoro and Rivlin [11] reported such a study on blends of Nomex with cotton and other flammable fibers, and they observed increased oxygen index values for the blends with increase in the percent Nomex in the blend.

Mechanism of Burning of Fabric

The mechanism of burning (combustion in air) is a complicated phenomenon but may be represented as shown in Figure 1. The process of ignition and burning of solid materials can be described schematically (Figure 2). If a flame attacks the surface of a flammable material, the latter will ignite if a sufficient supply of oxygen is available. The heat produced during burning causes convection of the surrounding atmosphere and/or diffuses into the solid. During the regression of the surface, ablation of solid particles into the flame occurs, thus acting as fuel for the flame. Hence, the rate of surface regression depends on the direct oxygen attack and the ablation. The heat which diffuses into the solid causes pyrolysis, an endothermic process, and results in the formation of volatile products. The latter diffuse out to the surface of the solid and form additional fuel, which increases the flame temperature. In contrast to surface regression, pyrolysis forms a solid residue (char)

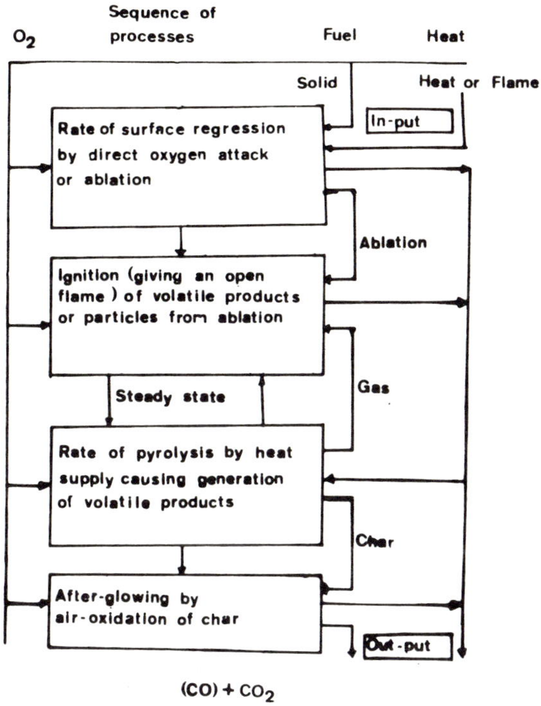

Figure 1. *Flow of processes involved in burning of solid materials like fibers.*

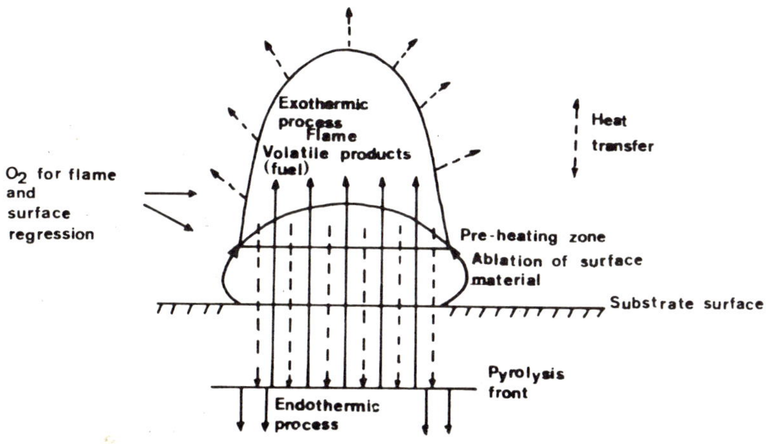

Figure 2. *Schematic representation of the burning process of solids like fibers..*

as an additional product of the process. Flaming combustion occurs through radical chain reactions, and the radicals are formed by the pyrolysis of the fabric. Glowing is simply oxidation of char (mainly carbon) by air to give carbon dioxide and some carbon monoxide.

Measurement of Fabric Flammability

The flammability of a fabric (untreated as well as treated) is usually measured in terms of ignitability and combustibility. Ignitability is a measure of the ease with which a fabric enflames while combustibility is related to the rate of flame travel along a fabric specimen under a given test procedure. A number of test methods have been developed for evaluating the flammability of untreated fabrics and the flame resistance of flame retarded fabrics [2, 12-15]. For untreated fabrics, the ease of ignition and the rate of burning, as measured by the AATCC 33-1962 [16] test method, are used to compare flammability. Tweedie et al. [17] using vertical strip method reported results of rate-of-burning measurements on a wide variety of fabrics, natural, synthetic and blends obtained from various practical sources.

In addition, the rate of flame travel along an ignited fabric specimen may also be used as a criterion for flammability and such measurements on a series of fabrics have been reported [9] The flammability of thermoplastic fabrics is evaluated by more or less the same methods. However, in order to prevent melt dripping of the fabric, the test specimen is given support either by the use of a backing or by sewing of threads (usually glass fiber) along the specimen. Such arrangements have been found to give reasonably good results of flammability measurement on thermoplastic fabrics [18]. Additional characteristics such as ignition temperature and the flame temperature (temperature inside the flame produced by a burning fabric) are also measured for evaluation of fabric flammability. Ignition temperature, "the maximum flame temperature" and the flammability [19-22] of common fibers are shown in Table 1. The maximum flame temperature of a fabric was found to depend on specimen dimension and on the concentration of oxygen surrounding the test specimen [20]. For flame retarded fabrics, the flame resistance is evaluated in terms of afterflame (duration for which a specimen flames after removal of the ignition source), char length (vertical length of the specimen charred) and afterglow (duration of glowing). Usually a test specimen is ignited by the application of a specified flame for a definite length of time and the above criteria are measured. The prescribed conditions vary from method to method and also depend on the requirements of the country concerned. The AATCC 34-1969 vertical char test method [16] has been widely used and has been recently modified to test both natural and synthetic fabrics; the flame application time has also been reduced from 12 sec to 3 sec [23]. Results obtained for flammability measurement by the test methods mentioned above have been critically discussed [12, 13]. It appears that the fabrics may rank in different orders, even if the same property is purportedly being measured. Suchecki [24] suggested an "Index of Merit" defined as the ratio

Table 1. Flammability Characteristics of Fibers (from Ref. 21)

Fiber	Ignition Temp. (°C)	Maximum Flame Temp. °C)	Flammability
Cotton	400	860	Burns readily with char formation and afterglow.
Rayon	420	850	Burns readily with char formation and afterglow.
Acetate	475	960	Burns and melts ahead of flame.
Triacetate	540	885	Burns readily and melts ahead of flame.
Nylon 6	530	875	Melts, supports combustion with difficulty.
Nylon 66	532	–	Melts, does not readily support combustion.
Polyester	450	697	Burns readily with melting and shoot.
Acrylic	560	855	Burns readily with melting and sputtering
Modacrylic	–	–	Melts, burns very slowly.
Polypropylene	570	839	Burns slowly.
PVC and Polyvinylidene chloride	–	–	Does not support combustion.
Wool	600	941	Supports combustion with difficulty.

of the burning time to the charred length for comparing the flammability of fabrics, but it has never been accepted and used for evaluation of flammability to any extent. In recent years, the Oxygen Index (OI) method has gained popularity for flammability measurement of untreated and flame retarded fabrics. Originally developed by Fenimore and Martin [25], the OI is defined as the minimum volume fraction of oxygen in a slowly rising oxygen-nitrogen mixture that will sustain candle like burning of the polymer. Tesoro et al. [10] found that OI measures flammability of fabrics as a function of chemical composition of the fiber. The OI values of some of the fabrics as measured by Tesoro et al. are shown in Table 2. Willard [26] found that the oxygen index is a simple function of the added nitrogen and phosphorus content of the fabric. Miller et al. [27] studied the effects of various physical parameters e.g fabric weight, specimen size, enclosure chimney, horizontal and vertical mounting and direction of gas flow on the OI and observed

Table 2. Oxygen Index Values for Some Fabrics (from Ref 10)

Fabric – 100%	Fabric weight oz/yd²	Oxygen Index
Acrilan	6.5	0.182
Arnel triacetate	6.5	0.184
Acetate	6.5	0.186
Polypropylene	6.5	0.186
Vinylon (PVA)	6.5	0.197
Rayon	6.5	0.197
Cotton	6.5	0.201
Nylon	6.5	0.201
Polyester	6.5	0.206
Wool	7.0	0.252
Dynel	6.5	0.267
Nomex N-4274	4.8	0.282
Phovyl (PVC) "55"	6.5	0.371

that they affect OI significantly. DiPietro [28] studied the effect of preignition temperature of the test specimen on the OI of flame retarded cotton and found that the OI decreases with increase in preignition temperature of the fabric. Hendrix and coworkers [29] studied the effect of moisture content on the oxygen index of cotton, polyester and their blends, both untreated and flame retarded, and found that a moisture content of less than 6% has very little effect. However, at a moisture content higher than 6%, the oxygen index values increase and this increase is more pronounced for flame retarded fabrics than for the untreated. The rapid and intense interest in the OI technique is due to the precision with which the OI values can be measured and the reproducibility of the results within a laboratory and among laboratories. Reproducibility of measurement by OI technique has been claimed to be 1-4% [30]. Recently, Miller et al. [31] developed apparatus and technique "A wheel-fed flame apparatus" for the determination of flame propagation rates by moving fabric specimen at an appropriate velocity while maintaining the flame at a fixed position. It is claimed that data with a high degree of precision can be obtained by this method for both non-fusible and fusible fabrics, the latter without any support grid.

TECHNOLOGY OF FLAME RETARDANCY

Development of Flame Retardancy

Man's desire and need to reduce flammability of readily combustible materials go back to the prehistoric era, but attempts to impart flame resistance to fabrics are of relatively recent origin. In 1638 plaster of Paris was used as flame retardant for canvas and in 1821 Gay-Lussac investigated a number of inorganic salts as flame retardant for textile fabrics. From then until about 1950, a limited amount of *ad hoc* research was done. Research and development has been increasing at a faster rate during the past 10-20 years. The early work was concerned primarily with cellulose fabrics and has been reviewed from time to time [30-36]. Ranney [37] has summarized patent literatures on flame retardant textiles. The introduction of synthetics and blends and their wide use necessitated the development of flame retardants for such fabrics. Some of the relevant literature have been reviewed [4, 21, 35, 38]. Nametz [21] provided a good survey of the techniques and flame retardants used to impart flame resistance to polyester, nylon, acrylics and a few other synthetic fibers. A large number of diverse compounds have been applied to fabrics, particularly to cotton. It appears that the majority of flame retardants contain phosphorus, nitrogen and halogen. In recent years some work has been done to explore potential of boron containing compounds as flame retardant for fabric [39, 40].

Techniques of Fabric Flame Retardation

Several different approaches have been used for flame retarding fabrics. These are: finishing fabrics with flame retarding chemicals, adding flame retardants to polymers before their ultimate extrusion, copolymerization and chemical modification of fiber. Chemical modification involves reaction of the polymer with some chemicals and the product is flame retardant. Examples include phosphorylation, phosphonomethylation, phosphoromydation, mesylation, tosylation and cyanoethylation of cellulose and introduction of chlorine into polyester. In the copolymerization process one of the monomers contains a flame retarding element, e.g., chlorine, and is copolymerized with a second monomer so that the polymer contains its flame retardant in the chain. Of the four approaches, finishing appears to be most widely used in textile field. Flame retardants used for textile finishing may be put into two general classes, unreactive and reactive. Apparently such a classification is based on whether or not the flame retardant reacts with the fabric (polymer) at some stage during application of the retardant to fabric. The unreactive class includes a host of inorganic and organic compounds, most of which are salts. Water or organic solvent soluble compounds are applied from solution while the insoluble compounds are formed *in situ* on the fabric, e.g., WO_3 is precipitated on cotton by the reaction of $NaWO_4$ and $SnCl_4$. The unreactive retardants are cheap, they are nondurable or semidurable to washing, depending on whether the retardant is soluble or insoluble in the washing solvent. Finishing of fabric with reactive type flame retardants offers fastness to standard home washing and the effect is known as durable. Durable flame retardants for cotton fabric includes THPC-polymers, APO and APO-THPC. Both APO and THPC are potential monomers and also cross-linking agents for cellulose. APO is capable of forming both homo and copolymer and such polymers have been formed on cotton by applying these compounds on cotton from water solution and curing at elevated temperature THPC forms cross-linked polymers with a number of polyfunctional molecules, e.g., amines, ammonia, acids, phenols and APO. It appears that both APO and THPC in addition to forming polymer react with cellulose. Most of the work on these two compounds as flame retardants for cotton have been done at the Southern Regional Research Laboratories (SRRL) of the U.S. Dept. of Agriculture (USDA). They have developed durable flame resistant cotton by the *in situ* polymerization of THPC-ammonia mixtures. The *in vitro* polymer chemistry of THPC and its derivatives have recently been studied [41]. Recent interest has been focused on another derivative of THPC, tetrakis (hydroxymethyl) phosphonium hydroxide (THPOH) and its *in situ* polymerization with ammonia, urea or melamine, to obtain durable flame resistant cotton [42-45] on a commercial basis. Other reactive flame retardants for cotton include cyanamide-phosphoric acid [46], methylol dicyandiamide, trimer of phosphonitrilic chloride $(PNCl_2)_3$ [47].

The principles of flame retarding blends are basically the same as those of single component fabrics. However, it is usually necessary to flame retard all the components of the blend. Since flame retardants are specific in their effect, it may be necessary to use more than one type of flame retardant for blends. Some of the recent studies in this area have been reported by Kruse [4].

Synergism in Flame Retardancy

Incorporation of nitrogen or halogen (F, Cl and Br) into a phosphorus containing flame retardant has been found to increase the flame retarding efficiency. Examples include brominated phosphates, $THPC\text{-}NH_3$ polymer, and phosphonitrilic chloride. Similarly, certain mixtures of compounds have been found to be more efficient flame retardants than the constituent compounds. Examples include borax-boric acid, organic phosphate-urea, antimony oxide-organic chloride, $THPC\text{-}NH_3$, THPC-diamine etc. [48]. Such increase in the flame retarding efficiency is an example of synergism and is an attractive field of research both from theoretical as well as practical view points.

MECHANISM OF FLAME RETARDANCY

General

The theoretical aspect of the mechanism of fabric flame retardancy is complicated both because of the complexity of the burning process itself, which varies from polymer to polymer, and because of the diverse nature of the compounds used as flame retardants. A generalized and coherent theory of flame retardancy applicable to all polymer is yet to be evolved although much effort is being given in recent years to understand flame retardancy. It appears [49] that much of the present knowledge is speculative. However, cellulose has been widely studied and a wealth of information is available. Several theories based on this information, some of which is empirical and intuitional, have been put forward, but a systematic study of the mechanism of flame retardancy in other polymers has begun only recently [49]. However certain generalized statements pertaining to the nature of the flame retardants can be made.

1. Flame retarding effect of a compound is "catalytic," although a "high minimum" (10 percent) add-on is required to obtain a "satisfactory" flame retardancy This minimum add-on varies from polymer to polymer e.g. for the same retardant viscose requires higher add-on than cotton.
2. The effectiveness in terms of add-on of flame retardants depends on many factors such as thermal stability of both retardant and the polymer and chemical reaction at elevated temperature of the flame retardant with the basic polymer [39].
3. Flame retardants are specific in their effect i.e. a compound may be a good flame retardant for cellulose but may not have any effect on polyester.
4. Organophosphorus compounds are especially effective for hydroxyl containing polymers e.g. cellulose, but their effectiveness does not appear to depend significantly on their structure. However, compounds decomposing above the flaming temperature (400°C) are ineffective.
5. The efficiency of halogen containing flame retardants increases in the order F, Cl, Br, I and halogens on aromatic structure are less effective than those on the aliphatics [50, 51].

The Function of Some Flame Retardants

In principle flame retardancy can be achieved by affecting any or all of the stages of burning of polymers. These stages are (Figure 2) ignition, pyrolysis and flaming. Derivation of the function(s) of flame retardants is a formidable piece of work because of the many interacting variables involved in flame retardancy. Nevertheless, some generalizations concerning the function of flame retardants pertaining primarily to cellulose (but perhaps applicable to other polymers) can be made.

1. Phosphorus containing flame retardants act predominately in the solid phase and affect pyrolysis of the polymer.
2. Halogen containing flame retardants act predominately in the gas phase and affect the flaming reaction by acting as free radical scavengers; they may also act in the solid phase or in both phases.
3. Differential thermal analysis (DTA) shows that flame retardants lower the decomposition temperature of certain polymers [28, 39, 52-55].
4. Thermogravimetric analysis (TGA) shows that many flame retardants affect the pyrolysis in such a way as to produce more char and less flammable gas; major weight loss of flame retardant fabric occurs at higher temperature compared to the untreated fabric [28, 53-55].

5. Flame retardants may lower the heat of combustion of the polymer because less combustible gases are produced; this in turn affects pyrolysis of the polymer. The non-combustible gases produced may reduce the exothermic surface regression because they dilute the oxygen necessary to attack the surface
6. Flame retardants may form a coating film around the fiber such that oxygen necessary for flaming cannot attack the surface directly. Flame retardants may, on decomposition, produce non-flammable gases which dilute the pyrolysis products to a concentration below that which is required to sustain flaming.
7. Flame retardants may act as inhibitors for the free radical flaming reaction e.g. halogens, acting as free radical scavengers in the flaming reaction.

Pyrolysis and Flame Retardancy

Pyrolysis of the polymer supplies fuel for flaming; hence it has been of utmost interest to study the effect of flame retardants on pyrolysis. With the exception of a few, the reported work is related to cellulose and an excellent review has been published [56] covering the basic aspects of cellulose pyrolysis which will not be discussed here. It appears that all work leads to a common conclusion that flame retardant cellulose, and probably most other fabrics, produces more char and less flammable gas. In addition, flame retardants catalyze the pyrolysis of a polymer [57, 58]. A recent mass spectrometric study [58] shows that the pyrolysis products from untreated cotton do not differ qualitatively from those of flame retarded cotton, although the proportion of the flammable gases in the latter is reduced. The effect of this increased or decreased availability of pyrolysis products (fuel) is to create an imbalance in the fuel to oxidizer ratio in the flaming reaction. A combustible mixture will undergo flaming reaction only over a narrow range of fuel to oxidizer ratio, called the lower and upper flammability limits [59]. The flame becomes unstable and may extinguish, lift off or flash back outside the flammability limits. The manner in which flame retardants affect the pyrolysis of polymers and the nature of interaction depends among others, on the chemical nature of both the polymer and the flame retardant and their thermal stability. In the case of cellulose and phosphorus containing flame retardants, it is speculated that the phosphorus compounds are converted to phosphoric or polyphosphoric acid which acts as dehydration catalyst for cellulose [49].

Synergism in Flame Retardancy

Basic aspect of synergism is a relatively new field of investigation and hence very little progress has been achieved. Only a limited number of systems have been studied. It is felt that the mechanism of synergism is depicted primarily by the thermal reaction of the constituent compounds. For the P-N synergism, Hendrix [60, 61] has shown that some organic phosphate in the presence of urea forms phosphoramide and the latter acts as flame retardant for cellulose. The antimony-halogen synergism is thought to be due to formation of antimony trihalide, which acts as flame retardant [50]. Similarly phosphorus-halogen synergism is thought to be due to formation of phosphorus -tri or -penta halide [50]. A detailed discussion of such synergistic effects can be found in the recent issue [49] of the Polymer Conference Series.

Dehydration Theory of Flame Retardancy

In principle a theory should be able to explain the mechanism of action of known flame retardants and in addition, should be able to predict new flame retardants. Such a theory has yet to be evolved. However several theories have been put forward for cellulose. These are: the coating, the gas, the thermal and the chemical theories and they have been described elsewhere [62] The former three are applicable only to a limited number of flame retardants. The chemical theory appears to be more general and applicable to a large number of compounds.

The chemical theory was developed by Schuyten [63]; it assumes that flame retardants are either Lewis acids or bases or their precursors and are available at or near the flaming temperature of cellulose to catalyze dehydration of cellulose through a carbonium or a carbanion as shown below. Such a dehydration process may continue until all the "available" hydroxyl groups of cellulose have been involved, leaving a residue of high carbon content (char).

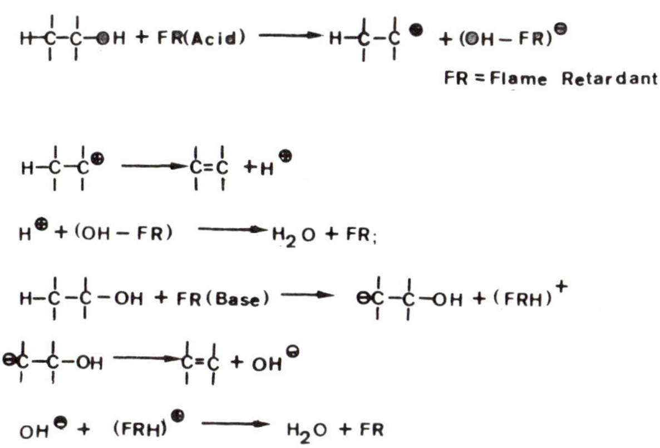

GLOW RETARDANCY

Glowing (flameless combustion) is due to the oxidation of carbon (char) to carbon dioxide. Thermolabile phosphorus and boron containing compounds and some metallic oxides have been found useful as glow retardants [64, 65]. Several postulations have been put forward to explain glow retardancy of cellulose. The oxidation of carbon to carbon monoxide is less exothermic than that to carbon dioxide and it has been proposed [66] that glow retardants divert oxidation of carbon to the less exothermic carbon monoxide so that the heat of oxidation is insufficient to sustain further oxidation of carbon. In addition, it has been proposed that glow retardants may coat the "active oxidation sites" in the char [64-66] so that oxygen can not diffuse into the char. Furthermore, boron and phosphorus containing glow retardants were proposed to act as inhibitors for the oxidation of carbon by interfering with the electron transfer process in the oxidation of carbon [65].

CONCLUDING REMARKS

The need and importance for flame retarding fabrics are increasing at a faster rate than ever before A large number of test methods are available for evaluating flammability of fabrics, but a method which can simulate the flammability of fabrics in actual use is far from a reality, although efforts are continuously being made to approach this goal. Nevertheless the current test methods provide data which may be used to classify fabrics as slow or fast burning, flammable, less flammable or more flammable under specified conditions.

Very little is known about the basic aspects of flammability and flame retardation of fabrics other than cellulose which shares most of the literature in this field and much is known about the basic and applied aspects of flammability and flame retardation of cellulose. Studies on synthetic fibers and their blends with natural fibers have been neglected in the past. In view of their increasing use, more efforts should be directed towards understanding basic aspects of flammability and flame retardation of synthetics and blends, and it appears that activity in this area is gaining momentum.

Although a large number of flame retardants are known particularly for cellulosic fibers, only a very few of them posses any reasonable fastness to normal home laundering. In addition, most flame retardant finishes affect both the physical properties and the aesthetic values of fabrics. Only from the results of systematic studies, may it perhaps be possible to tailor flame retardants which would meet most if not all the requirements asked for by the consumers.

Flammability and flame retardation of polymeric materials, particularly fabrics, is an interesting, challenging and fruitful field for investigation and it remains up to scientists to explore and develop this rather unsystematic and unorganized branch of science and technology.

ACKNOWLEDGMENTS

Thanks are due to the Sun Chemical Corporation (USA) for financial support and permission to publish this paper and to Prof. H. Zollinger for discussion.

REFERENCES

1. E. P. Martin, Ciba Rev , No. 4 (1969).
2. H. R. Richards and D. M. Wiles, Can. Text. J. *86* (9), 47 (1969).

3. J. E. Clark, Sources and Resources *4* (1), 3 (1971).
4. Textile Flammability and Consumer Safety, gdi occasional publication No. 45, 1969, Ruschlikon – Zurich.
5. G. Stamm, Textilveredlung *6*, 631, 656 (1971).
6. Chem. & Eng. News, 16, Oct. 18 (1971).
7. M. J. Korosky, Sources and Resources *4*, (1971); also Amer. Dyest. Rep. *60*, 49 (1971).
8. E. V. Thompson, Polymer Letters *4*, 361 (1966).
9. E. B. Nielson and H. R. Richards, Text. Chem. Color. *1* (12), 27 (1969).
10. G. C. Tesoro et al., Text. Res. J. *40*, 430 (1970).
11. G. C. Tesoro and J. Rivlin, Text. Chem. Color., *3* (7), 27 (1971).
12. H. R. Richards, Can. Text. J. *80* (20), 41 (1963).
13. M. J. Korosky, Amer. Dyest. Rep. *58* (6), 15 (1969).
14. H. R. Richards, 12th Canadian Textile Seminar (International), August 24-26, 1970, Queens University, Kingston, Ontario.
15. R. Winstein, Text. Chem. Color., *3* (2), 55 (1971).
16. AATCC Tech. Mannual, 1971, p 204-205.
17. A. S. Tweedie et al Can. Text. J., *88* (11), 73 (1971), also *88* (12), (1971).
18. E. B. Nielson and H. R. Richards, Text. Chem. Color., *2* (4), 78 (1970).
19. L. W. Sayers, Text. Inst. Ind., *3*, p. 168 (1965).
20. B. Miller et al , Proceedings of the Third Annual Meeting, Information Council on Fabric Flammability, New York, Dec. 1969, p. 116.
21. R. C. Nametz, Ind. Eng Chem., *62* (3), 41 (1970).
22. U. Einsele, Textil-Industrie, *72*, 984 (1970).
23. R. B. Leblanc, Text. Chem. Color., *2* (3), 47 (1970).
24. S. M. Suchecki, Text. Indust., *133* (2), 69 (1969).
25. C. P. Fenimore and F. J. Martin, Mod. Plastics, *44*, 141, November 1966.
26. J. J. Willard and R. E. Wondra, Text. Res. J. *40*, 203, (1970).
27. B. Miller and C. H. Meisen, Jr., Text. Chem. Color., *2* (12), 35 (1970).
28. J. DiPietro et al , Text. Chem. Color., *3* (2), 45 (1971).
29. J. E. Hendrix et al., Text. Res. J. *41*, 854 (1971).
30. J. L. Isaacs, J. Fire and Flammability, *1*, 36, Jan. 1970.
31. B. Miller and C. H. Meisen, Jr., Text. Chem. Color., *3* (5), 55 (1971).
32. E. M. Akin et al., Amer. Dyest. Rep., *29*, p. 418, 445 (1940).
33. C. J. West et al., Flameproofing, Bibliography series No. 185, The Institute of Paper Chemistry, Wisconsin 1959.
34. Anon, Textile World, *119*, 102 (1969).
35. H. Hantge, Textilveredlung *6*, 637 (1971).
36. L. H. Chance et al., Melliand Textilbar., *10*, 1211 (1971).
37. M. W. Ranney, Flame Retardant Textiles, Noyes Data Corporation, 1970.
38. Anon, Amer. Dyest. Rep. *57* (10), 40 (1968).
39. M. A. Kasem, and H. R. Richards, Ind. Eng. Chem., Product R&D, *11*, 114-133 (June 1971).
40. Anon, Cotton Research Notes, Oct. 1964, p. 4.
41. M. A. Kasem et al., J. Appl. Polym. Sci., *15*, 2237 (1971).
42. J. J. Duffy, Proceeding of the Third Annual Meeting, ICFF, New York, 1969.
43. J. V. Beninate et al., Textile Indust. *131* (11), 110 (1967).
44. J. V. Beninate et al., Text. Res. J., *38*, 267 (1968).
45. W. A. Reeves and J. D. Guthrie, U. S. Pat. 2,809,941 (1957).
46. S. J. O'Brien, Text. Res. J., *38*, 256 (1968).
47. S. J. O'Brien, Text. Chem. Color , *2* (12), 31 (1970).
48. G. C. Tesoro, Text. Res. J., *39*, 180 (1969).
49. Flammability Characteristics of Polymeric Materials, Polymer Conference Series, 1970, The University of Utah, Salt Lake City (USA).

50. Encyclopedia of Polymer Science and Technology, H. Mark et al., Ed., Interscience, Vol. 7, 1967, p. 16.
51. J. G. Frick, Jr., Text. Res. J. *26*, 525 (1956).
52. J. E. Parker and A. E. Lipska, Final Report, OCD Work Unit No. 2531C, NRDL-TR-69, May 1969, USNRDL, San Francisco, California.
53. J. W. Schuyten et al., Ind. Eng. Chem., *47*, 1433 (1955).
54. R. M. Parkins et al., J. Appl. Polym. Sci., *10*, 1041 (1966).
55. W. K. Tang and W. K. Neill, J. Polym. Sci., Part C, No. 6, 65 (1964).
56. F. Shafizadeh, Adv Carbohy Chem., *23*, 419 (1968).
57. P. Hofmann and F. Raschdorf, Textilveredlung *5* (6), 486 (1970).
58. A. E. Lipska and A. Wodley, J. Appl. Polym. Sci., *13*, 851 (1969).
59. J. N. Bradley, Flame and Combustion Phenomenon, Mathuen and Co., Inc., 1969.
60. J. E. Hendrix et al., J. Appl. Polym. Sci., *14*, 1701 (1970).
61. J. E. Hendrix, Ph.D. Dissertation, Clemson University, South Carolina, 1970.
62. Kirk-Othmer Encyclopedia of Chemical Technology, H. Mark et al., Ed., 1966, Vol. 9 300.
63. J. W. Schuyten et al., Advances in Chemistry Series, No. 9, 1953, p. 7.
64. J. M. Sather and H. R. Richards, Amer. Dyest. Rep. *58* (3), 21 (1970).
65. J. F. Rakszawski and W. E. Parker, Carbon *2* (1), 53 (1964).
66. R. W. Little, Text. Res. J. *21*, 901 (1951).

Fire Studies in Oxygen-Enriched Atmospheres

V. A. Dorr

Research and Development Department
Ocean Systems, Incorporated
Tarrytown, New York

(Received January 14, 1970)

ABSTRACT

Studies of the effects of environmental parameters upon combustion in hyperbaric environments and evaluations of selected materials for flammability in oxygen-enriched atmospheres have been conducted. Burning rate data for a standard material in oxygen-nitrogen and oxygen-helium mixtures are presented and a scale of fire resistance for measuring flammability in oxygen-enriched atmospheres is discussed. Tentative recommendations for some of the more fire safe materials are included.

INTRODUCTION

The ever-increasing application of oxygen-enriched atmospheres* in hospital hyperbaric chambers, oxygen tents and divers' decompression chambers has pointed up to the vital need for research into the combustion hazards associated with the use of these environments. Although fires in hyperbaric and hypobaric chambers are rare, they have occasionally occurred with astonishing suddenness and often fatal results [6, 7, 8, 9]. In order to reduce the probability of further accidents in oxygen-enriched environments, one must strive to eliminate any one of the three fundamental prerequisites for a fire: an ignition source, a fuel and an atmosphere capable of supporting combustion. In practice, it is often difficult or impossible to remove any one of these requirements with absolute certainty. Attempts should therefore be made to reduce the contributing effects of all three essential factors, i. e., electrical items that may serve as potential ignition sources should be avoided, the quantity of flammable materials in the chamber should be reduced, and the absolute oxygen concentration and the oxygen partial pressure within the chamber should be minimized as much as possible.

In order to evaluate the influence of the above factors upon combustion in oxygen-enriched environments, a study has been conducted to (1) assess the effects of environmental parameters upon the combustion rate of a standard material, (2) determine the minimum oxygen concentration required for the combustion of solid flammables in the presence of nitrogen and helium, and (3) categorize the flammability hazards of cloth fabrics, elastomers and insulations in hyperbaric environments.

* An oxygen-enriched atmosphere (OEA) is generally defined as an atmosphere which contains greater than 21 mole-% oxygen and/or contains a partial pressure of oxygen greater than 0.21 atmosphere.

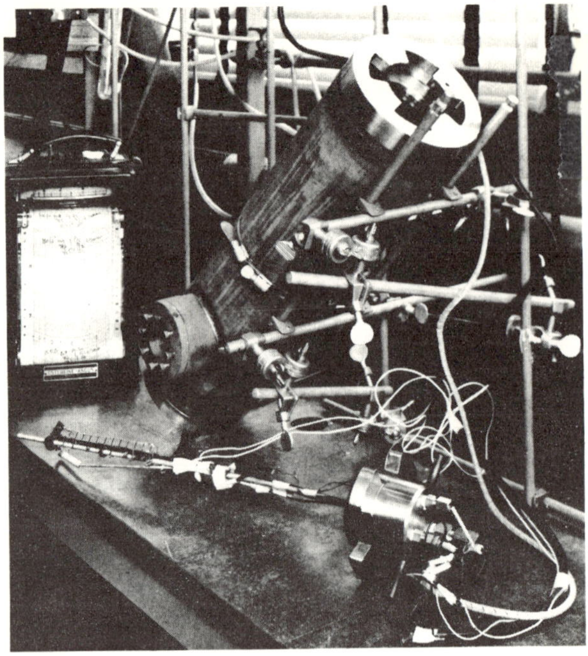

Figure 5. *Fifteen centimeter ID pressure vessel, its closure and sample holder.*

EFFECTS OF ENVIRONMENTAL PARAMETERS UPON COMBUSTION

Small-scale tests and rate determinations are conducted in a 15 cm. ID by 51 cm. long chamber with a working pressure of 40 atm. abs. (Figure 5). A strip of test material is mounted on a series of brass pins and the lower end is held firmly in an igniter grid of chromel wire which may be heated to 1400° F. Thermocouples welded to the igniter wire and mounted on the sample holder allow measurement of ignition temperature and burning rate.

The burning rates of filter paper have been measured in oxygen-nitrogen and oxygen-helium mixtures containing 15% to 100% oxygen over the pressure range of 1 to 10 atm. abs. [3, 4]. Figure 1, a plot of the burning rates of filter paper in oxygen-nitrogen mixtures at an angle of 45° versus the total pressure, shows that the burning rate is accelerated for all gas mixtures as the total pressure is increased. The rate is also accelerated as the oxygen content is increased at any given pressure. In the case of air, 21% oxygen, the combustion rate of filter paper is increased from 1.1 cm./sec. to 2.9 cm./sec. as the pressure is raised from 1 atm. abs. to 10.1 atm. abs. The same material burns at a rate of 2.1 cm./sec. in a 50% oxygen/50% nitrogen mixture at 1 atm. abs., yet this rate decreases to 1.2 cm./sec. as the total pressure is increased to 3.3 atm. abs. by adding nitrogen without changing the oxygen partial pressure of 0.5 atm. The burning rates of filter paper in oxygen-helium mixtures are analogous to the preceding, as seen in Figure 2. The burning rate is accelerated by either an increase in the total pressure of any given gas mixture or an increase in the oxygen concentration at any given total pressure.

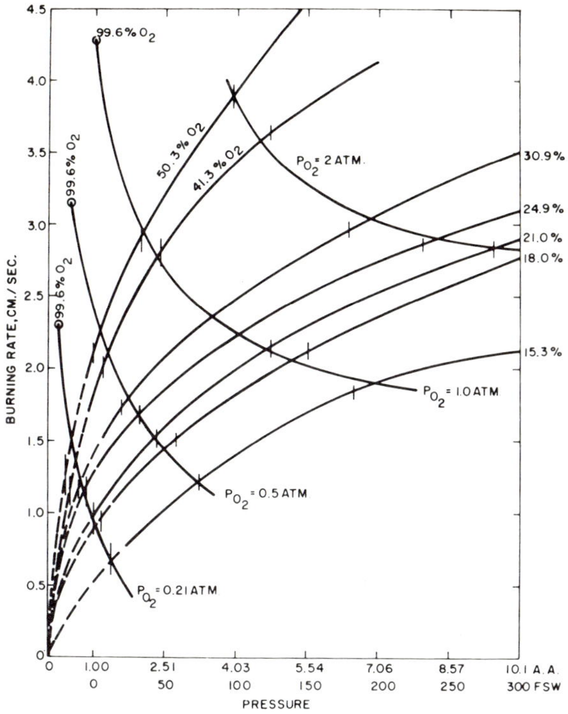

Figure 1. *Burning rates of filter paper at an angle of 45° in oxygen-nitrogen mixtures.*

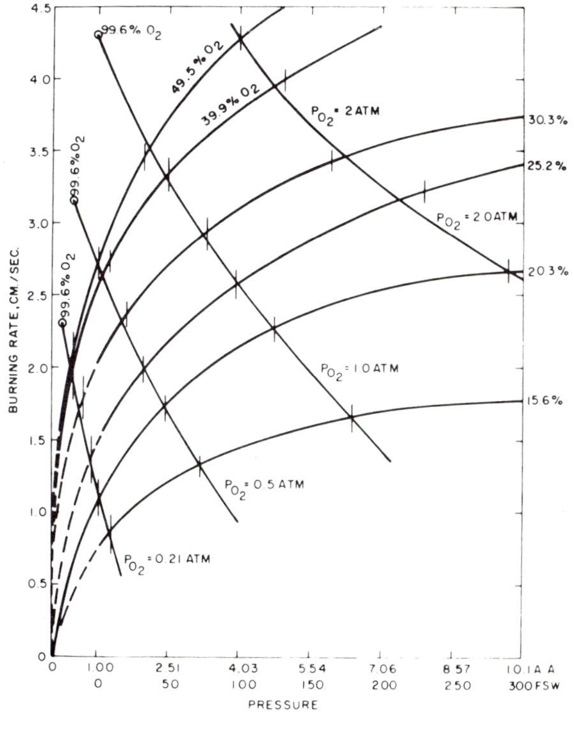

Figure 2. *Burning rates of filter paper at an angle of 45° in oxygen-helium mixtures.*

A comparison of the burning rates in helium versus nitrogen (with identical oxygen concentration and total pressure) shows that the combustion process tends to be accelerated in the helium environment. This effect has been observed to be most pronounced at atmospheric pressure in a 30% oxygen atmosphere. In the helium mixture, the burning rate of filter paper at an angle of 45° is measured as 1.9 cm./sec. whereas the burning in the nitrogen mixture is measured as only 1.4 cm./sec., a 26%decrease. The difference in burning rates tends to diminish (in the oxygen concentration range of 15% to 50%) as the total pressure of the mixture is increased. A reduction of the difference in the densities of oxygen-helium and oxygen-nitrogen mixtures at higher pressures probably accounts for this.

MINIMUM OXYGEN CONCENTRATIONS NECESSARY FOR COMBUSTION IN HYPERBARIC ENVIRONMENTS.

The risk of fire is not always increased in hyperbaric environments. At high pressures, the oxygen concentration in a mixture of oxygen and inert gas is normally reduced in order to avoid the danger of oxygen toxicity. This decrease in the oxygen percentage may also render the atmosphere incapable of supporting combustion under certain carefully defined conditions [1, 5]. Three zones representing complete combustion, incomplete combustion and no combustion at all are shown in Figure 3, a plot of oxygen concentration versus total pressure for vertical strips of filter paper in oxygen-nitrogen mixtures. Isobars showing oxygen partial pressures of 0.21 and 1.5 atm. are also shown as these oxygen concentrations represent minimum and maximum partial pressures normally used for a respirable environment in hyperbaric chambers. It is seen that the area denoted ABC lies both within the region of noncombustion and between the two oxygen isobars, thus these conditions will maintain life yet will not support the combustion of ordinarily flammable materials such as paper and cotton cloth. The narcotic effect of nitrogen at high pressure, however, makes its use as a diluent gas unadvisable at partial pressures exceeding approximately 5.5 atm. Therefore, only the area denoted ADE is both physiologically acceptable and free from all risk of accidental fire. The absence of narcosis in oxygen-helium mixtures (up to pressures of at least 35 atm. abs.) produces a much larger and more practical range of gas composition and pressure (Figure 4) that is physiologically satisfactory and still devoid of combustion hazards.

FLAMMABILITY OF MATERIALS IN HYPERBARIC ENVIRONMENTS

When it is impractical to maintain the hyperbaric environment within the region of noncombustion and if all possible sources of ignition within the chamber cannot be precluded with absolute certainty, great care should be exercised in the selection of materials to be placed in the chamber. The results of standard flammability tests in the ambient atmosphere are not valid indications of the burning behavior the same material exhibits under hyperbaric conditions. In our laboratory, we have devised a scale of fire resistance for flammability under oxygen-enriched atmospheric conditions [2, 4, 5]. Materials are tested for combustion in air at atmospheric pressure, compressed air and in oxygen-nitrogen mixtures containing increasing amounts of

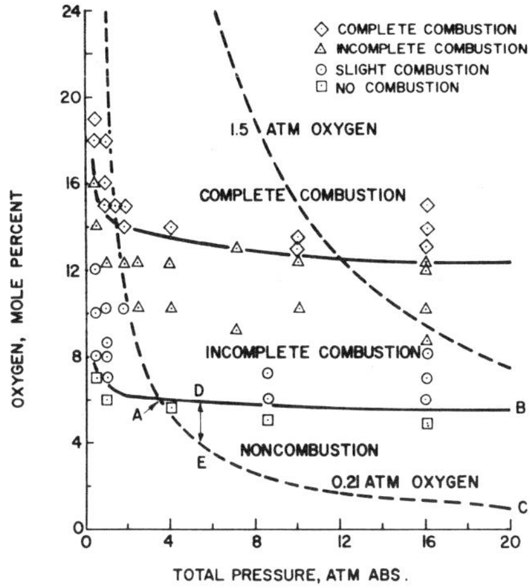

Figure 3. *Three combustion zones for vertical filter paper strips in oxygen-nitrogen mixtures.*

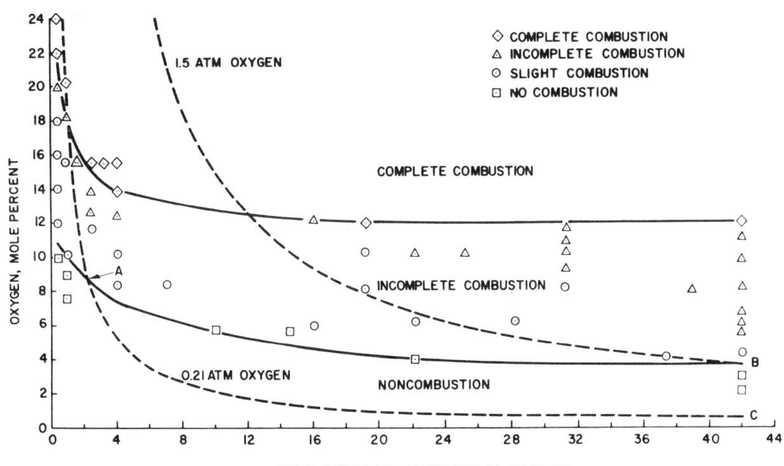

Figure 4. *Three combustion zones for vertical filter paper strips in oxygen-helium mixtures.*

oxygen at a total pressure of one atm. abs. The material is then rated in one of ten classifications described below:

Scale of Fire Resistance

Class 0. Burns readily in air at atmospheric pressure

Class 1. Has an appreciably higher ignition temperature and/or burns at an appreciably lower rate in air at 1 atm abs pressure than cotton cloth or paper. An example of a Class 1 material is wool.

Class 2. Non-flammable or self-extinguishing in air at one atm abs pressure.

Class 3. Self-extinguishing or burns slowly in air at a pressure of 100 feet of sea water (4.03 atm abs).

Class 4. Self-extinguishing or burns slowly in air at a pressure of 200 fsw (7.06 atm abs).

Class 5. Self-extinguishing or burns slowly in a mixture of 25% oxygen and 75% nitrogen at a pressure of 1 atm abs.

Class 6. Self-extinguishing or burns slowly in a mixture of 30% oxygen and 70% nitrogen at a pressure of 1 atm abs.

Class 7. Self-extinguishing or burns slowly in a mixture of 40% oxygen and 60% nitrogen at a pressure of 1 atm abs.

Class 8. Self-extinguishing or burns slowly in a mixture of 50% oxygen and 50% nitrogen at a pressure of 1 atm abs.

Class 9. Non-flammable in 100% oxygen at a pressure of 1 atm abs.

Although over one hundred materials have been tested for flammability and classified [4, 5], there are very few that possess enough intrinsic flame resistance to be considered truly acceptable for use in oxygen-enriched hyperbaric chambers. Test results and comments on several of these materials are presented below.

FABRICS AND CLOTHING

Owens-Corning Beta Fiberglas is a completely nonflammable Class 9 fabric which, perhaps, has the greatest application in oxygen-enriched atmospheres. The material does not burn in pure oxygen at atmospheric pressure even when ignition is attempted with the intense flames of burning cellophane tape. The material does, however, melt under these conditions and small, brittle beads of fused glass remain. The material has very low abrasion resistance (about ⅛ that of nylon) and consequently Beta garments have poor durability. We feel that this is a minor inconvenience in comparison with the greater safety Beta fabric affords over conventional materials. One-piece coveralls made from Beta 4190B (B. Welson Company, Style BW-1027-001) are being used successfully in the Ocean Systems manned diving research currently underway in Tarrytown, N. Y. Another Beta garment, a two-piece suit with parka made from a heavier gauge fabric, is the Fyrepel, Model 320-30-G9.

Although minor skin irritation has been reported in conjunction with the use of glass garments, we have experienced no discomfort when wearing the Welson suit. The abrasion resistance and, presumably, comfort of the fabric may be increased by coating it with Teflon. A sample of the Teflon coated Beta Fiberglas 4190, obtained from B. Welson, was considerably stiffer than the uncoated material. When this was tested in oxygen at atmospheric pressure, the coating did not burn when in contact with the igniter wire alone but when cellophane tape was used to provide a flaming ignition source, the coating burned almost completely off. The DuPont Armalon P.S. 6439 and 63339AS which are also Teflon coated glass fabrics as well as Armalon 98-101 which is Teflon coated Nomex have been rated as Class 7 as the coating burns off in a mixture of 40% oxygen and 60% nitrogen at atmospheric pressure. An additional material obtained from B. Welson, "super Beta", is made by coating the glass fibers with Teflon before weaving and the resultant fabric is strong and more pliable than the coated fabric. The coating did not burn in oxygen even with the flaming ignition source and the material is Class 9. This technique appears to overcome the major limitations of regular Beta Fiberglas but it is still under development and to our knowledge the material is not yet being produced commercially.

Beta may also be obtained in the form of webbing of several different widths from the Carolina Narrow Fabric Company. We have used this material with limited success (due to breakage) for interwoven supports in the bunks and chairs used in our chambers.

There are several other types of fabrics in the Class 8 and Class 9 categories which are currently under development and which may have future application. Ordinary asbestos fabrics contain only about 80% asbestos and 20% of a carrier fabric added for strength. This carrier is usually cotton or a polyester which will burn off intensively in a hyperbaric environment. Uniroyal Fiber and Textile, in their D67504 and D67505 series, has developed a 79% asbestos, 14% Beta Fiberglas and 7% Nomex composite which we rate as Class 9. Although the samples did not burn in oxygen, there were, however, periodic small flashes of light from within the fabric as ignition was attempted. This stems presumably from the interwoven Nomex. The material also is produced with several types of finishes which reduce linting and scuffing of the asbestos. Although the material is rather stiff and heavy for clothing applications, it seems plausible that an adequate, nonflammable blanket (when several layers are sewn in between a cover, e.g. Beta Fiberglas) could be fabricated.

Fiberfrax alumina-silica ceramic fiber from Carborundum is a Class 9 material which is produced as a fabric, felt, board and rope (note that several types must be heat treated at 800°F. for several hours by the customer to remove organic residues before qualifying for Class 9). The material is heavy, coarse and will irritate the skin with prolonged contact. It is a very good insulation material and might also serve as a blanket when placed inside a suitable cover. It might also serve as a good fire blanket to smother fires as it is rated for service up to 2300°F. (this, however, has been shown to be a very ineffective method to extinguish chamber fires [10]). The fiberfrax rope, also nonflammable has a very low tensile strength but this may be increased by using a flexible steel cable in the center.

Several types of inorganic fabrics under development by Union Carbide Corporation have been tested. Zirconia cloth and boron nitride cloth* are Class 9 materials and Thornel carbon fabric is Class 8. Of the three, boron

* Boron nitride cloth is also under development by The Carborundum Company.

Table 1. Burning Rates of Mine Safety Appliances Fire Retardant Wool Blanket and Union Carbide Dynel Blanket.

MSA Fire Retardant Wool Blanket No. 04-12487	UCC Dynel Blanket
0.39 cm/sec	0.35 cm/sec
0.45	0.29
0.47	0.28
Average = 0.44[2]	Average = 0.31

1. All samples burned at an angle of 45° in 30% oxygen-70% nitrogen at 15 psia.
2. Wool strips were self-extinguishing in several runs whereas Dynel burned steadily in all runs. Wool, however, always burned with a stronger flame and longer flame front than Dynel.

nitride most closely resembles the weave and appearance of normal cloth. The Carbon Products Division of Union Carbide produces several styles of Ucar electrically conductive cloth in which strands of conductive carbon yarn are woven in a carrier fabric. Style 9966, carbon yarn in E-glass carrier, is Class 8 and style 8020, carbon yarn in B-glass, is Class 5. These fabrics seem particularly well suited for use in heated diving clothing. Two other styles employ cotton and olefin fabrics as the carrier and are readily flammable.

Union Carbide's Dynel fabric, a copolymer of vinyl chloride and acrylonitrile, is a Class 5 material so rated due to its ability to rapidly shrink away from a heat source and therefore prevent ignition. When it does burn, the rate is slow and the heat of combustion appears to be low. Strips of 5 oz./sq. yd. Dynel were burned in 25% oxygen at atmospheric pressure at an angle of 45° and the burning rates were 0.18 and 0.26 cm/sec. In five other attempts, the material was self-extinguishing and rates could not be determined.

Samples of Monsanto Chemical Corporation's X-400 and X-410 fabrics have been supplied to us by the U.S. Navy Clothing and Textile Research Unit. The accompanying technical data sheet from Monsanto describes the X-400 fabric as flameproof in that it "will not burn or produce noxious fumes when subjected to an 1800°F. Meker burner flame." We have verified this statement although it was noted that the fabric initially produced a yellow flame while in contact with the burner flame and the resultant material after this treatment was brittle and would crumble to the touch. In so far as testing these materials in our standard manner, we have rated both the X-400 and X-410 as Class 6. In compressed air at 100 fsw and 200 fsw, both materials smoldered only slightly before self-extinguishing. They also smoldered slightly past the igniter grid in 25% and 30% oxygen at atmospheric pressure but burned completely in 40% oxygen. In addition to possessing improved flame resistance, both materials have the feel and wearability of ordinary textiles and appear to be good candidates for diving clothing as they become more available.

Comparative burning rate tests were conducted on a Mine Safety Appliances fire retardant wool blanket No. 04-12487 and a Union Carbide Dynel blanket and the results are shown in Table 1. MSA informs us that the fire retardant treatment is not resistant to laundering and will wash out in water. Although the burning rate of the wool averaged 0.44 cm/sec versus 0.31 cm/sec for the Dynel, the wool strips were self-extinguishing in several runs whereas the Dynel burned steadily in all runs. Wool, however, always burned with a stronger flame and longer flame front (a nap effect possibly) than Dynel. Of the two blankets, the Dynel appears to be a better choice from a flammability point of view, however, from the same point of view, neither is particularly suited for actual chamber use. Blankets would be safest if custom made in the manner previously described.

An organic material of excellent flame resistance is Pluton heat resistant fabric from the Minnesota Mining and Manufacturing Company (3M). This material appears to be impervious to all ordinary heat sources. When it was exposed to the flame of a natural gas-oxygen torch at ambient conditions, the fabric heated to red hot but cooled upon removal and did not appear to suffer any damage. In a pure oxygen environment, however, the material glowed intensely and disintegrated after contact with the igniter wire. In a mixture of 50% oxygen and 50% nitrogen at atmospheric pressure, the glowing (smoldering) slowly progressed up the strip but was self-extinguishing before the end; Pluton is therefore in Class 8 (style B-1 seems to be slightly more fire resistant than style H-1). To our knowledge, Pluton fabrics have not been used in clothing and bedding applications. The 3M Company states that its susceptibility to damage by abrasion is high and that automatic laundering is not recommended. Otherwise Pluton seems to be a good choice for a compressed air environment from the fire safety point of view.

A new fiber made from polybenzimidazole (commonly referred to as PBI) has been produced by the Celanese Chemical Corporation under contract to the U.S. Air Force. Samples of PBI fabric supplied by the Air Force Materials Laboratory were tested and placed in Class 7. Proposed uses for this material include space capsule parachutes that have to withstand the extreme heat of re-entry and aviation flight suits.

Samples of DuPont's woven brown and white Teflon cloth obtained from Stern and Stern Textiles were basically similar in combustion characteristics under the variety of conditions tested with the exception of pure oxygen at atmospheric pressure in which the white Teflon cloth is significantly less flammable. This difference may be attributed to the approximately 7% carbonaceous residue, formed in the polymerization process, that is left in brown Teflon. The white Teflon has been purified to remove this but is 75% more expensive as a result.

ELASTOMERS

To date, we have located only two nonflammable, Class 9, elastomers. Neither Raybestos-Manhattan's Refset fluoroelastomer nor Thiokol Chemical Corporation's CNR Nitroso Rubber ignites or burns in oxygen at one or two atmospheres absolute pressure. The CNR rubber merely vaporizes in contact with the igniter and the Refset forms a brittle ash. Unfortunately, it appears at this time that Thiokol has discontinued the production of CNR and Raybestos-Manhattan manufactures Refset only as a specialty item.

Other highly nonflammable fluorinated elastomers are currently marketed. While none may be rated as truly nonflammable as the CNR or Refset, they can, however, presently be purchased as regular commercial products. Minnesota Mining and Manufacturing Company's Fluorel brand elastomers are sold to several other companies who complete the processing and sell the finished rubber. One such company, the Mosites Rubber Company has supplied us with samples of Fluorel numbers 1059 and 1071 which we have tested and rated at Class 8. In pure oxygen at one atm abs, the rubber strip burned slowly and completely. A large amount of hydrogen fluoride was apparently released during the combustion as the glass sight port quickly fogged over and was permanently etched. In 50% oxygen-50% nitrogen at one atm abs as well as air at 200 fsw, no sustained burning of the rubber was observed even when the flaming cellophane tape ignition source was utilized.

DuPont's Fairprene 80-080 (Viton), a fluoroelastomer, is a Class 6 material that is not as flame resistant as the Fluorel rubbers discussed previously but is still an improvement over natural rubber. The Fairprene 85-001 (Viton on glass cloth) is only Class 3. The ease of ignition and burning is greater with the thin Viton layer on the coated material than with the thicker, solid slab. Other DuPont elastomers tested such as Fairprene 70-001, Hypalon (Class 3) and straight neoprene (Class 2) are not flame resistant enough for safe use in the chamber.

ELECTRICAL INSULATION

We previously tested a braided copper wire (approximately 1 mm in diameter) insulated with a 7 mil thickness of DuPont's Kapton polyimide film (designated 019-919). The insulation was rated as Class 9 as there was no combustion in 1 atm abs oxygen, however, a Kapton sheet alone is rated as Class 7. It appears that the heat sink effect of the wire plays an important part in the rating of the insulation. One should always take this into consideration when selecting an insulation.

Another flame resistant insulation, tested during the first year, is ITT's Poly/Kynar. Samples of the WTE 1932A (20 gauge) and WTE 1930A (18 gauge) wires were tested and rated as Class 8. Again the heat sink effect of the wire may appear to make the insulation look more nonflammable than it actually is.

We strongly recommend the use of Teflon insulated wire for all chamber wiring. This Class 8 material is commonly being used on wiring for high temperature applications and is readily available on all ordinary sizes and types of wiring. We have used Teflon insulated hook up wiring (Mil-W-16878/D type E, 18 gauge) as leads to the igniter wire and thermocouples during our full scale tests. We have never observed it to sustain combustion during any of the tests although it has flamed and melted off the wire when exposed to direct flames. It is an absolute requirement that additional electrical and mechanical insulation be provided when Teflon insulated wire is used. The abrasion resistance of Teflon is low and the insulation might well wear through when subjected to continuous chafing. The probable cause of the fatal fire in the space environment simulator at Brooks AFB on 31 January 1967 was a short circuit occurring when one of the occupants stepped on an abraded Teflon wire and pressed it against the metal floor [8]. Each Teflon wire should be placed inside a second Teflon or fiberglas sheath and the bundle of doubly insulated wires should be contained inside a flexible metal armor or solid conduit. This scheme should provide ample protection against short circuits and fire.

An excellent choice for all permanent hard wiring in the chamber is General Cable's Mineral Insulated Cable. The conducting wires, imbedded in a mineral insulation, are enclosed in a copper tube. The initial installation effort is greater than for normal wiring due to the care that must be exercised during bending of the cable and sealing of the ends. Moisture penetration may also be a problem and certain techniques have to be employed for its elimination. Once the MI cable is in place, however, it is absolutely noncombustible and virtually impervious to accidental damage.

SUMMARY

Studies of the effects of environmental parameters upon combustion in hyperbaric environments and evaluations of selected materials for flammability in oxygen-enriched atmospheres have been conducted. Test results show that the burning rate of filter paper strips is substantially accelerated as the total pressure of an oxygen-inert gas mixture is increased to pressures to 10 atm abs. Rates are also increased as the oxygen content in the mixture is increased. A comparison of nitrogen and helium as oxygen diluents shows that the combustion process generally proceeds faster in the helium environment, all other factors being equal.

Tentative recommendations for some of the more fire safe materials as well as fabricated end-item products that may have use in hyperbaric chambers are as follows:

Fabrics
- (a) Beta Fiberglas—Owens Corning
- (b) Teflon—DuPont

Elastomers
- (a) Refset—Raybestos—Manhattan
- (b) Fluorel—3M Company
- (c) CNR—Thiokol

Electrical Insulation
- (a) Kapton—DuPont
- (b) Teflon—DuPont

Diver Clothing
- (a) Beta fiberglas coveralls—B. Welson

This work has been supported by Contract N00014-66-CO149 with the Office of Naval Research, Washington, D. C.

REFERENCES

1. G. A. Cook, V. A. Dorr and B. M. Shields, "Region of Noncombustion in Nitrogen-Oxygen and Helium-Oxygen Atmospheres," I and EC Process Design and Development *7,* 308 (1968).
2. G. A. Cook, R. E. Meierer and B. M. Shields, "Combustibility Tests on Several Flame-Resistant Fabrics in Compressed Air, Oxygen-Enriched Air, and Pure Oxygen," Textile Research Journal *7,* 591 (1967).
3. G. A. Cook, *et al,* "Effects of Gas Composition on Burning Rates Inside Decompression Chambers at Pressures up to 300 Feet of Seawater," Under-Ocean Technology (Proceedings of the 54th Annual Meeting of the Compressed Gas Association, January 17, 1967).

4. G. A. Cook, R. E. Meierer and B. M. Shields, "Screening of Flame-Resistant Materials and Comparison of Helium with Nitrogen for Use in Diving Atmospheres," Defense Documentation Center No. AD-651583 (March 31, 1967).
5. V. A. Dorr and H. R. Schreiner, "Region of Noncombustion, Flammability Limits of Hydrogen-Oxygen Mixtures, Full-Scale Combustion and Extinguishing Tests and Screening of Flame-Resistant Materials," Defense Documentation No. AD-689545 (May 1, 1969).
6. J. V. Harter, "A Review of the Navy Chamber Safety Program" Proceedings of Fire Hazards and Extinguishment Conference, Defense Documentation Center No. AMD-TR-67-2.
7. J. V. Harter, "Fire at High Pressure" Proceedings of the Third Symposium on Underwater Physiology, The Williams and Wilkins Company, Baltimore, 1967.
8. A. G. Swan, "Two Man Space Environment Simulator Accident," Proceedings of Fire Hazards and Extinguishment Conference, Defense Documentation Center No. AMD-TR-67-2.
9. Report of Apollo 204 Review Board to the Administrator, National Aeronautics and Space Administration.
10. L. Segal, *et al*, "Fire Suppression in Hyperbaric Chambers," Fire Journal, May 1966, p. 87.

Victor A. Dorr

Victor A. Dorr is a development chemist in the Research Department of Ocean Systems, Inc. at Tarrytown, New York. He received his B.A. degree in Chemistry from Harpur College of the State University of New York at Binghamton. He worked for Union Carbide Corporation at Tonawanda, New York, before joining Ocean Systems, Inc. in 1968. His experience includes polymer synthesis and development, automotive lubricants, chemical and instrumental analysis, and combustion safety in diving atmospheres.

APPENDIX A

CLASSIFICATION OF MATERIALS ACCORDING TO SCALE OF FIRE RESISTANCE

Class 0

Burns readily in air at atmospheric pressure

Material	Name	Manufacturer
Fabric	Cotton and cotton terrycloth	—
Fabric	Orlon, Style 20543F	Uniroyal
Fabric	Wondershield coated with Fasslon G	J. P. Stevens
Fabric	Ucar Electrically Conductive Fabric Style 9955A (conductive carbon yarn in cotton fabric)	Union Carbide
Fabric	Ucar Electrically Conductive Fabric Style 1103 (conductive carbon yarn in olefin fabric)	Union Carbide
Yarn	Tungsten textile yarn	Union Carbide
Elastomer	Natural Rubber	—
Paint	Heat-resistant aluminum primer	Federal Stock No. 8010-815-2692

Class 1

Has an appreciably higher ignition temperature and/or burns at an appreciably lower rate in air at 1 atm abs pressure than cotton cloth or paper.

Material	Name	Manufacturer
Fabric	Wool	—
Fabric	X-101	Monsanto Chemical
Paint	White cover paint (enamel)	Federal Stock No. 8010-577-4738

Class 2

Nonflammable or self-extinguishing in air at one atm abs pressure.

Material	Name	Manufacturer
Fabric	Roxel [cotton treated with tetrakis (hydroxymethyl) phosphonium chloride (THPC)]	Hooker Chemical
Fabric	Asbeston S/5670	Uniroyal
Fabric	Asbeston S/6555 (asbestos/polyester composite)	Uniroyal
Fabric	Plasticon	Continental Hospital Industries
Fabric	Silastic 2316 coated on 128 glass	Dow Corning
Elastomer	Neoprene	Du Pont
Elastomer	Special Rubber No. 59356	B. F. Goodrich
Elastomer	Silastic S-23416U	Dow Corning
Elastomer	SE-9029 and SE-9044	General Electric
Foam	Experimental urethane foam No. 584RD69	Union Carbide

Class 3

Self-extinguishing or burns slowly in air at a pressure of 100 fsw (4.03 atm abs).

Material	Name	Manufacturer
Fabric	Nomex	Du Pont
Fabric	Fairprene 85-001 (Viton coated fiberglass)	Du Pont
Fabric	Fairprene 72-020 (Hypalon coated nylon)	Du Pont
Fabric	Flameproof Jean cloth	Capital Cubicle
Fabric	Verel V-304	Eastman
Fabric	Brattice cloth	American Brattice Cloth
Fabric	Cotton sateen treated with tris (1-aziridinyl) phosphine oxide (APO)	Dow Chemical
Fabric	50% Nomex/50% THPC treated cotton	Hooker Chemical
Fabric	Green cotton No. 60	Wheeler Protective Apparel
Fabric	Submarine flameproofed mattress cover	Supplied by U. S. Navy Experimental Diving Unit
Elastomer	Fairprene 70-001 (Hypalon)	Du Pont
Felt	Nomex batt	Du Pont

Class 4

Self-extinguishing or burns slowly in air at a pressure of 200 fsw (7.06 atm abs).

Material	Name	Manufacturer
Fabric	Fairprene 5710 (neoprene coated nylon)	Du Pont
Fabric	Fairprene 88-003 (butyl coated Nomex)	Du Pont
Fabric	Dynel	Union Carbide
Fabric	Wool treated with THPC	Hooker Chemical
Fabric	Fire-Retardant Wool Blanket	Mine Safety Appliances
Fabric	Fabric F	David Clark Inc.
Plastic	Hetron 92	Hooker Chemical

Class 5

Self-extinguishing or burns slowly in a mixture of 25% oxygen and 75% nitrogen at a pressure of 1 atm abs.

Material	Name	Manufacturer
Fabric	Silverprened Fiberglass No. 35	Wheeler Protective Apparel
Fabric	Asbeston Cloth (17% flameproof cotton)	Wheeler Protective Apparel
Fabric	Cotton Khaki No. 61	Wheeler Protective Apparel
Fabric	Rovana, patterns 5700, 5800 and 6400	Dow Badische
Fabric	Wondershield	J. P. Stevens
Fabric	Dynel Tent Material	Uniroyal
Fabric	Dynel Blanket	Union Carbide
Fabric	Ucar Electrically Conductive cloth style 8020 (conductive carbon yarn in fiberglass fabric)	Union Carbide
Fabric	CP-2198 (Aluminum coated asbestos)	Johns-Manville
Fabric	Terry Knit Nomex	Du Pont
Fabric	Panel No. 221-158-1 and 2 (hypalon coated fiberglas)	Du Pont
Film	Kapton film type 100H	Du Pont

Class 6

Self-extinguishing or burns slowly in a mixture of 30% oxygen and 70% nitrogen at a pressure of 1 atm abs.

Material	Name	Manufacturer
Fabric	Asbeston S/1445 (asbestos/dynel composite)	Uniroyal
Fabric	Asbeston S/3620-IAR	Uniroyal

Fabric	Navy brattice cloth	American Brattice Cloth
Fabric	X-400	Monsanto Chemical
Fabric	X-410	Monsanto Chemical
Elastomer	Fairprene 80-080 (Viton)	Du Pont

Class 7

Self-extinguishing or burns slowly in a mixture of 40% oxygen and 60% nitrogen at a pressure of 1 atm abs.

Material	Name	Manufacturer
Fabric	PBI (polybenzimidazole)	Celanese Corporation under contract to U. S. Air Force.
Fabric	XF-91 coated fiberglas	Reliable Rubber Products
Fabric	Armalon P.S. 6439 and 63339AS (Teflon coated fiberglass)	Du Pont
Fabric	Armalon 98-101 (Teflon coated Nomex)	Du Pont
Elastomer	XF-91	Reliable Rubber Products
Film	Kapton film types 300H and 200F919	Du Pont

Class 8

Self-extinguishing or burns slowly in a mixture of 50% oxygen and 50% nitrogen at a pressure of 1 atm abs.

Material	Name	Manufacturer
Fabric	Teflon	Du Pont
Fabric	Teflon coated Beta Fiberglas	Supplied by B. Welson
Fabric	Pluton H-1 and B-1	3M
Fabric	Ucar Electrically Conductive Fabric Style 9966 (conductive carbon yarn in fiberglass fabric)	Union Carbide
Fabric	Thornel (carbon fabric)	Union Carbide
Fabric	Avceram CS	FMC
Fabric	100% Rhovyl Knit* (Polyvinylchloride Cloth)	John J. Ryan
Elastomer	Fluorel 1071	Mosites Rubber
Electrical Insulation	Type WTE electric wire insulation	ITT
Plastic	Hetron 92TG + Sb_2O_3	Hooker Chemical

* Conditional classification because the fabric shrinks away from flame and is self-extinguishing. The material is not intrinsically flame resistant.

Class 9

Nonflammable in 100% oxygen at a pressure of 1 atm abs.

Material	Name	Manufacturer
Fabric	Beta Fiberglas	Owens-Corning
Fabric	Beta Fiberglas	Supplied by B. Welson
Fabric	"Super Beta" (Glass fibers coated with Teflon before weaving)	Supplied by B. Welson
Fabric	Boron Nitride Cloth	Union Carbide
Fabric	Zirconia Cloth	Union Carbide
Fabric	D67504 and D67505 series (asbestos/fiberglass composites)	Uniroyal
Fabric, felt, board and rope	Fibergrax Ceramic Fiber (many styles)*	Carborundum
Webbing	Beta Fiberglas webbing	Carolina Narrow Fabrics
Elastomers	CNR Nitroso Rubber	Thiokol
Elastomers	Refset	Raybestos-Manhattan
Paper	Nonflammable Paper (Fiberglass/asbestos composite)	Dynatech
Electrical insulation	Kapton electric wire insulation type 019-919	Du Pont

* Several types must be heat treated (800°F for several hours) to remove organic residues before qualifying for Class 9.

Flammability Measurements and Thermal Decomposition of Textiles

J. E. Hendrix, T. K. Anderson,
T. J. Clayton, E. S. Olson, and
R. H. Barker

Clemson University
Clemson, South Carolina 29631

(Received March 16, 1970)

INTRODUCTION

When considering flammability, it must be remembered that virtually all materials are flammable in certain atmospheres and conditions. Even metals and ceramics flame when exposed to extreme conditions. Therefore, flammability becomes a relative, rather than an absolute, matter. There is no such thing as a fireproof fabric,but there are different degrees of fire resistance.

A survey of the technical literature related to flame-retardant compounds has led to the generalization that "most or all flame-retardant finishes reported to be effective consist of insolubilized organophosphorous polymers containing nitrogen" [1]. Such generalizations have led to common acceptance of the theory that nitrogen exerts a synergistic effect on the flame-retardant properties of phosphorous-containing compounds.

HISTORICAL

Application of flame retardants to textile materials had its beginning about 1640 when theater canvas was treated with clay and plaster of Paris. By 1740, alum was found to give textiles flame-retardant properties, as was ammonium phosphate by 1786. Gay Lussac completed the first systematic investigation of the use of flame retardants in 1820. From this investigation he concluded that the most effective salts were those which were either low melting and capable of covering the material with a glassy layer or those which would decompose into nonflammable vapors on heating [2]. Since Gay Lussac's investigation, extensive literature pertaining to flame retardants has become available. Unfortunately, much of this information is of more practical than theoretical value, and consequently it is of little application to a description of fundamental mechanisms of pyrolytic degradation and flame retardation. In recent years, with the use of chromatographic and thermal analytical techniques to the study of thermal decomposition, fundamental data have been collected.

Four mechanistic theories pertaining to flame retardation and two pertaining to reduction of afterglow received widespread acceptance by 1947 [3]. Strong alkali, mineral acids, and oxidizing agents are effective flame retardants for cellulose but degrade the fabric substrate. Some derivatives of these substances do not degrade the textile and produce flame-retardant substances when decomposed at or near flaming temperatures. According to chemical theories, the flame-retardant substance acts as a catalyst by promoting degradation of the substrate in a direction other than that taken by untreated cellulose. Substances which produce strong alkali at high temperatures include the alkali metal salts of weak acids. None of the neutral salts of strong acids and bases decompose to a large extent at flaming temperature; thus, to produce strong acids the acid salt of a weak base is preferred. Ammonium salts of strong acids are among those most easily decomposed. Magnesium and zinc chlorides have low thermal stability and usually exhibit pronounced flame-retardant properties. Easily reducible oxides have proven to be effective in both the hydroxide and hydrated forms [3].

The coating theory proposed by Gay Lussac is still widely used to explain the flame-retardant properties of low-melting inorganic salts. It is postulated that upon contact with a flame, a glassy film of fused salt is formed on the surface of the fabric. The film excludes oxygen from the system thus preventing propagation of the combustion reaction. Compounds which melt with the evolution of gases to form stable foams also coat the substrate surface. The flame retardation by a mechanism similar to that of the low-melting salts is accomplished.

Certain materials decompose at elevated temperatures yielding inert or difficultly oxidizable gases. Supporters of the gas theories hypothesize that application of such compounds to textile materials decreases flammability by either diluting the flammable gases produced during pyrolysis or by blanketing the substrate with an inert atmosphere. Thus, exposure of the substrate to an oxidizing atmosphere should be reduced or eliminated. Such changes in the fuel to air ratio are presumed to raise the ignition temperature of the substrate, thereby retarding flaming. However, if the quantity of inert gas produced by the retardant is compared with that of the gaseous products resulting from the thermal decomposition of cellulose, none of the known flame retardants produces inert gases in quantities substantially greater than those of volatile gases produced by cellulose [4]. Despite this fact, the theory has been widely accepted [3].

Supporters of the thermal theories of flame-retardant mechanisms hypothesize that effective retardants should be capable of maintaining fabric temperatures below the minimum combustion temperature for cellulose. According to such theories, the retardant must either be able to dissipate large quantities of energy internally or possess an ability to conduct this energy away from the flame front at a rate comparable to or greater than that at which it is supplied. Numerous compounds undergo energy changes during pyrolysis that are comparable to those of known retardants, but are completely ineffective. It has been shown by differential thermal analysis (DTA) that the number and magnitude of endothermic changes in no way correlate with the effectiveness of a compound in flame-retardant applications [3].

Afterglow retardation in cellulose substrates is directly related to interference with normal high-temperature oxidation of charred surfaces resulting from lower-temperature pyrolysis. Many metallic oxides exert catalytic effects on the oxidation of carbon which result in lower ignition temperatures and increased burning rate [5]. Metallic oxides appear to enhance afterglow in the case of cellulose, presumably by a similar catalytic path. Advocates of the chemical theory of glow-retardance postulate that if the catalytic influence altered the course of oxidation to a less exothermic path, producing carbon monoxide rather than carbon dioxide, the heat thus liberated should be insufficient to sustain combustion after the igniting flame was removed. Advocates of physical theories have been unable to support their hypotheses with experimental data [3].

No well organized effort has successfully standarized a definition of terms describing flammability in textiles. The following definitions, however, have been accepted through general usage [6]:

Flame-retardant textile: refers to a fabric which does not support combustion after the igniting flame has been removed. Charring of the areas in contact with the flame is expected. A fabric may be classified as flame-retardant by a particular test method but not so when measured by other tests. Therefore, a more precise definition should include the test method employed.

Flame-resistant textile: a term which can be used interchangeably with flame-retardant textile.

Flame-proof textile: a term which has no place in the study of flame-retardant cellulosic fabrics. It should be applied only to those fabrics which undergo no change when exposed to any flame.

Afterglow: the flameless combustion of a fabric after the igniting flame has been removed. It may occur only in the charred area, or may consume the entire specimen.

Nondurable flame-retardant finish: refers to a chemical system which imparts to textiles flame-retardancy but which requires a replacement after each laundering of the textile.

Semidurable flame-retardant finish: refers to a system which imparts to textiles flame-retardancy lasting through a few mild washings.

Durable flame-retardant finish: refers to a chemical system which retains its ability to impart flame-retardancy through repeated launderings.

Many test methods have been designed to determine the flammability characteristics of textile materials. The majority of these test methods do not generate data clearly defining the properties being tested. Difficulty in reproducing data with and between laboratories is also inherent when these tests are used. Despite these drawbacks, and until improved methods are developed, current test methods must of necessity continue to find application in defining fabric flammability.

One of the most widely used flammability tests is to ignite a fabric sample with a lighted match and observe its burning characteristics. This simple match test is obviously inapplicable to evaluate flame retardants properly, but it does offer a quick check on flammability properties.

Test methods used most frequently as the basis for evaluation of flammability characteristics of cellulose are AATCC 34-1966, Fire Resistance of textile Fabrics and AATCC 33-1962, Flammability of Clothing Textiles [7].

AATCC 34-1966, the vertical char test, is probably the most widely accepted test available. Samples being tested are exposed to a conditioned atmosphere for a period of eight hours and are then suspended over a propane flame for twelve seconds in a draftfree cabinet. Times of afterflame and afterglow, plus char length, are recorded as the basis for evaluations. This method, due to the severity of the test, is used primarily for fabrics known to be flame-resistant. It is generally reserved for fabrics weighing at least six ounces per square yard [6] because lighter-weight flammable fabrics often appear nonflammable by this method [7].

AATCC 33-1962 is designed to indicate which textiles ignite easily, and which, when once ignited, tend to burn with sufficient rapidity and intensity as to be hazardous when used as clothing. Samples to be tested are dried, desiccated, and mounted at forty-five degrees with the horizontal in a ventilated chamber. Flame from a butane-charged microburner is directed on the lower portion of the sample for a period of one second and the time required for the flame to travel up the fabric for a distance of five inches and sever the stop-cord is recorded as a measure of the intensity of burning. Observations of flaming characteristics and examination of the charred sample may provide further pertinent information [7].

To answer the need for improved flammability tests, efforts have recently been directed toward designing test methods by which the flammability properties of textiles may be characterized through numerical designations. One such method involves mounting the sample on a substrate of ashless filter paper and then igniting the substrate. A flammability index, based on the minimum number of layers of filter paper necessary to consume the sample during burning of the substrate, provides a measure of the ease of ignition [8]. Another method has recently been developed by the General Electric Company in an attempt to determine the relative flammability of materials through precise and reproducible measurement of the minimum concentration of oxygen required to support combustion of the sample [9, 10].

Flammability testing, by itself, offers little insight into either the mechanisms by which cellulose pyrolytically degrades, or into the manner in which these mechanisms are altered through the action of flame-retardant substances. To determine these mechanisms, investigators must have data to include not only the relative flammabilities but also the thermodynamics and kinetics of the degradation and the qualitative and quantitative information concerning the degradation products. To obtain this additional information, recent investigators have applied thermal gravimetric analysis (TGA) [11, 12, 13], differential thermal analysis (DTA) [12, 13] and differential scanning calorimetry (DSC) [13] to the study of the pyrolytic degradation of cellulose. Determination of the intermediate and final products of pyrolysis or combustion has been made possible through use of infrared spectroscopy [12, 14, 15], paper chromatography [16], gas chromatography [17, 18], and mass spectrometry [19].

Very little has been published with regard to the thermal decomposition of fibrous polymers other than cellulose. Early investigations of the pyrolysis of cellulose dealt with the relative amounts of gases, chars, and tars produced, and with observations of how flammability varied with the relative proportions of these fractions. These studies established that the production of large proportions of tars favors high flammability [3, 20].

As early as 1918, Pictet and Sarasin [21] isolated 1-6-anhydro-β-D-glucopyranose (levoglucosan) from the products resulting from the destructive distillation of cellulose in vacuo. In addition to levoglucosan pyrolysis of cellulose yields mainly carbon monoxide, carbon dioxide, methane, hydrogen, acetic acid, formic acid, and various ketones, phenols, and furans. Schwenker and Pacsu [16, 22], using gas chromatography, obtained chromatograms showing the presence of at least thirty-seven different volatile products. They were able to identify about half of these by retention volume data and functional group analysis. Essentially the same products resulted whether or not the environment was an oxidizing or a nonoxidizing atmosphere, thereby indicating that the dominant mechanism is nonoxidative in nature. Higgins [15], however, concluded from infrared spectroscopic data that the degradation was oxidative in nature. The data indicated that carbonyl absorption increased with time of heating cellulose in air.

In 1921, Irvine and Oldham [23] proposed that the first step in pyrolysis was hydrolysis to glucose, followed by dehydration of the glucose to form levoglucosan. Essentially this same mechanism was proposed by Heuser [24] in 1944.

Tamaru [20], in 1951, proposed that the mechanism of cellulose pyrolysis involved depolymerization of the macromolecules to a low molecular weight species. He contended that some of the species underwent dehydration and polymerization with continued and increased heating to yield a high polymeric material, while the remainder of the species underwent further decomposition. The polymeric material was believed to be degassed by heat with resulting aromatization and evolution of carbon monoxide, carbon dioxide, methane, and hydrogen. Esteve and coworkers [25] proposed a similar mechanism to that of Tamaru. They believed that depolymerization of cellulose to levoglucosan through scission of 1, 4-glycosidic linkages followed by internal isomerization to the glucosan, was the first and rate-determining step in pyrolytic degradations. It was thought that part of the levoglucosan polymerized and aromatized to form chars while the remaining levoglucosan underwent destructive distillation to form flammable tars, volatile gases, and water. Ivanov et al. [26], in 1956, independently proposed a similar mechanism by which cellulose pyrolytically degraded in vacuo through splitting of the 1,4-glycosidic linkages with the formation of levoglucosan.

A free radical mechanism was suggested by Martin [27] as a possibility for the pyrolytic degradation of cellulose. Day and coworkers [28] also proposed a homolytic mechanism involving stabilization of the intermediate free radical by delocalization but were unable to find any evidence to support their theory [29]. Esteve [25] calculated the energy of activation for the pyrolysis of cellulose to be substantially lower than that required for a free radical process.

Madorsky et al. [14], in 1956, proposed a mechanism based upon random dehydration along the cellulose chain occurring simultaneously with scission to produce levoglucosan, water, carbon monoxide, and carbon dioxide. He based this theory on the fact that C-O bonds are thermally less stable than C-C bonds. In 1960, Madorsky revised his theory to involve thermal scission of the C-O linkages between rings, followed by transfer of a hydrogen from the primary alcohol to the oxygen of the broken glycosidic linkage and formation of an oxygen linkage between C-1 and C-6. He deduced that when a similar scission took place at a point one structural unit removed from that of the original scission, levoglucosan would be formed. Scissions at points more than one unit removed from the original should result in nonvolatile multiple-unit chain fragments [30].

Schwenker and Pacsu [16, 18] proposed that if, in the pyrolysis of cellulose, a first order depolymerization to levoglucosan was the first and rate-determining step, then modifying cellulose at the C-6 position should minimize levoglucosan formation and decrease flammability. In support of their proposal, they oxidized approximately half of the primary hydroxyls of a cellulose sample and found that when pyrolyzed it yielded only half as much levoglucosan as did the untreated sample. The sample also exhibited reduced flammability.

In 1957, Golova et al. [31] investigated the effects of the packing density of cellulose chains on the pyrolytic decomposition of cellulose in a vacuum. They found normal cellulose to give approximately 60% levoglucosan, while mercerized cellulose yielded only 38%, hydrocelluose from cuprammonium solution about 15%, cellulose hydrate from unoriented viscose staple about 4%, and cellulose hydrate from oriented viscose silk about 5%. This indicates that orientation did not significantly increase packing density of the chains. With this information they concluded that a change in the density of the cellulose chains caused a change in the courses of the chemical processes occurring during pyrolytic decomposition. Decreased packing density was believed to cause a decrease in levoglucosan formation, while increased packing density hindered the side processes of hydrolysis and dehydration, thus increasing levoglucosan formation.

Golova proceeded to investigate Irvin's and Oldham's theory [23] which considered levoglucosan to be formed through an intermediate stage in which cellulose was hydrolyzed to glucose and then dehydrated. Golova found that pyrolyzed cellulose gave a levoglucosan yield of 60% under vacuum while β-D-glucose gave only 5-6%. Decreasing the degree of polymerization (D.P.) of cellulose to 150 did not significantly affect the yield of levoglucosan; yet cellobiose was found to yield 3-4% levoglucosan. From this data, Golova et al. [33] concluded that the glucose intermediate theory was incorrect. They further proposed that in regions of high packing density, levoglucosan formation is favored and side reactions are secondary. In regions of different physical structure, dehydration and hydrolysis are the primary reactions with the production of levoglucosan less favored. Based on this information, Golova and coworkers proposed a theory, similar to that proposed by Madorsky [31], in which levoglucosan was thought to be formed through scission of the 1, 4-glycosidic linkages followed by hydrogen abstraction and isomerization to levoglucosan. They further concluded that there is some limiting D.P. below which levoglucosan formation is not favored.

In an effort to gain further support of their cellulose pyrolysis theory, Golova et al. [33] prepared cellulose samples through mild hydrolysis which represented a range of D.P. They thermally degraded the samples at constant temperature for successively increasing time increments and found that all of the samples rapidly degraded to a limiting D.P. of about 200 which remained unchanged during further decomposition. They then concluded that the mechanism of pyrolysis involved initial breakdown to a D.P. of 200, followed by complete depolymerization of the chains through internal isomerization yielding levoglucosan. Golova [34] later showed that levoglucosan production was low during the initial stages of pyrolysis and that a large amount of volatile products was produced. Later, during pyrolysis, the yield of levoglucosan increased and remained constant as degradation continued.

In reporting their work, Golova and coworkers did not speculate about the reason for the apparent stability of the cellulose molecule having a 200 D.P., nor did their results establish whether this number represented an average over a broad spectrum or whether a 200 D.P. was actually a critical size for the cellulose chains. They did, however, surmize the latter. From their observations, it might be further hypothesized that the 200 D.P. represents the average length of the crystalline regions, where the chain packing density would be greatest. Thus, one might expect the initial breakdown, to be that of the amorphous areas where side reactions would be favored. Breakdown of the amorphous areas could be followed by complete depolymerization of the chains within the crystalline areas, where levoglucosan formation would be favored. However, it is of some interest to speculate at this point on the seeming inconsistency of these conclusions with the theory of flame propagation by levoglucosan and the known ability of some phosphorus compounds to act as flame retardants. Since it seems highly unlikely that these reagents penetrate the crystalline regions of the cellulose fibers, it is difficult to see how they can be effective in inhibiting flaming by reducing levoglucosan formation. For these conclusions to be compatible, it would seem to be necessary to propose that phosphorus-containing flame retardants exert their effect by acting on the levoglucosan after it is formed and escapes from the inaccessible regions of the polymer matrix. This reasoning would be consistent with the recent demonstration by Shafizadeh [46] that $ZnCl_2$, acting as a Lewis acid, catalyzes the dehydration of both cellulose and levoglucosan under pyrolytic conditions.

Honeyman [35[has investigated cellulose pyrolysis in an effort to gain some information applicable to the improvement of flame retardant treatments for cellulose. He pointed out that, in order to form levoglucosan from the glucosan units of the cellulose chain, a thermodynamically unfavored interconversion from the all equatorial to the all axial conformation must occur. Chatterjee and Conrad [36] and several other groups]39, 43] have followed the kinetics of cotton cellulose pyrolysis and interpreted their results in terms of Honeyman's conclusions. The initial stages of the reaction, which Tang and Neill found to follow pseudo zero-order kinetics, was quite complex, as would be expected from the random scission of glycosidic linkages in the amorphous areas. The latter stages of the decomposition conformed to a first order rate equation which was interpreted as evidence to support Honeyman's postulated chain unzipping with concurrent formation of levoglucosan. Using this data, along with the postulate by Kilzer and Broidie [47] that cellulose pyrolysis consists of competing dehydration (in the temperature region 200-380° C), and tar formation (in the temperature region 280-340° C). Mack and Donaldson studied the effect of added inorganic bases in attempts to reduce levoglucosan formation. By catalyzing the decomposition in this manner, they were able to markedly affect the thermal behavior of the cellulose and increase the percentage of residual char obtained. Their interpertation of these results was that by catalyzing the degradation so that it would occur at a temperature below the minimum required for the conformational interconversion necessary for levoglucosan formation, tar formation and flaming should be reduced. This is basically the type of approach which most workers in the area of flame retardant cellulose have taken in recent years.

The majority of the investigations of cellulose pyrolysis and its relation to flaming have been carried out near or below the cellulose combustion temperature. Holmes and Shaw [37] investigated the pyrolysis of different cottons in vacuum and dry air mainly at 418° C. Berkwitz-Mattuck and Noguchi [38] investigated the effect of one-second exposure of cellulose to high flux levels of 5-25 cal./cm.^{2}-sec. Both investigators concluded that the same mechanism observed in lower-temperature pyrolysis operated in these investigations. Reductions in tar formation with increased char yields were reported to be the resultant effects of flame retardants in these investigations. This was also the case in lower-temperature degradations.

The mechanism of cellulose pyrolysis after the formation of levoglucosan still remains vague. Berkowitz-Mattuck and Noguchi [38] proposed that fragmentation of levoglucosan through a free radical mechanism accounted for the observed gases, aldehydes, ketones, and unsaturated products. This mechanism, although logical, has not been substantiated.

Tang and Neill [39] used TGA and DTA in their studies concerning the effect of flame retardants on the pyrolysis and combustion of α-cellulose. Their TGA studies showed that flame retardants lowered the threshold temperature for pyrolysis and raised the char yield of the products. DTA studies in oxygen showed that, although flaming was reduced, flame retardants often extended glowing. They explained that the heat of combustion is concentrated in a narrow temperature range and is redistributed by flame retardants over a wider temperature range and at a lower maximum intensity. Results of depolymerization of cellulose and of polymerization and aromatization of pyrolysis products gave reduced heat of pyrolysis.

In their most recent work, Tang and coworkers [40] postulated four major types of reactions occuring in the pyrolysis from observations related to their DTA studies on wood cellulose. At 220° C, cellulose was observed to undergo a decomposition involving loss of water to yield a partially dehydrated species. Loss of water was confirmed by mass spectrometric thermal analysis. Depolymerization of the dehydrated species and volatilization of the resulting product was reported to begin about 280° C. In this process, previously untreated cellulose broke into smaller fragments which volatilize via an endothermic process competitive with the dehydration. The remaining reactions involve decomposition of the dehydrated cellulose via exothermic reactions into a number of gaseous products and residual char, and condensation of simpler substances from the decomposition products. As in the previously discussed investigations, effective flame retardants were reported to minimize the depolymerization, producing water and char.

.Perkins and coworkers [2] used DTA and TGA for their investigations of the decomposition of cellulose fabrics treated with various flame retardants. They discovered that less than ½% of selected flame retardants in the cellulosic material was frequently sufficient to alter the course of reaction from endothermic to exothermic. Additionally, they found that decomposition under nitrogen was exothermic rather than endothermic. TGA was found useful in explaining endotherms related to volatilization of degradation products, as was infrared spectroscopic data in characterizing resulting char.

Numerous compounds containing both phosphorus and nitrogen have been successful in retarding flaming in cellulose. To date the most effective permanent flame retardant treatments for cotton cellulose are those in which phosphorus and nitrogen-containing monomers are polymerized within the cellulose fiber matrix. The use of tris (1-aziridinyl-phosphine oxide) (APO) has been studied extensively since it is effective in producing both flame resistance and chemical crosslinking in a single treatment. Thus it offers the obvious advantage of imparting durable press characteristics at the same time as flame retardance. However, APO treatments have not been widely accepted by the industry due to their high cost and toxicity. More commonly, the industry has gone to the use of tetrakis (hydroxmethyl) phosphonium chloride (THPC) or, just recently, tetrakis (hydroxmethyl) phosphonium hydroxide (THPOH). The exact chemical nature of the latter compound is still somewhat in doubt. Both THPC and THPOH are usually copolymerized with such nitrogen-containing compounds as urea and melamine resins. The resulting fabrics show the expected pyrolytic behavior when examined by DTA and TGA [12] and generally give good flame resistance. However, they present the concurrent disadvantages of requiring high add-ons, frequently severe fabric tendering, and poor aesthetic qualities. The properties of these finishes have been reviewed widely [44]. Jones and Noone [41] have investigated the permanency of the flame retardant properties of cellulose treated with various phosphorus and nitrogen compounds and then at a later time subjected to alkaline laundering conditions. They studied compounds containing phosphorous bound to nitrogen and carbon. The compounds containing P-N bonds were found to be effective in reducing or eliminating flaming but were unsuitable for use as permanent treatments, while the compounds containing P-C bonds were much more promising as permanent flame retardants. The polar nature of the phosphorous-nitrogen bond encourages nucleophillic attack at the phosphorous atom and, therefore, allows its hydrolysis and removal from the fabric during washing. Such an effect can be minimized by strongly electron-donating groups substituted on the nitrogen atom.

Tesoro [42] has recently developed a durable flame-retardant finish for cellulose based on the *in situ* insolubilization of a polyphosphonate having a high nitrogen content. In this work and in a later investigation of the flame-retardancy of cellulose reacted with N-methylolacetamides [1], she observed that increasing the nitrogen-to-phosphorous ratio in the compound with constant phosphorous add-on resulted in improved flame-retardancy. She described this phenomenon as resulting from a synergistic effect of nitrogen on the flame-retardant effectiveness of the organophosphorous compounds. Since the origin of such an effect is not immediately apparent, it is being intensively investigated in this and other laboratories.

A similar synergisitic effect was recently proposed for halogen-containing compounds [45] but this has not been generally accepted. It is thought that since many halogen compounds are flame retardants themselves, their effect when applied in conjunction with phosphorus compounds is probably simply additive rather than synergistic. This would seem particularly logical since the flame retardant ability of the halogens is thought to be attributable to their ability to scavenge free radicals. Although the reasons for the ability of halogens to act as flame retardants has, at times, been ascribed to their ability to act as free radical chain inhibitors. At least one group has recently presented evidence which they interpret as showing that the pyrolytic liberation of bromine free radicals catalyzes the decomposition of the cellulose below the ignition point of the liberated gases [12]. However, considerably more work will be required before this point is fully clarified.

EXPERIMENTAL

The basic data came from tests using a vertical flame tester, muffle furnace, General Electric Candle Tester, Du Pont 950 Thermogravimetric and Du Pont 900 Differential Thermal Analyzers.

Vertical Flame Testing

The vertical flame tester was used to determine the extent of burning or charring of the test samples by AATCC Test Method 34-1966 [7].

Muffle Furnace Tests

A standard Hevi-Duty, Type 62 muffle furnace (Basic Products Corp.,

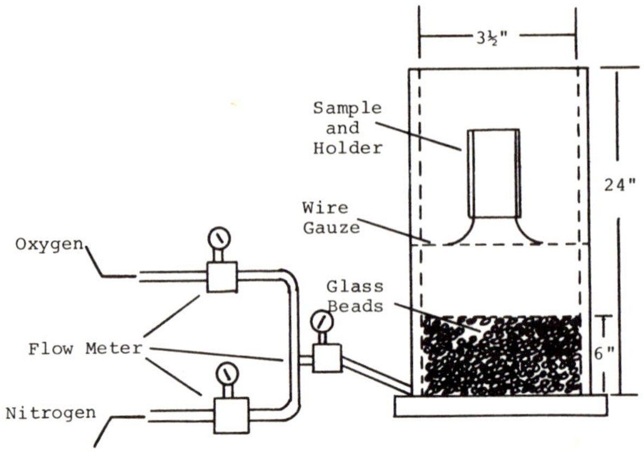

Figure 1. *Candle Tester for Oxygen Demand.*

Watertown, Wisc.) equipped with a glass opening for sample viewing was used in these tests. The tests were conducted by exposing untreated and treated cotton fabrics in 300°C and 400°C atmospheres. A two-inch square fabric sample was placed on a specially constructed copper wire stand, and then the assemblage was rapidly placed in the furnace. A stop watch was started as the furnace door was being closed and the sample was viewed for 30 seconds, or until flaming began, whichever was the shorter time.

Experiments indicated that when samples were not pyrolyzed within 30 seconds, using a given temperature, they could be regarded flame resistant. On the other hand, if the sample ignited before 30 seconds of exposure, the time to ignition was recorded.

Oxygen Demand Determination

The candle tester [9, 10] was used to determine the relative flammability of selected textiles. This test produces negligible convective heating, because the flame burns downward from the top of the sample. The sample is not preheated by hot vapors released during the combustion of the sample and the flame geometry is simple. The data produced by these tests give the minimum amount of oxygen required to support complete combustion of the sample. The reported data resulted from the conversion of volume flow rates to mass flow rates and then conversion to oxygen demand using the formula:

$$\text{Oxygen Demand} = \frac{O_2}{O_2 + N_2} \times 100.$$

The top of a 2.5-inch vertical sample of fabric was ignited using a 0.25-inch hydrogen flame mounted in the apparatus shown in Figure 1. The hydrogen flame was kept in contact with the sample until combustion had taken place. The nitrogen/oxygen atmosphere was adjusted so that flaming, sufficient to consume the sample, was low but constant. Tests were repeated and oxygen flow rates adjusted until the replication of results was achieved using minimum oxygen.

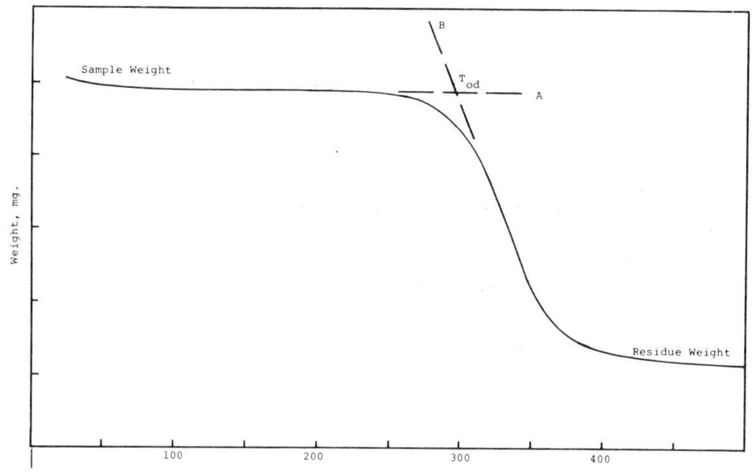

Figure 2. *TGA curve illustrating a method for the determination of the extrapolated decomposition temperature (T_{od}). The intersection of tangents A and B Is the T_{od}.*

Thermogravimetric Analysis

The samples were prepared by being pulverized in a Wiley Mill. The measurements were made in a dynamic nitrogen atmosphere using a Du Pont 950 Thermogravimetric Analyzer with a constant heating rate of 5°C per minute and a scale sensitivity of 4 mg per inch. The data collected were the initial sample weight, weight of residue after decomposition, extrapolated onset decomposition temperature (T_{od}) and completed decomposition temperature (T_d).

The initial sample weight and residue weights were used to determine percentage residue by the following formula:

$$\% \text{ Residue} = \frac{\text{Residue (mg)}}{\text{Sample (mg)}} \times 100$$

The extrapolated onset decomposition temperature (T_{od}) was determined by extending the temperature curve a prior to decomposition, and the decomposition curve b until they intersected, Figure 2. The temperature at the intersection of a and b is the onset decomposition temperature.

Completed decomposition temperature (T_d) was then determined by extending the residue curve c so that it intersected with the major decomposition curve b.

Differential Thermal Analysis

The fabric specimens were pulverized in the same manner as for the TGA. A Du Pont 900 Differential Thermal Analyzer was used to obtain the thermal data and was equipped with 2-mm bore capillary tubes using glass as the reference material. The heating rate, scale sensitivity, slope and other necessary data are included in the appropriate figures.

Table 1—Identification of Test Fabrics Shown in Table 6.

Sample	#1 —	Cotton Printcloth
	#2 —	Vat Tan Gabardine
	#3 —	" " "
	#4 —	Soft Filled Sheeting Napped
	#5 —	" " " "
	#6 —	" " " "
	#7 —	White Printcloth
	#8 —	Green Printcloth
	#9 —	Soft Filled Sheeting Napped
	#10 —	" " " "
	#11 —	" " " "
	#12 —	" " " "
	#13 —	" " " "
	#14 —	" " " "
	#15 —	Printcloth—Printed
	#16 —	" "
	#17 —	Soft Filled Sheeting Napped—Printed
	#18 —	" " " " "
	#19 —	" " " " "
	#20 —	" " " " "
	#21 —	Purfied cellulose—CP
	#22 —	Type 1 Army Twill Prepared for Dyeing but not Dyed
	#23 —	" " " "
	#24 —	Type 1 Army Twill
	#25 —	" " " "
	#26 —	" " " "
	#27 —	" " " "
	#28 —	" " " "
	#29 —	Type 1 Army Twill Vat Tan Dyed
	#30 —	" " " " " " "
	#31 —	" " " " " " "
	#32 —	" " " " " " "
	#33 —	" " " " " " "
	#34 —	100% Wool—No Finish
	#35 —	100% Wool + 15% H_3PO_4
	#36 —	Nomex
	#37 —	100% Cotton
	#38 —	" "
	#39 —	" "
	#40 —	" "
	#41 —	" "

Fabric Data

Laboratory and mill production samples were used. The sulfur dyed samples were different mill dye lots. These samples were later tested as a comparison of the untreated and treated samples using a polyethylene emulsion. The tan vat-dyed samples were taken from a mill dye lot and subsequently treated with different percentages of THPOH. Two other fabrics were selected, Nomex and wool flannel, to represent the class of fabrics which are believed to be flame resistant in the untreated state.

Supplemental data for all specimens tested are shown in Table 1.

DISCUSSION OF RESULTS

A basis for the understanding of the mechanism whereby phosphorus-containing flame retardants act on cellulose has recently been laid [48] by the study of several series of model flame retardants. From these investigations it can be concluded that at least two mechanisms are operative in the solid phase and at least one in the vapor phase. The vapor phase reaction is apparently a free radical termination of the oxidation process itself. In the solid phase, both a phosphorylation process and a dehydration and rearrangement seem to be operative. Both of these result in lower levoglucosan production and thus a decrease in the fuel supply for the flame. Since these processes are competing with the chain unzipping of cellulose to form levoglucosan, it is necessary to catalyze the desired reactions to decrease the energy of activation so that the process proceeds more rapidly at a lower temperature.

The lowering in temperature of decomposition is particularly significant and easily followed by thermalanalysis. This can be demonstrated in the simplest case by the treatment of cotton fabric with phosphoric acid or diammonium phosphate. These two compounds are extremely effective in imparting self-extinguishing characteristics to the fabric. Differential thermalanalysis (DTA) and thermogravimetric analysis (TGA) of the treated fabrics show that the decomposition occurs well below 300°C, whereas, untreated cellulose decomposes in the 350°-400°C region. Thus, the decomposition temperature of the treated cellulose seems to be a simple and rather effective criterion for establishing the effectiveness of a particular flame retardant treatment. As a corollary to this, increased residues are almost inevitably found when the flame retardant treatments are effective. This is to be expected since the flame retardant acts in the solid phase to decrease the amount of gaseous products formed and to increase, therefore, the remaining unvolatilized carbonaceous material.

An example of a situation in which this mode of investigation was extremely successful is the investigation of the chemical nature of the phosphorus-nitrogen synergistic effect [49]. Treatment of cellulose with nitrogenous bases, such as urea or guanidine carbonate, has been found to result in very little change in the temperature of decomposition and residue formation. Similarly, treatment with triphenyl phosphate, a poor flame retardant, shows only a minor alteration in the energistics of decomposition. However, when the phosphate is mixed with one of the nitrogenous bases a significant decrease is observed in the decomposition temperature as determined by thermogravimetric analysis. Greatly increased residues are also seen. This effect is more pronounced when the more nucleophilic nitrogenous bases are used. This has been explained in terms of the formation of P-N bonds during the heating of the treated fabric. Those systems containing P-N bonds are apparently very effective flame retardants for cellulose.

A correlation between these studies on the mechanism of thermal decomposition of the cellulose with the actual burning characteristics can be es-

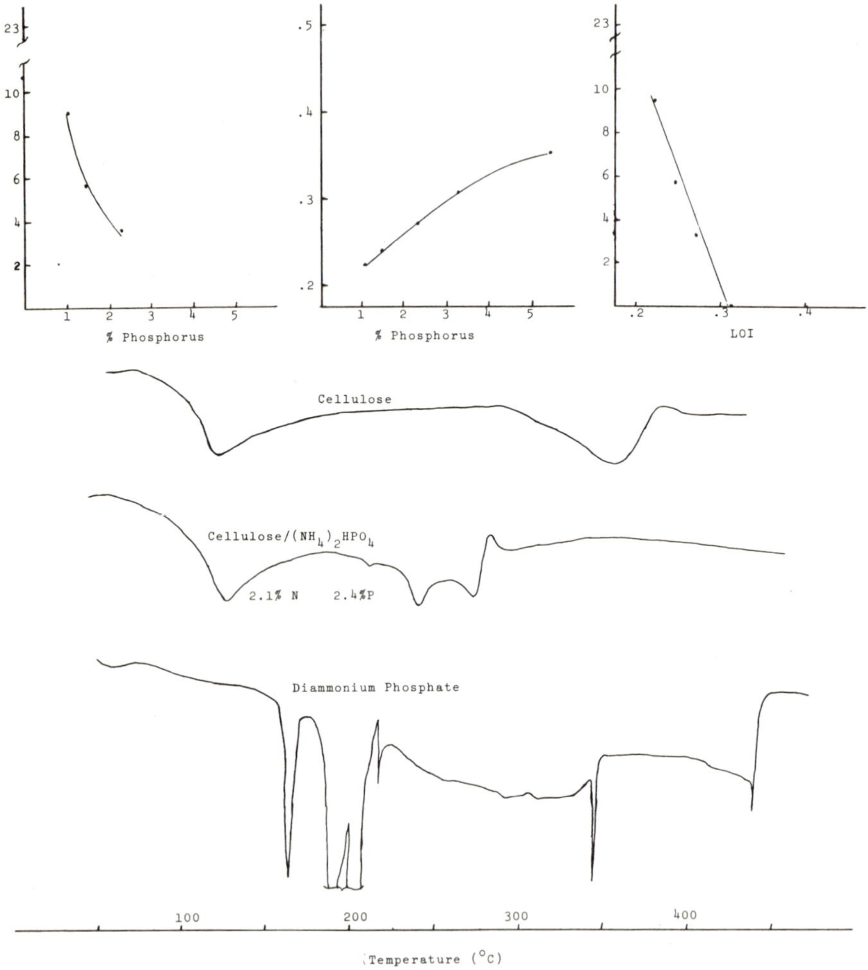

Figure 3. *DTA thermograms of $(NH_4)_2HPO_4$ applied to a cotton fabric and purified cellulose. The Rate of Burning (inches per minute) and LOI are compared with each other and as a function of percent phosphorus add-on.*

tablished [48]. Fabric treated with diammonium phosphate exhibits rates of combustion typical of most simple flame retardants acting in the solid phase. As can be seen in Figure 3, the rate of combustion decreases very rapidly up to approximately 2% phosphorus. Beyond this, the incremental decreases in burning rate with increases in phosphorus contents become considerably smaller. This type of burning rate curve is similar to that which is observed in studies of the heat of combustion or the heat of pyroylsis of treated fabrics as determined by differential scanning calorimetry [50, 51]. The thermal decomposition of cellulose is known to follow a first order rate law and should produce this type of curve. The heat and burning rates both tend to approximate a linear relationship with the logarithm of the phosphorus content. On the other hand, LOI values vary linearly with the phosphorus content itself. Perhaps of

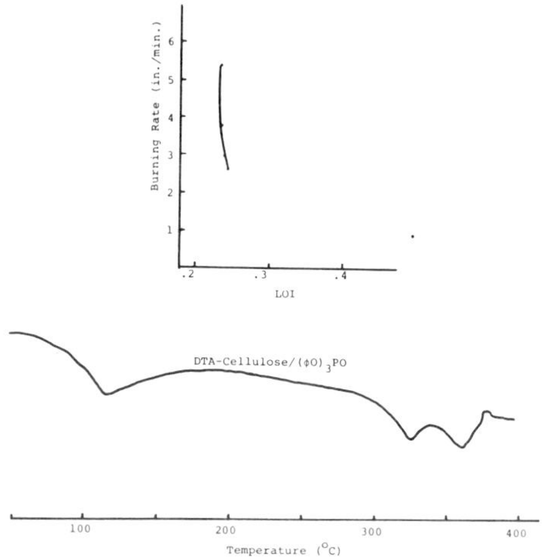

Figure 4. *DTA thermogram of triphenyl phosphate applied to a cotton fabric. The Rate of Burning (inches per minute) is shown as a function of LOI.*

greater significance is the relationship between the rate of burning and the LOI values which is observed to be essentially linear for fabrics treated with flame retardants which operate by a single simple mechanism. This type of linear relationship between burning rates and LOI values seems to be a rather good criterion for establishing that a single mechanism is operative.

The use of burning rates and LOI values in conjunction with the thermal data has lead to confirmation of the conclusions reached previously regarding the nature of the phosphorus-nitrogen synergistic effects. Using triphenyl phosphate alone, the DTA curves indicate only a transesterification and no catalysis of any dehydration or rearrangement processes. As expected, a plot of burning rate *vs* LOI is essentially linear. When urea is added to the system along with triphenyl phosphate and cellulose, another linear relationship between burning rate and LOI is found, as seen in Figure 4. Similarly the DTA curve indicates some catalysis of the solid phase reaction and the possibility of a slight exothermic process occuring above 300°C. The linearity of the burning rate *vs* LOI curve shows that this high temperature exothermic reaction does not contribute significantly to the flame retardant process. But in the case of guanidine carbonate, a contribution of the higher temperature exothermic process can be seen in Figure 5. With low nitrogen add-on, linearity is observed between LOI and burning rate. However, as the nitrogen content increases the DTA shows much more pronounced exothermic activity and the burning rate *vs* LOI curve begins to deviate from linearity indicating that a second mechanism is now operative. This is essentially what would be expected on the basis of previously published information on mechanisms of decomposition [52]. The polymer formed from guanidine carbonate and triphenyl phosphate should operate by essentially a dehydration-rearrangement mechanism. However, the excess guanidine carbonate, being considerably more basic than urea, should act by a mechanism similar to that proposed by Mack and Donaldson [13] for the catalyzed degradation of cellulose at high temperatures. This second process becomes important only at very high nitrogen contents and with very basic nitrogenous systems.

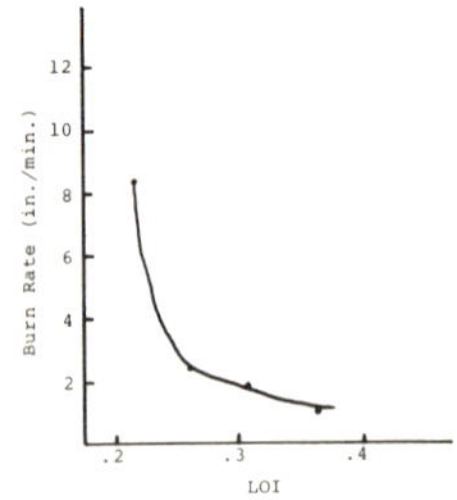

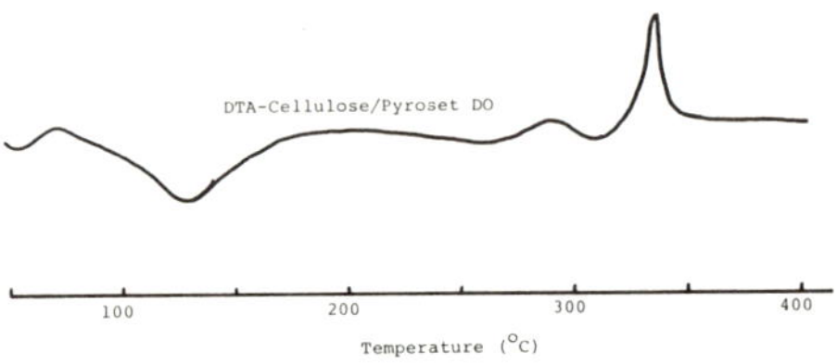

Figure 5. *DTA thermogram of Pyroset DO applied to a cotton fabric. The Rate of Burning (inches per minute) is shown as a function of LOI.*

When this type of treatment is extended to commercial flame retardants similar effects are very often seen. In the case of fabric treated with Pyroset DO, the DTA curve, Figure 6, indicates both the endothermic activity typical of the acid catalyzed phosphorylation-dehydration-rearrangement process and the base catalyzed degradation which is seen as a large exotherm above 300°C. Two separate mechanisms are active in the substrate as indicated by the marked curvature in the plot of burning rate *vs* LOI. Further, it can be demonstrated that the curvature is a function of the dual mechanisms rather than an artifact from the base catalyzed process. Fabric treated with Arko AS demonstrates this, Figure 7. In this system, the DTA curve exhibits a large exotherm centered about 350°C which is typical of the base catalyzed decomposition reaction. However, a plot of burning rate *vs* LOI exhibits close adherence to linearity, indicating a single mechanism and demonstrating the validity of this type of approach to base catalyzed degradation reactions.

THPC treated fabrics show further the usefulness of this approach. The DTA curve is quite uninformative. The burning rate *vs* LOI plot, however, is nearly linear indicating a simple mechanism which, in this particular system, must involve the phosphorus catalyzed dehydration-rearrangement reaction. Thus, the use of the burning rate *vs* LOI plots is of particular value when the DTA curves fail to give the complete picture of the degradation.

Other dramatic portrayals are found in cases involving competing or complementary solid phase and vapor phase reactions. The DTA curves of fabrics treated with Flamort U, Figure 8, exhibit a single sharp endotherm with a nadir at about 250°C indicating that the solid phase decomposition reaction has been catalyzed. Apparently this is not the only mode of reaction exhibited by this compound. The burning rate *vs* LOI curve shows marked deviation from linearity indicating that two mechanisms are invalid. Since the DTA curve shows little evidence of other solid phase reaction it is probable that this material is active in the vapor and possibly acts as a free radical

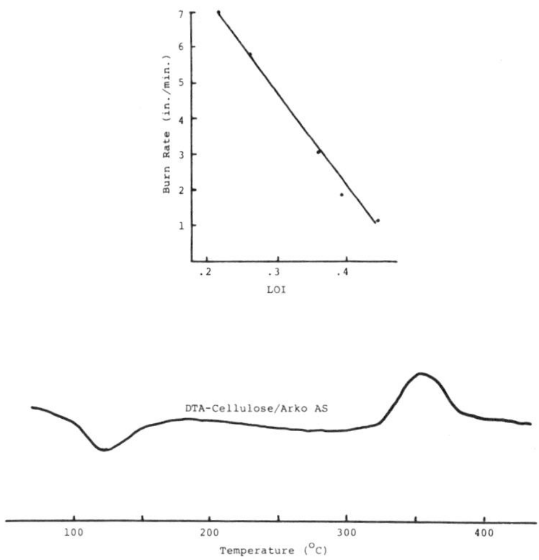

Figure 6. *DTA thermogram of Arko AS applied to a cotton fabric. The Rate of Burning (inches per minute) is shown as a function of LOI.*

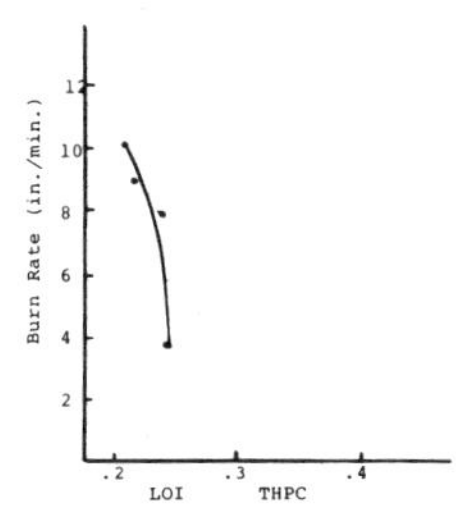

Figure 7. *The Rate of Burning (inches per minute) of a cotton fabric treated with THPC is compared with LOI.*

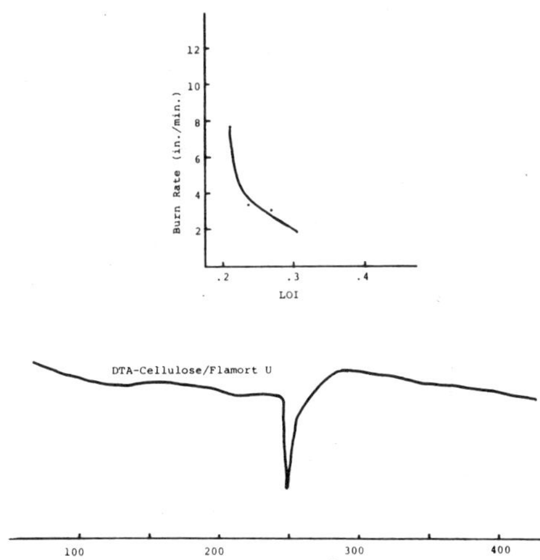

Figure 8. *DTA thermogram of Flamort U applied to a cotton fabric. The Rate of Burning (inches per minute) is shown as a function of LOI.*

terminator. Chemical analysis of Flamort U confirms this by indicating the presence of halogen, probably in the form of bromide. Bromides are known to act as free radical terminators in the vapor phase.

It is apparent that many commercial flame retardants exert their effect by more than one mechanism. The predominant mechanism for non-thermoplastic polymers, however, appears to be by direct action on the decomposition reaction of the substrate. This solid phase reaction should be more amenable to mechanistic investigations and correlation with actual flammability tests. Therefore, the primary effort has been concentrated on a study of the solid phase degradation reactions.

The correlation between the extrapolated onset temperature for thermal decomposition and the ability of many of these systems to act as flame retardants has been demonstrated. Therefore, some evaluation of the flame retardant activity of various samples should be possible by a simple investigation of the charring reaction in a muffle furnace.

Muffle Furnace Tests

Conventional test methods, such as the vertical and 45° flame tests, usually indicate that a true flame retardant finish becomes increasingly difficult to ignite with successive additions of the finish. At some concentration it is usually found that tendencies to flame are quenched. Tests indicated, however, that precisely the opposite effect can occur when elevated temperatures are used in a muffle furnace.

The initial tests using the muffle furnace were to determine a satisfactory set of two temperatures. The lower temperature should not ignite the untreated cotton sample, but the higher temperature should produce ignition after 20-30 seconds exposure. Successive replications showed that when the untreated samples were exposed to a temperature of 300°C the samples were scorched. When exposed at 400°C, pyrolysis occurred after 20 seconds. The results from these tests are shown in Table 2.

The muffle furnace table of results shows that when a temperature of 300°C was used, no sample, either treated or untreated, flamed within the 30-second test. However, sample appearances and apparent strengths were all changed, except for the Nomex. Only the untreated cotton showed slight scorching and good strength retention. All THPOH samples became black, while those treated with Pyrovatex were browned. Oddly enough, the 10% Pyrovatex-treated sample was lightest in color, while the 25% and 30% treated samples were darkest. In addition, the strength characteristics of the treated cotton samples were, in all cases, drastically lowered.

When the 400°C exposure tests were applied, Nomex was once again found to be unchanged in both appearance and strength. All cotton samples, treated and untreated, ignited and were either totally destroyed, or were without significant strength when the fabric structures remained.

Table 2 also shows that ignition times changed when untreated and treated fabrics ignited and were either totally destroyed, or were without significant strength when the fabric structures remained.

Table 2 also shows that ignition times changed when untreated and treated cotton samples were tested at 400°C in the muffle furnace. The cotton twill 30% THPOH treated fabric required only 7.5 seconds exposure time to ignite, while the 15% THPOH treated sample required 10 seconds. However, the untreated cotton twill control required 20 seconds. On the other hand, the 30% Pyrovatex treated sample required only 9 seconds to ignite. The other samples

Table 2. Muffle Furnace Test Results When Selected Untreated and Flame Retardant Finished Fabrics Were Exposed at 300°C and 400°C Temperatures.

Sample	Ignition Time In Seconds 300°C	Ignition Time In Seconds 400°C	Sample Appearance 300°C	Sample Appearance 400°C	Apparent Sample Strength 300°C	Apparent Sample Strength 400°C
Dyed Cotton Twill						
Untreated	—	20	L.S.	T.D.	Good	T.D.
30% THPOH	—	7.5	Black	Black	None	None
25% THPOH	—	8.0	Black	Black	None	None
20% THPOH	—	9.0	Black	Black	None	None
15% THPOH	—	10.0	Black	Black	Slight	None
THPC	—	20.0	Tan	Black	Fair	None
White Napped Cotton						
Untreated	—	18.0	L.S.	T.D.	Good	T.D.
30% P	—	9.0	Dk. Brown	Black	Little	None
25% P	—	—	Dk. Brown	Black	Little	None
20% P	—	—	Brown	Black	Weak	None
15% P	—	—	Brown	Black	Weak	None
10% P	—	11.0	Dk. Tan	Black	Weak	None
Resin and Catalyst	—	11.0	Brown	T.D.	Weak	T.D.
Nomex	—	—	U.C.	U.C.	U.C.	U.C.

* L.S. = Light Scorching
T.D. = Total Destruction
U.C. = Unchanged

smouldered, turned black and lost all strength characteristics. Because flaming did not occur, the ignition times were not recorded.

The results produced by the muffle furnace tests cast some doubt on the ability of a flame retardant finish (FRF) to retard burning in a heated environment such as in a burning room, or while a fabric is exposed to a hot stove or similar situations. This observation may be explained by the fact that while a FRF lowers the quantity of combustible gases that can be produced, it does not produce combustible or self-supporting flammable gas mixtures at room temperature, when attempts are made to ignite it with a flame. When the same fabric sample is placed in a heated environment, the decomposition products may readily burn because of the higher temperature.

The following hypothetical cases, in which A represents a fabric treated with FRF and B an untreated reference fabric, may help to explain the point.

Consider what happens when A is exposed to a flame at room temperature. It will decompose into approximately 40% carbon and other nonburnable inorganics, 40% CO_2, H_2O and other nonburnable gases and 20% burnable gases. However, because of the low temperature and dilution by other gases, no combustion results.

If A is exposed to a temperature of 400°C, it will produce the same decomposition products. However, it will produce a combustible mixture in this environment more quickly than the untreated fabric. While the total amount of combustible products may be lower with A than with B, they are produced more quickly with A because the finish catalyzes the decomposition of the fabric.

B, at both room temperature and at 400°C, will produce approximately 10% carbon and nonburnable inorganics, 20% CO_2, H_2O and nonburnable gases and 70% burnable gases. The burnable gases from B at 400°C are produced more slowly than those from A at the same temperature. This is shown by the quicker ignition of A at 400°C. Clearly this result means that one has to be very careful when defining flame retardance. Some obvious factors which may influence flame retardance and make it difficult to measure are:

—Chemical composition of fabrics, dyes, finishes and blends. Each of these factors changes a fabric's ease of burning and the temperature at which spontaneous flaming occurs.

—Construction of the fabric. This may determine a fabric's ease of burning, because of accessibility of heat to the fiber surface or presence of a combustible atmosphere.

—End use. The final use of a fabric, that is, the temperatures to which it may be subjected, will influence its flammability.

This test indicates that no single test will effectively define a fabric's resistance to flaming. The results of any single test would be difficult to correlate with end use.

DTA, TGA And Oxygen Demand

Thermal analysis techniques can determine decomposition temperatures and residues during the combustion of a given sample. Every cotton sample analyzed shows that the decomposition temperature of cotton is lowered when an adulterant is present in the fabric. The adulterant may be a softener, dyestuff, flame resistant finish or residues not removed during the preparation process. Differential Thermal Analysis (DTA) studies show that not only is the temperature differential important, but that the mode of decomposition of cellulose is changed when resistance to flaming is achieved. Figure 9 illustrates the results obtained when three different cotton samples were studied by DTA. The lowest point of the 100% cotton endotherm is at 360°C, while the lowest point of the dyed cotton endotherm has been shifted to 320°C because of the effects of sulfur dye. However, the analysis indicates that these two modes of decomposition remain essentially the same. On the other hand, analysis of the flame retardant finished sample indicates the presence of two exotherms, one minor and one major, at 313°C and 267°C, respectively. It should also be noted that the evidence for the presence of cellulose is totally changed during pyrolysis because the entire mode of decomposition has been altered.

Thermogravimetric Analyzer (TGA) data support the change of temperature relationship and provides the residue data which is of importance. Figure 10 shows the progression of residues and temperatures when five different fabrics were analyzed, and Table 3 provides the consolidated data for these same fabrics.

These results show that the addition of products lowers the onset decomposition temperature and that the percentage residue varies with residual materials. A major residue of the sulfur dyed fabric is associated with a temperature deflection of 65 degrees below that of pure cotton cellulose, even

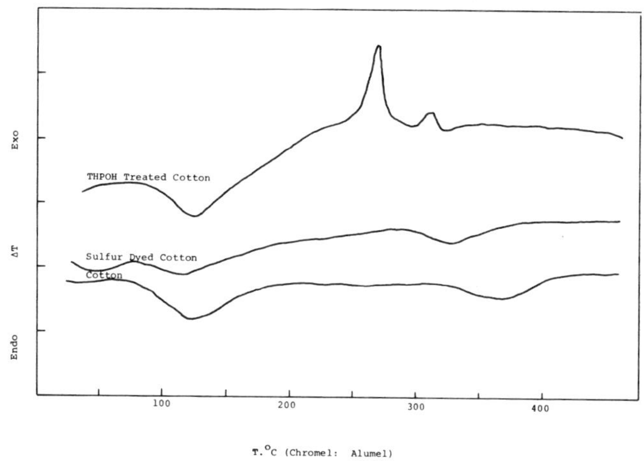

Figure 9. *DTA comparing THPOH treated cotton and sulfur dyed cotton fabrics with bleached cotton in an air atmosphere using a 2 mm capillary tube, glass reference, and a 20°C/minute heating rate starting at 25°C. Base line slope is +0.2 and ΔT is 0.5°/division.*

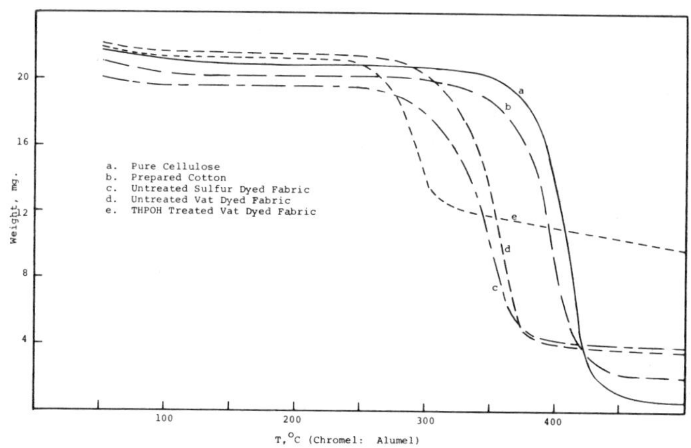

Figure 10. *TGA of pure cellulose, prepared cotton fabric, sulfur dyed cotton fabric, vat dyed cotton fabric and THPOH treated vat dyed fabric using ½ liter/minute nitrogen atmosphere, 0 shift, 5°C/minute heating rate and 4 mg/division vertical scale.*

Table 3. Onset Decomposition Temperatures, Percent Residues and Oxygen Demand Data for Selected Cotton Fabrics.

Fabric	T_{od} °C	%R	O.D.
Pure Cotton Cellulose	380	0.93	—
Prepared Cotton Fabric	366	9.18	20.27
Untreated Sulfur Dye Control	315	16.51	19.67
Untreated Vat Dye Control	320	13.90	21.17
THPOH Vat Dye (30%)	275	43.46	41.69

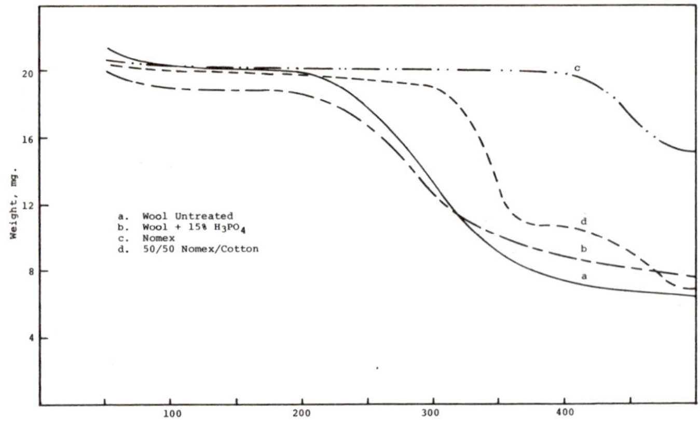

Figure 11. ***TGA** of untreated wool fabric, wool treated with 15% phosphoric acid, Nomex and 50/50 Nomex/Cotton mixture using ½ liter/minute nitrogen atmosphere, 5°C/minute heating rate and 4 mg/division vertical scale.*

Table 4. Onset Decomposition Temperatures, Percent Residues and Oxygen Demand Data for Wool and Nomex Fabrics.

Fabric	T_{od} °C	%R	O.D.
Untreated Wool	244	29.82	25.53
Wool + 15% H_3PO_4	241	39.30	47.05
Nomex	420	71.90	36.36
50/50 Nomex/Cotton	427/320	37.14	—

though the mode of decomposition is unchanged. However, when the mode of decomposition is changed, as with the THPOH treated samples, the onset temperature is deflected from 380°C to 275°C. Oxygen Demand data show that only the THPOH treated sample required atmospheres of high oxygen to support combustion.

The decomposition mode associated with temperature change does not seem to be a criteria for other fibers, such as wool and Nomex. Both of these fibers are considered to be fire resistant without further treatment. Figure 11 and Table 4 show that the T_{od} for untreated wool was 244°C, while that of the wool which had been treated with 15% H_3PO_4 was 241°C. The residues are important, because the residues of the wool samples are 29.82% and 39.30% respectively. Oxygen Demand data for both wool samples show that an increased quantity of oxygen was required to support the combustion of the treated wool. This agrees with the residue increase hypothesis. Untreated wool required more oxygen than was contained in the ambient atmosphere for complete combustion. It should be noted that the 29.82% residue is well above that deposited by cotton, unless the cotton had been treated to limit its flammability.

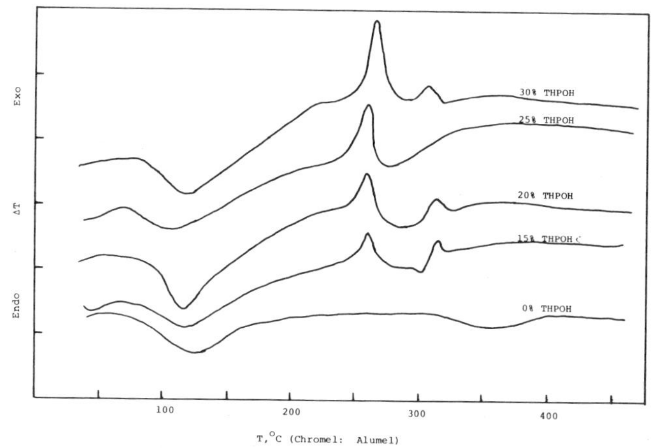

Figure 12. *DTA of cotton fabrics treated with different percents of THPOH thermalized in an air atmosphere, 4mm glass capillary, glass reference, 20°C/minute heat rate and +0.2 base line slope. ΔT is 0.5°C/division.*

Table 5. Onset Decomposition Temperatures, Percent Residues and Oxygen Demand Test Results when THPOH was Applied to THPOH Treated Fabrics.

% THPOH	T_{od} °C	%R	O.D.
0	320	13.90	21.19
15	272	35.85	25.13
20	278	37.56	25.53
25	275	41.59	26.51
30	274	43.46	41.69

Tan vat dyed cotton twill samples, commercially treated with 0, 15, 20, 25 and 30% THPOH, were subjected to DTA, TGA and Oxygen Demand tests. The DTA results in Figure 12 show the change in mode of decomposition between cotton and the four THPOH treated samples, plus the step-wise transition change of temperature in the exothermic reaction recorded in the 270°C range. Table 5 shows the influence of increasing THPOH concentration on T_{od}, percentage residue and oxygen demand. Because sufficient flame retardant had been applied to alter the thermal degradation at or below the 15% concentration, no further temperature change is noted. Of particular interest are the increases in percentage residues, which parallel the increases in the concentration of THPOH. The oxygen demand data show that the untreated vat dyed fabric required a slight increase of oxygen over ambient atmosphere conditions. The 15-25% additions produced only slight increases in oxygen demand commensurate with increasing concentration. However, an apparent threshold value was achieved when 30% THPOH was applied.

The muffle furnace tests can be related to the DTA and TGA data shown in Figures 9, 12, 13 and 14. All THPOH and Pyrovatex treated cotton fabrics charred or darkened in a shorter period of time than did the untreated con-

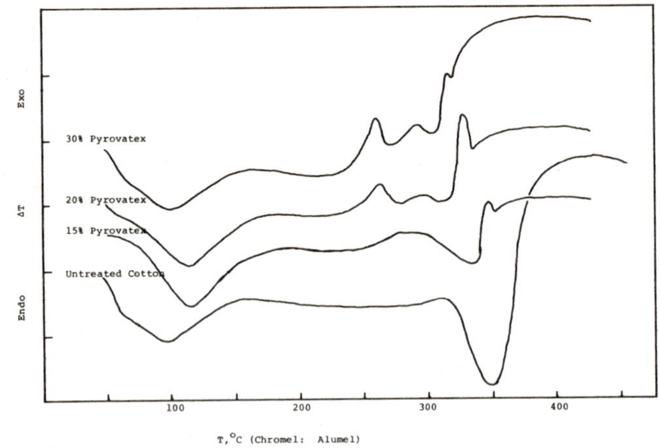

Figure 13. *DTA comparing untreated bleached cotton with different percents of Pyrovatex treated fabric when thermalized in an air atmosphere, 4 mm glass capillary, glass reference and 20°C heating rate. The base line slope is +0.2 and 0.5°C ΔT/ division.*

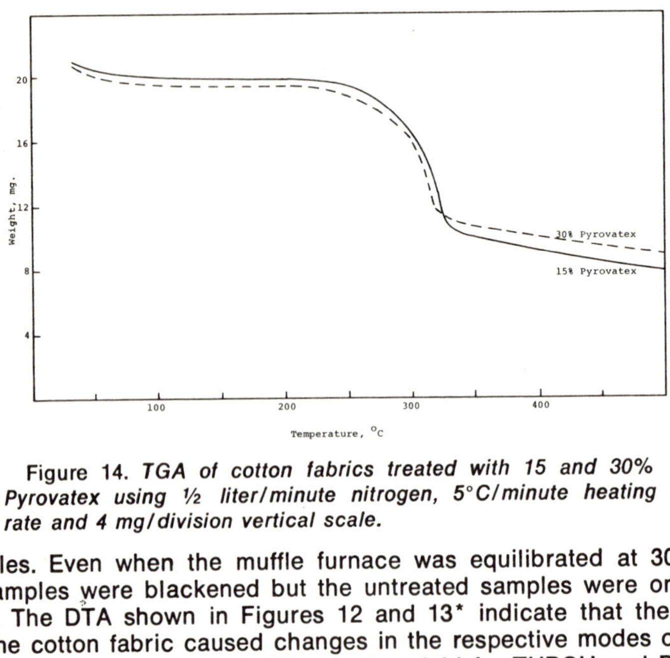

Figure 14. *TGA of cotton fabrics treated with 15 and 30% Pyrovatex using ½ liter/minute nitrogen, 5°C/minute heating rate and 4 mg/division vertical scale.*

trol samples. Even when the muffle furnace was equilibrated at 300°C, the treated samples were blackened but the untreated samples were only lightly scorched. The DTA shown in Figures 12 and 13* indicate that the FRF applied to the cotton fabric caused changes in the respective modes of decomposition. The TGA data shown in Figures 9 and 14 for THPOH and Pyrovatex, respectively, effectively present the evidence for the change in onset of decomposition temperature. All figures, therefore, substantiate the muffle furnace finding, that the FRF treated fabrics decompose more quickly and at lower temperatures than do the untreated samples. Untreated cotton had a

* Sample weights used for this part of the study were approximately four times greater than those used for the other DTA studies, therefore changes in the magnitude of endothermic action of cellulose occurred.

Table 6—Oxygen Demand, TGA and Vertical Flame Test Data.

	Sample	OD	T_{do} C	% Residue	Inches Burned Vertical Flame Test	$\sqrt{\frac{R^3}{T}}$
1	Cotton, Prepared	20.27	366	9.18	∞	1.10
2	THPC Treated	43.24	279	41.28	3.5	11.31
3	THPC Washed	41.66	278	40.35	3.5	10.89
4	9396 Print Treated	29.99	280	38.73	4.9	10.23
5	" Print 50 washes	27.58	283	38.42	4.6	10.10
6	THPOH-NH_3	39.50	282	42.93	2.8	11.97
7	9283 White	24.61	262	36.53	6.5	9.56
8	" Green	27.58	284	40.38	∞	10.91
9	9282 Flannel Outing	25.53	284	40.00	5.4	10.73
10	" Flannel Outing	27.58	263	38.87	∞	10.48
11	7763 Print Treated	26.51	277	42.08	4.2	11.64
12	" " 50 Washes	26.51	282	41.91	4.0	11.27
13	" " Treated	25.53	286	41.33	3.6	11.28
14	" " 50 Washes	26.51	280	41.74	4.5	11.46
15	9047 Flannel Outing Treated	27.58	284	39.29	5.6	10.45
16	" " 50 Washes	23.76	282	36.07	∞	9.21
17	Plaid Treated—9A	34.89	293	38.10	4.5	9.89
18	" 50 Washes	32.21	292	35.53	3.8	8.91
19	" Treated—4A	38.08	290	38.42	4.4	10.03
20	" 50 Washes	32.21	292	35.90	4.8	9.05
21	Purified Cellulose	—	380	.93	—	.04
22	Sul. Dye Control	19.67	315	16.51	∞	2.77
23	#1 Sul. Dye	21.52	308	20.36	∞	3.81
24	#1 Sul. Dye + P/E	22.22	310	17.12	∞	2.93
25	#2 Sul. Dye	21.52	306	14.85	∞	2.38
26	#2 Sul. Dye + P/E	22.22	302	19.23	∞	3.52
27	#3 Sul. Dye	21.52	311	19.30	∞	3.51
28	#3 Sul. Dye + P/E	22.22	304	13.11	∞	1.98
29	Tan Control	21.19	320	13.90	∞	2.13
30	15% THPOH	25.13	272	35.85	4.3	9.23
31	20% THPOH	25.53	278	37.56	3.4	9.86
32	25% THPOH	26.51	275	41.59	3.7	11.47
33	30% THPOH	41.69	275	43.46	4.1	12.31
34	Wool Control	25.53	244	29.82	3.4	7.17
35	Wool + 15% H_3PO_4	47.05	241	39.30	3.0	10.86
36	Nomex	36.36	420	71.90	None	23.16
37	5% Pyrovatex	22.9	295	29.6	∞	6.68
38	15% Pyrovatex		293	36.5	4.5	9.27
39	20% Pyrovatex		283	40.9	4.0	11.09
40	25% Pyrovatex		289	43.1	3.8	11.93
41	30% Pyrovatex		292	42.0	4.0	11.58

T_{od} of 365-380°C while the FRF treated samples had T_{od} of 278°C and 292°C. Therefore, these FRF samples should decompose at temperatures slightly less than 300°C, while the untreated samples are relatively untouched. However, the 400°C muffle furnace testing should cause combustion of all samples.

From the above results and observations it is readily apparent that some relationship between T_{od}, residue and oxygen demand should exist. Additional tests were made and compared with the vertical flame test results made on each sample. These results are shown in Table 6.

The oxygen demand data show that three cotton fabrics required more than 40% oxygen, five required more than 30%, 4 required more than 25% and 11 required less than 25%. The oxygen demands of the printed samples were in the mid to high twenties, while the four plaid samples, 9A and 4A, were in the thirties. The flannel outing samples were all in the mid-range twenty percentages. Surface effects apparently contributed to the combustibility of these latter samples during the vertical flame and oxygen demand testing. The T_{od}s were lowered appropriately and residues remained high even after washing.

All residues of the flame resistant finishes were greater than 30%. When the samples were washed, residues decreased slightly except for the 7763-38C sample. The decreases were attributed to the lack of fixation of the fire resistant finish and its subsequent removal by washing.

The sulfur dyed samples, after-treated with a polyethylene softener, produced consistent oxygen demand results of 22.22%. The untreated sulfur dyed samples also produced uniform results, requiring 21.52% oxygen for three samples. Variations in the residue data were attributed to variations of additives during applications. The T_{od}s were in the 300°C-315°C range. None of these samples was flame resistant when tested in the vertical flame tester.

From the data presented and discussed, it should be obvious that graphic plots, linear and semilogarithmic, are probable as functions of oxygen demand, T_{od}, residue and inches burned.

Oxygen demand appears to be an additive rather than an exponential function. During the course of this study, attempts were made to relate oxygen demand to other apparent functions, but these were either unsuccessful or did not improve the data presentation. T_{od} indicated that temperature was a matter of some consequence and should be considered as a function of the flammability of the fabric. Flame retardants were shown to lower the temperature 60-100 degrees below that of prepared cotton and, therefore, the onset decomposition temperature should receive some attention. Great importance was not attached to this quantity since common adulterants may reduce the onset decomposition temperature. Miller, et al. [53] have shown that there is a relationship between the maximum flaming temperature of different fabrics and flame retardant finishes with respect to the percentage of oxygen used in the oxygen demand test. The percentage of oxygen has a linear relationship to flame temperature for a given fabric or finish. Since temperatures used for radiation work depend upon absolute values, all plots and calculations were converted to absolute temperatures.

The higher residues produced during the TGA cotton studies were, with only two exceptions, found to be nonflammable when the samples were subjected to the vertical flame test. For this reason, different weightings for the residue relations were believed to be necessary. A GE 350 Computer was programmed to convert the residue and temperature data to exponential values as $R^2, R^3, R^4, \frac{R}{T}, \frac{R^2}{T}, \frac{R^3}{T}, \frac{R^4}{T},$

$$\sqrt{\frac{R^2}{T}}, \sqrt{\frac{R^3}{T}}, \sqrt{\frac{R^4}{T}}, \sqrt[3]{\frac{R^2}{T}}, \sqrt[3]{\frac{R^3}{T}}, \sqrt[3]{\frac{R^4}{T}}, \text{ and } \sqrt[4]{\frac{R^4}{T}}.$$

The resultant data were then plotted on linear, semi-log and log/log graph paper as functions of char length, temperature and oxygen demand.

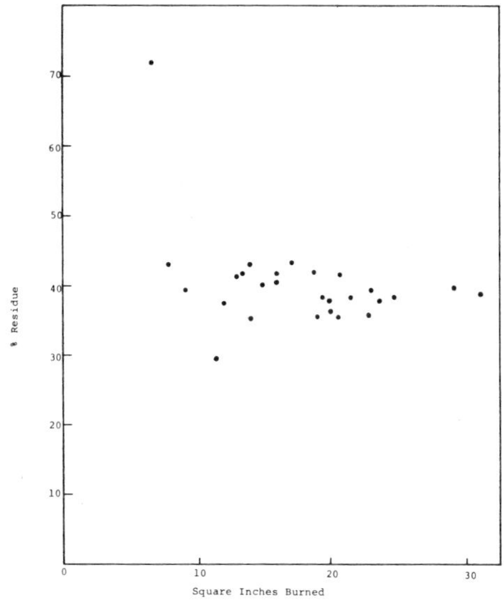

Figure 15. *A linear plot of the square inches burned using the vertical flame test versus % residue.*

Selected linear and semi-log plots are shown in Figure 15-21. The charred inches vs. percentage residue, Figure 15, shows that no valid interpretation exists by the correlation of this data. The plot of temperature vs. percentage oxygen, Figure 16 shows the agglomeration of test results produced by adulteration rather than the relative flame resistance. Hence the temperature vs. percentage oxygen demand was eliminated. The residue, residue-T_{od} and oxygen demand date indicate that relationships do exist. These are plotted in Figures 17-21. The semi-log plot of percentage residue vs. oxygen demand, Figure 17, and the linear plot of $\sqrt{\frac{R^3}{T}}$ vs. percentage oxygen demand, Figure 19, make reasonable presentations of the data. It is interesting that both data plots are exponential functions. The similarity between the two plots tends to indicate that the oxygen demand-residue exponential relation is a valid assumption. The semi-log plot of the functions of $\sqrt{\frac{R^3}{T_A}}$ vs. percentage oxygen demand, Figure 18, indicates that the data are of a discontinuous nature. On the other hand, the semi-log plot of $\sqrt{\frac{R^2}{T_A}}$ vs. oxygen demand, Figure 20, tends to compress the data into either vertical or horizontal relationships. Figure 21 presents the $\sqrt{\frac{R^2}{T_A}}$ data in an expanded linear form. Without the expansion, the plot was so compressed that analysis was difficult.

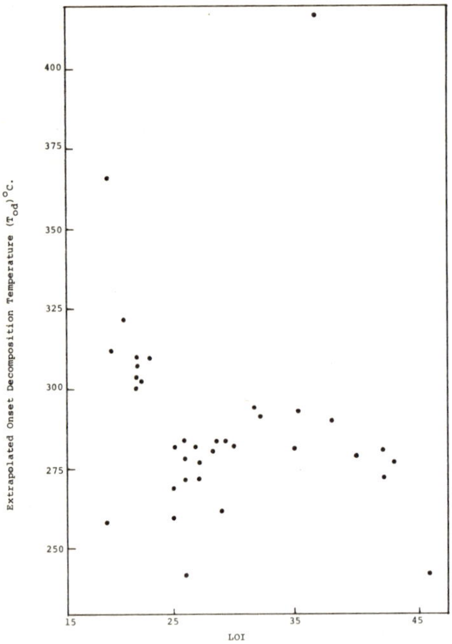

Figure 16. *A linear plot of the extrapolated onset decomposition temperature °C (T_{od}) versus LOI (% oxygen demand).*

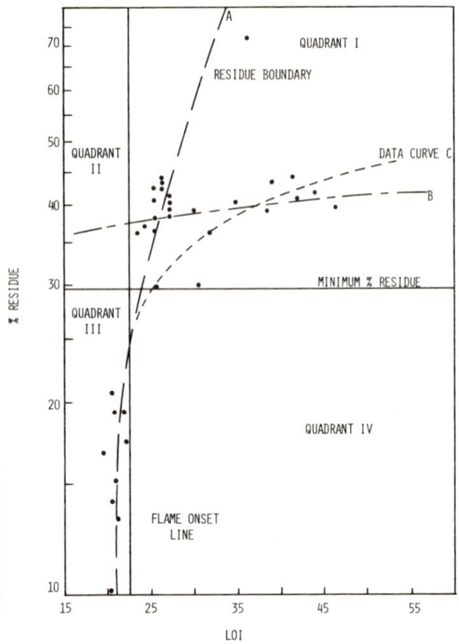

Figure 17. *A semi-logarithmic plot of % residue versus LOI (% oxygen demand) and the four hypothesized quadrants.*

Figure 17 also presents a hypothetical solution to the flammability dilemma. A vertical line was drawn so that all of the oxygen demand values for untreated cotton fell to the left of this line, at 22.5% oxygen demand. The horizontal axis, R, was then drawn at a value slightly less than that of the wool control, because wool has partial fire resistance and an appreciable residue. The intersection of the two axes produced four quadrants I-IV, with quadrants I and III believed to be more important. Quadrant I was regarded as being the flame resistant one because it is in the high oxygen demand and high residue part of the graph. Wool treated with H_3PO_4 required an atmosphere of 47.05% oxygen and produced a mid-range residue. Nomex had a lower oxygen demand than did the treated wool, but it produced the largest residue of any sample tested. In quadrant III all samples burned in an air atmosphere and were excluded from the fire resistant class. Quadrants II and IV do not present so clear-cut a case. Though a glow and smoulder hypothesis may be advanced, this group was not included in the work presented.

Curves A and B in Figure 17 are the assumed data curves for the points plotted. Since there are two intersecting lines for the points plotted instead of one, as shown by curve C, the average data curve, the assumptions previously used are of importance in this hypothesis. Curve A is regarded as being the residue boundary on which an increase in percentage residue shows increased flame-resistant characteristics. The points along this line are of fabrics whose main dependency is on the percentage residue. The lower values

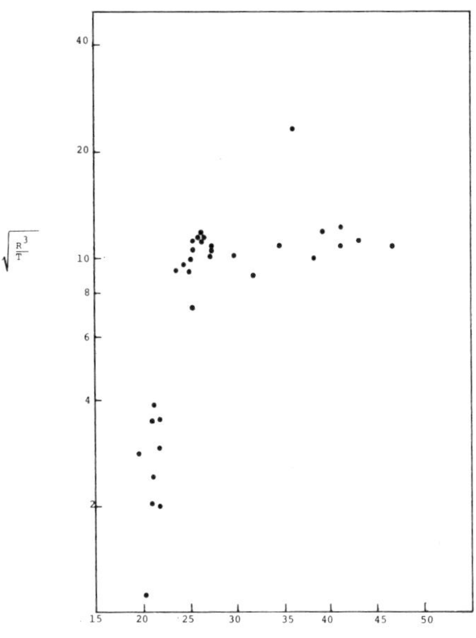

Figure 18. *A semi-logarithmic plot of the function* $\sqrt{\frac{R^3}{T}}$ *versus LOI (% oxygen demand) where R is % residue and T is temperature (°K).*

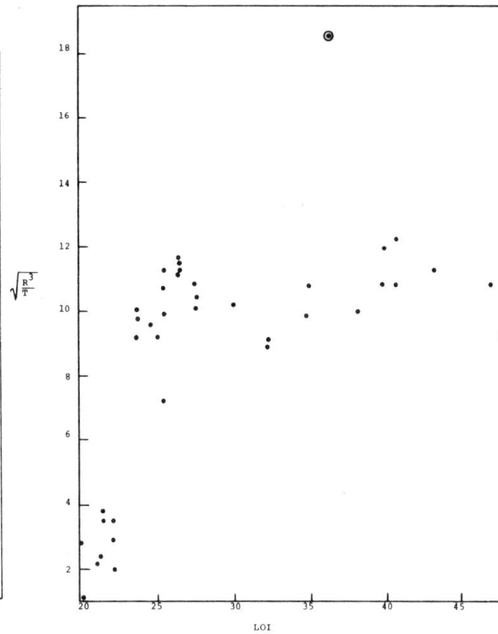

Figure 19. *A linear plot of the function* $\sqrt{\frac{R^3}{T}}$ *versus LOI (% oxygen demand) where R is % residue and T is temperature (°K).*

of this index indicate little resistance to flaming. Curve B is the oxygen demand curve on which any resistance to flame is primarily a function of the oxygen demand of the sample.

From the plotted physical data and the intersection of curves A and B, four quadrants are established as previously hypothesized. Quadrant I is termed the flame resistant quadrant; quadrant III, the flame quadrant; and quadrants II and IV, the probability quadrants. The probability quadrants are defined as containing that sample which may be hazardous when exposed to selected conditions.

The A-B quadrant, when placed on the percentage residue vs. oxygen demand semi-log plot, appears to produce a good fit. However, when these quadrants are superimposed on the $\sqrt{\frac{R^3}{T}}$ vs. oxygen demand linear plot, the fit includes only two flammable samples in the fire resistant category.

CONCLUSIONS

Effective flame resistant cotton fabrics decompose at temperatures approximately 100°C lower than pure cellulose and 80°C lower than prepared cotton fabrics. The evidence for temperature lowering was produced by TGA

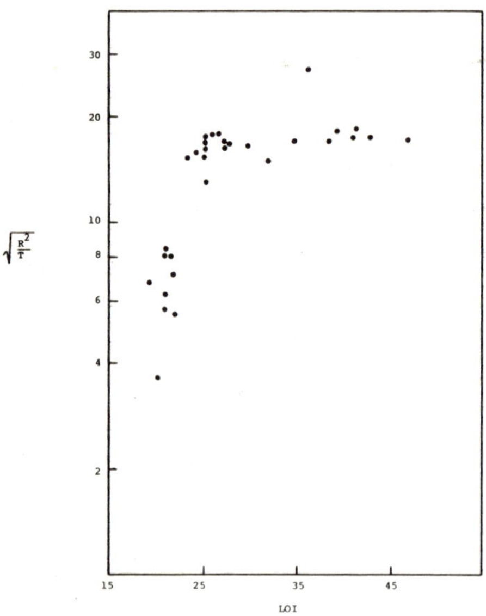

Figure 20. *A semi-logarithmic plot of the function* $\sqrt{\frac{R^2}{T}}$ *versus LOI (% oxygen demand) where R is % residue and T is temperature (°K).*

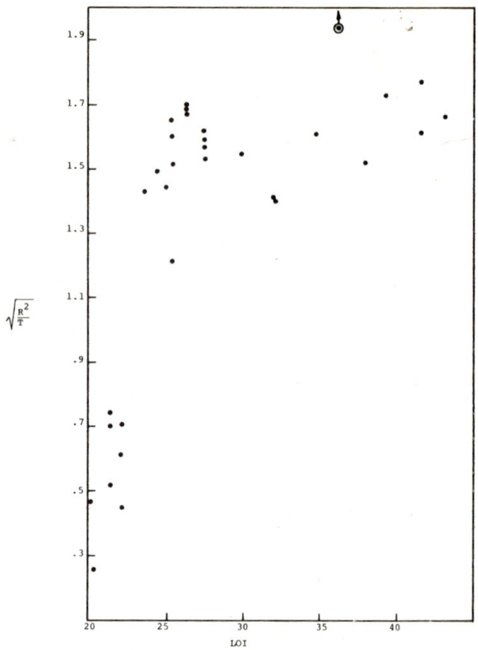

Figure 21. *A linear plot of the function* $\sqrt{\frac{R^2}{T}}$ *versus LOI (% oxygen demand) where R is % residue and T is temperature (°K).*

tests in nitrogen atmospheres and supported by muffle furnace tests with ambient air. While decrease in decomposition temperature is a characteristic of these nonflammable fabrics, percentage of residue and oxygen demand are of major significance. Tests have shown that fabrics with residues and oxygen demands of less than 30% and 25%, respectively, are not appreciably fire resistant.

An apparent exponential relationship exists between the onset decomposition and residue data produced by TGA and oxygen demand. This relationship may be plotted. The data plotted as $\sqrt{\frac{R^3}{T}}$ vs. oxygen demand may then be evaluated by a system such as the quadrant system.

Since decomposition of the phosphorus compounds applied to cotton fabrics takes place at reduced temperatures, their propensity to flame may be determined by the muffle furnace test. Because this property is usually unforeseen and not tested, the muffle furnace test method should be examined further.

REFERENCES

1. G. C. Tesoro, et al., "Flame Properties of Phosphonate Derivatives of Cotton Cellulose," *Tex. Res. Jour. 38* (3), 245-55 (1968).

2. J. T. Marsh, *An Introduction to Textile Finishing,* Chapman and Hall Ltd., London, 1957, p. 525-39.
3. R. W. Little, *Flameproofing Textile Fabrics,* Reinhold Publishing Corp., New York, 1947.
4. H. D. Tyner, "Fire Extinguishing Effectiveness of Chemicals in Water Solution," *Ind. Eng. Chem. 33,* 60-5 (1941).
5. W. A. Kramer, "The Catalytic Oxidation of Carbon," Thesis Ph.D., Ohio State University (1927).
6. *Encyclopedia of Polymer Science and Technology,* Vol. 7, Interscience Publishers, New York, 1967, p. 1-64.
7. American Association Textile Chemists and Colorists (AATCC). *Technical Manual,* 1967, p. B-135-139.
8. H. K. Townsley, "Flammability Index Test Method to Determine the Ease of Ignition of a Fabric," Paper presented before the AATCC Committee on Fire Resistance Test Methods, February 15, 1968.
9. C. P. Fenimore, and F. J. Martin, "Candle-type Test for Flammability of Polymers," *Modern Plastics,* November, 1966.
10. C. P. Fenimore, and F. J. Martin, "Flammability of Polymers," *Combustion and Flame 10* (2), June, 1966.
11. A. Cerniana, "The Comparative Thermal Decomposition of Some Monosaccharides, Oligosaccharides, and Polysaccharides," *Ann. Chim. (Rome) 41,* 455-64.
12. R. M. Perkins, et al., "DTA and TGA Studies of Flame-Resistant Fabrics," *J. Appl. Poly. Sci. 10,* 1041-66 (1966).
13. C. H. Mack, and D. J. Donaldson, "Effects of Bases on the Pyrolysis of Cotton Cellulose," *Tex. Res. Jour. 37* (12), 1063-71 (1967).
14. S. L. Madorsky, et al., "Pyrolysis of Cellulose in a Vacuum," *J., Res. N.B.S. 56* (6), 343-54 (1956).
15. H. G. Higgins, "Degradation of Cellulose in Air as Shown by Infrared Spectroscopic Examination," *J. Poly. Sci. 28,* 645-8 (1958).
16. R. F. Schwenker, Jr., and E. Pacsu, "Pyrolytic Degradation Products of Cellulose," *Chem. Eng. Data Ser. 2,* 83-8 (1963).
17. C. T. Greenwood, et al., "Analysis of the Thermal Decomposition Products of Carbohydrates by Gas Chromatography," *Chem. Ind. (London),* 1878-9 (1961).
18. R. F. Schwenker, Jr., and L. R. Beck, Jr., "Pyrolytic Decomposition of Cellulose by Gas Chromatography," *J. Poly. Sci. Pt. C, 2,* 331-40 (1960).
19. A. Broido, and S. B. Martin, "Effects of Potassium Bicarbonate on the Ignition of Cellulose by Radiation," *Fire Res. Abs. and Rev. 3,* 193-201 (1961).
20. K. Tamaru, "Pyrolysis and Combustion of Cellulose in the Presence of Inorganic Salts," *Bull. of the Chemical Society* (Japan) 24, 164-8 (1951).
21. A. Pictet, and J. Sarasin, "Distillation of Cellulose and Starch in Vacuo," *Helv. Chim. Acta 1,* 87-96 (1918).
22. E. Pacsu, and R. F. Schwenker, Jr., "Pyrolytic Degradation of Cellulose and Chemical Modification for Flame and Glow Resistance," Final Report on the Oxidation of Flame Resistance Treated Cellulose to Research, Development, and Technical Services, Dept. of the Army. Contract No. DA 19-129-QM-122, Textile Res. Inst., Princeton, N. J. (1957).
23. J. C. Irvine, and J. W. H. Oldham, *J. Chem. Soc. Transactions 119,* 1744 (1921).
24. E. Heuser, *The Chemistry of Cellulose,* John Wiley and Sons, Inc., New York, 1944.
25. R. M. Esteve, et al., Progress Reports Nos. 12 and 13, QM Contract No. DA 44-109-1414, University of Rhode Island Kingston, R. I.
26. Ivanov, et al., *Izvest. Akad. Nauk. S.S.S.R., Otdel, Khim.* No. 10, 1266 (1956).
27. S. Martin, "The Mechanism of Ignition of Cellulose Materials by Intense Thermal Radiation," Research and Development Technical Report USNRDL-TR-102 (1956).
28. A. Day, et al., Progress Report No. 1, QM Contract No. DA 19-129-QM-211, University of Penn., Philadelphia, Pa. (1954).

29. A. Day, et al., Final Report, QM Contract No. DA 19-129-QM-211, University of Penn., Philadelphia, Pa.
30. S. L. Madorsky, et al., "Thermal Degradation of Cellulose Materials," *J. Res. N.B.S. 60* (4), 343-9 (1958).
31. O. P. Golova, et al., "New Data on the Bonds of Polysaccharide (Cellulose) Structure and the Course of Chemical Reactions Occurring During Thermal Decomposition," *Doklady Akad. Nauk. S.S.S.R. 112,* 430-2 (1957).
32. O. Golova, et al., "On the Mechanism of the Thermal Decomposition of Cellulose Under Vacuum and the Formation of 1, 6-Anhydro-1, 5-Glucopyranose-Levoglucosan," *Doklady Akad. Nauk. S.S.S.R. 115,* 1122-5 (1957).
33. O. P. Golova, et al., "Thermal Decomposition of Cellulose and Its Structure," *Doklady Akad. Nauk. S.S.S.R. 116,* 419-21 (1957).
34. O. P. Golova, et al., "Mechanism of Thermal Decomposition of Cellulose in Vacuum," *Vysokomole-Kulyarnyi Soedineniya 1,* 1295-1304 (1959).
35. J. Honeyman, "A Fundamental Study of the Pyrolysis of Cotton Cellulose to Provide Information Needed for Improvement of Flame-Resistant Treatments for Cotton," Final Progress Report, Contract No. UR-E29-(20)-9, Shirley Institute, Didsbury, Manchester, England, April 1964, p. 103-4.
36. P. K. Chatterjee, and C. M. Conrad, "Kinetics of the Pyrolysis of Cotton Cellulose," *Tex. Res. Jour. 36* (6), 489-94 (1966).
37. F. H. Holmes, and C. G. Shaw, "The Pyrolysis of Cellulose and the Action of Flame-Retardants. I Significance and Analysis of the Tar," *J. Appl. Chem. (London) 11,* 210-6 (1961).
38. J. B. Berkowitz-Mattuck, and T. Noguchi, "Pyrolysis of Untreated and APO-THPC Treated Cotton Cellulose During One-Second Exposure to Radiant Flux Levels of 5-25 cal./cm. 2-sec.," *J. Appl. Poly. Sci. 7,* 709-25 (1963).
39. W. K. Tang, and W. K. Neill, "Effect of Flame Retardants on Pyrolysis and Combustion of α-Cellulose," *J. Poly. Sci. Pt. C (6),* 65-78 (1964).
40. W. K. Tang, et al., "Effect of Inorganic Salts on Pyrolysis of Wood, Cellulose, and Lignin Determined by Differential Thermal Analysis," U. S. Forest Service Research Paper FPL 82, Forest Products Laboratory, Madison, Wisconsin, January 1968.
41. D. M. Jones, and T. M. Noone, "Some Approaches to the Permanent Flameproofing of Cotton: System Containing Phorphorus," *J. Appl. Chem. 12,* 397-405 (1962).
42. G. Tesoro, "Flame Retardants for Cotton Fabrics," *Texilveredlung 2* (7), 435-40 (1957).
43. N. Grassie, *Trans. Faraday Soc. 48,* 379 (1952).
44. J. V. Beninate, et al., *Text. Res. J. 38* (3), 267-72 (1968).
45. G. Tesoro, et al., Abstracts of Papers Presented, 155th National Meeting American Chemical Society, San Francisco, California.
46. F. Shafizadeh, et al., Abstracts of Papers Presented, 156th National Meeting, American Chemical Society, Atlantic City, New Jersey.
47. F. J. Kilzer, and A. Broidio, *Pyrodynamics 2,* 151 (1965).
48. J. E. Hendrix, "Thermal Degradation of Cellulose and Related Carbohydrates in the Presence of Selected Organic Phosphates," Ph.D. dissertation, Clemson University, 1970.
49. R. H. Barker, et al., "The Effect of Selected Nitrogen and Phosphorus Containing Compounds on the Thermal Degradation of Cellulose-I," Paper presented before the Cellulose Division, American Chemical Society, April 1969.
50. K. Yeh, and R. H. Barker, unpublished results.
51. E. Klein, et al., "Enthalpy Effects in Cotton Flame Retardants," Paper presented at 1969 Chemical Finishing Conference, Washington, D. C.
52. J. E. Hendrix, et al., "Aspects of the Thermal Decomposition of Cellulose," Paper presented to National Technical Conference, AATCC, Charlotte, N. C., November 1969.
53. B. Miller, et al., "Flammability of Textiles," *Progress Report 4,* Textile Research Institute, Princeton, N. J., July 1969, p. 8.

J. E. Hendrix

J. E. Hendrix is a cellulose research chemist with FMC Corporation at Princeton, New Jersey. He received his B.S. degree in Textile Science from Auburn University in 1966, his M.S. degree in Textile Chemistry from Clemson University in 1968, and his Ph.D. in Chemistry with a major in Textile Chemistry from Clemson University in 1969.

T. K. Anderson

Terry K. Anderson is pursuing his M.B.A. degree in Business Administration at Wharton School of Business. He received his B.S. degree in Textile Chemistry from Clemson University in 1969.

T. J. Clayton

Thomas J. Clayton is a chemist with Hystron Fiber Corporation, Spartanburg, South Carolina. He received his B.S. degree in Textile Chemistry from Clemson University in 1969.

E. S. Olson

Edward S. Olson is an Associate Professor of Textile Chemistry at Clemson University, Clemson, South Carolina. He received his B.S. and M.S. degrees in Textile Chemistry from Clemson College in 1938 and 1960 respectively. Before joining the faculty at Clemson, he was employed by Deering Milliken at Pendleton, South Carolina, by Dan River Mills at Danville, Virginia, and by North Carolina Finishing Company at Salisbury, North Carolina.

R. H. Barker

R. H. Barker is a Professor of Textile Chemistry at Clemson University, Clemson, South Carolina. He received his B.S. degree in Textile Chemistry from Clemson College in 1959 and his Ph.D. degree in Chemistry with a major in Organic Chemistry from the University of North Carolina in 1963. Before joining the faculty at Clemson, he was Assistant Professor of Chemistry at Tulane University.

Taken in part from the B.S. Thesis of T. K. Anderson and T. J. Clayton and M.S. Thesis of J. E. Hendrix. Presented in part as the Palmetto Section Contest Paper at the AATCC National Technical Convention, Charlotte, N. C., Nov. 1969.

Environmental Temperatures and Oxygen Index (OI) Values for Textile Fabrics

James E. Hendrix, John V. Beninate, George L. Drake, Jr., and Wilson A. Reeves

*Southern Regional Research Laboratory**
New Orleans, Louisiana

(Received September 20, 1971)

ABSTRACT

The flammability properties of selected natural, synthetic, and blended textile fabrics were investigated using the OI technique. Environmental temperatures during testing were varied over the –50 C to 260 C range. Under conditions of controlled environmental temperatures, thermal activities within the samples or interactions of the samples with oxidizing atmospheres cause variations in OI values. DTA and TGA thermograms of the samples in oxidizing atmospheres illustrate these thermal interactions and activities, and constitute a basis for explanation of the origin of the observed OI variations.

INTRODUCTION

One is not necessarily correct in citing an oxygen index (OI) value of 0.180 for cotton or 0.214 for polyester, although the values may have been carefully determined. It has been demonstrated that such statements are true only for a particular sample of specified weight, construction, degree of purity [1] and moisture content [2], with the determination made at given environmental temperatures and on a sample of a given size [3]. Thus OI values are not simple numbers describing sample flammability, but represent complex parameters dependent on sample preparation and testing environment.

Initial investigations indicated that environmental temperatures have dramatic effects on the flammability properties of untreated and flame-retarded cottons, as determined by the OI technique [4]. OI values decrease with increasing environmental temperatures, while environmental temperatures fluctuate during testing. Consequently, careful control of environmental temperatures during testing is necessary to avoid erroneous results.

---

*One of the laboratories of the Southern Marketing and Nutrition Research Division, Agricultural Research Service, U.S. Department of Agriculture.

DiPietro and co-workers recently reported that OI values for plastics or cellulose decrease with increasing sample temperature over the 25-175 C range [5]. They reported also that preheating flame-retarded samples to prescribed temperatures and cooling them again to room temperature before testing often results in lowered OI values. This they attributed to degradation of the flame retardant during preheating.

This study reports the effects of environmental temperature on the OI values for cotton, cotton-polyester blend and synthetic textile fabrics. Environmental temperatures ranged from sub-zero to the scorch temperatures for cotton. Responses of OI values to environmental temperature changes are discussed in terms of thermal analysis parameters.

EXPERIMENTAL

Data for this investigation were obtained using a Michigan Chemical Oxygen Index Apparatus, a DuPont 900 Differential Thermal Analyzer equipped with a high-temperature cell and a DuPont 950 Thermogravimetric Analyzer.

Fabric Samples

Fabrics utilized were obtained from commercial sources and treated with flame-retardant finishes in this laboratory. Fabrics containing cotton had been desized, scoured, and bleached. The fabrics included:

1. Cotton print-cloth (3.4 ox/yd^2)
   THPC-APO* (19% solids add-on)
   THPC-Amide* (23% solids add-on)
   THPOH-NH_3* (24% solids add-on)
2. 70:30 Cotton-polyester blend (3.4 oz/yd^2)
   THPC-APO (18% solids add-on)
   THPC-Amide (22% solids add-on)
   THPOH-NH_3 (22% solids add-on)
3. 50:50 Cotton-polyester (3.4 oz/yd^2)
   THPC-APO (18% solids add-on)
   THPC-Amide (20% solids add-on)
   THPOH-NH_3 (24% solids add-on)
4. Cotton sateen (8.6 oz/yd^2)
   THPC-APO (9.3, 11.1 and 16.0% solids add-ons)
   THPOH-Amide (25% solids add-on)
   THPOH-NH_3 (30% solids add-on)
   THPC-Cyanamide (13% solids add-on)
   PNE-$CHBr_3$ adduct* (22% solids add-on)
5. Greige industrial cotton duck (24.3 oz/yd^2)
6. Cotton knit (6.8 oz/yd^2) treated commercially with phosphonate flame retardant
7. Nomex (2.1 oz/yd^2)
8. Nylon 6 (0.6 oz/yd^2)

9. Polyester (1.2 oz/yd^2)
10. Wool (6.3 oz/yd^2)
11. Cellulose acetate (1.6 oz/yd^2)

*Tetrakis (hydroxymethyl)phosphonium chloride – tris(aziridinyl)phosphine oxide adduct [THPC-APO]; THPC – trimethylolmelamine – urea adduct [THPC-amide]; neutralized THPC-ammonia adduct [THPOH-NH_3]; Bromoform adduct of the allyl ester formed from phosphonitrilic chloride and allyl alcohol [PNE-$CHBr_3$ adduct].

Oxygen Index (OI) Determinations

OI values were determined according to ASTM procedures [6]. Variations in chimney temperature were achieved by wrapping that portion of the chimney surrounding the sample with a heating tape connected to a variable power supply or with a sleeve containing solid carbon dioxide. Chimney temperatures recorded represent readings from a thermometer placed between the chimney and the heat source at bulb level one inch below the top of the metal sample holder.

The relationship between chimney and sample temperatures was approximated by suspending a thermometer in the center of the chimney with the bulb level equal to that of the external thermometer. Readings were taken after five-minute equilibration periods under normal operating conditions. The relationship between chimney and approximated sample temperatures (Table 1) proved to be very close to the relationship of recorded temperature to sample temperature in differential thermal analysis. Thus OI data are discussed in terms of chimney temperatures, bearing this relationship in mind. OI data are reported as volume fractions of oxygen in the oxygen-nitrogen atmosphere, rather than per cent oxygen.

Differential Thermal Analysis (DTA)

The fabric specimens were pulverized in a Wiley mill. All curves were obtained from 10 mg samples under dynamic (2 liters/min) air atmospheres. The reference material was glass in all cases. A constant heating rate of 20 C/min and a ΔT sensitivity of 0.08 mv/in (20 C/inch) were employed. One must realize that high rates of heating increase ΔT sensitivity, but also shift all thermal activity to higher recorded temperatures [7]. The shift results from many complex factors affecting kinetics and thermodynamics of each sample, thus no exact heating rate-temperature relationship was derived. It was observed, however, that weight loss in TGA thermograms (5 C/min heating rate) occurs 30-50 C lower than onset of thermal activity in DTA

Table 1. Relationship Between Sample and Chimney Temperatures

	Temperatures (°C)				
Chimney	−57	25.5	103	202.5	272
Sample	−28	25	78.5	161.5	225

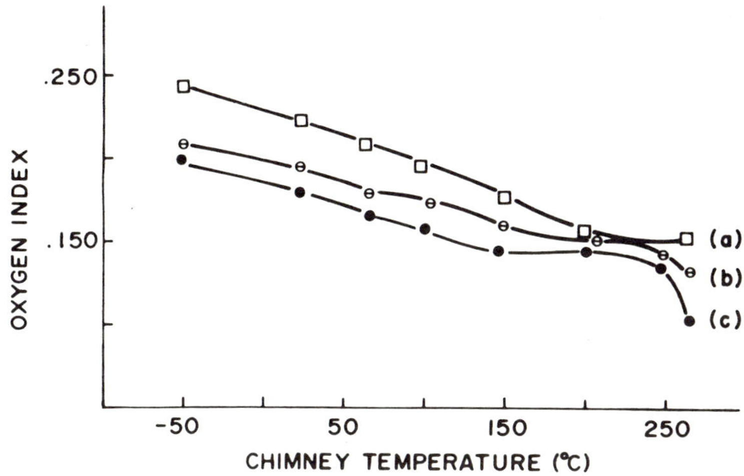

Figure 1. *Temperature-OI relationships for cotton cellulose. (a) Greige industrial duck (24.3 oz/yd²); (b) Sateen (8.6 oz/yd²); (c) Print cloth (3.4 oz/yd²).*

thermograms for activity near 300 C. This allows a direct comparison to be made between chimney temperatures and thermal activity observed in corresponding DTA thermograms.

Thermogravimetric Analysis (TGA)

All thermograms were obtained from 10 mg fabric specimens under dynamic (2 liters/min) air atmospheres. A constant heating rate of 5 C/min and a scale sensitivity of 2 mg/inch were employed.

RESULTS AND DISCUSSION

Cotton Cellulose exhibited OI values ranging from 0.133 to 0.243 (Figure 1), depending on extent of purity, fabric weight and construction, and environmental temperatures during testing. The reasons for the wide variations in values may be better understood if one considers the processes involved in burning. For propagation of flaming, the gas phase oxidation of cellulose decomposition products must be sufficiently vigorous to maintain the fabric substrate at temperatures high enough to effect pyrolytic degradation. Also, the rate of degradation must be great enough to maintain an adequate fuel supply to the gas phase. Thus, the further the sample and environment are below the temperatures necessary for substrate decomposition, the more vigorous the gas phase oxidation must be.

Consider, for instance, a cotton fabric sample at its normal moisture regain and placed in a sub-zero environment. Before the sample reaches its decomposition temperature, the following must occur: (a) heating of the sample and retained water to vaporization temperatures, (b) vaporization of retained water, and (c) heating of the

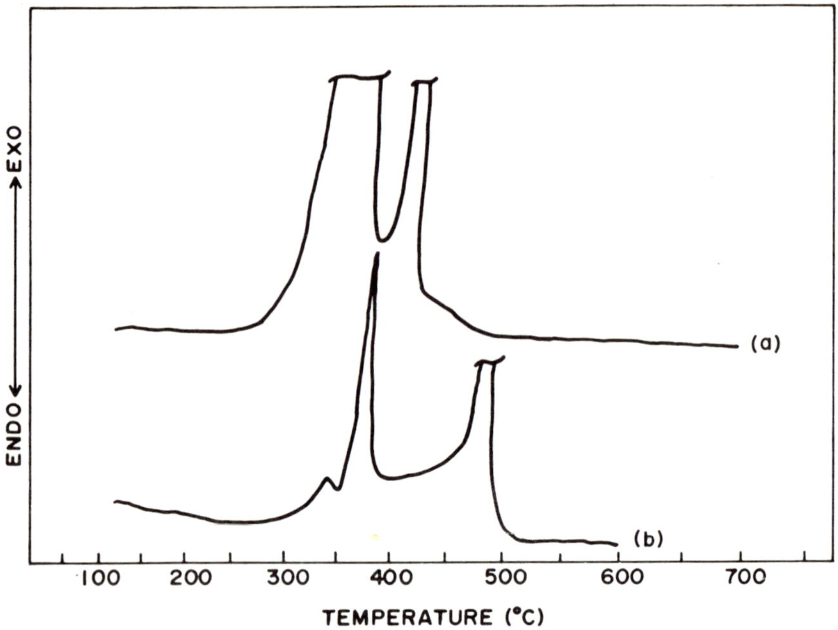

Figure 2. *DTA thermograms for cotton fabrics. (a) Greige industrial duck (24.3 oz/yd^2); (b) Print cloth (3.4 oz/yd^2).*

dry sample to decomposition temperatures. Since these processes are endothermic, lowered environmental temperatures necessitate a more vigorous combustion to maintain the fabric substrate at its decomposition temperature. Additionally, transfer of heat from the gas phase to the fabric substrate is an inefficient process, utilizing only a fraction of the total heat produced for substrate decomposition. Small increases in heat required for substrate decomposition, therefore, necessitate enormous increases in heat which must be produced in the gaseous phase.

Temperature-OI relationships for actual cotton samples (Figure 1) illustrate the reality of the processes discussed. OI values for cotton fabrics in sub-zero environments lie well above corresponding values at room temperature, and decrease essentially linearly with temperature over the −50 C to 100 C range. At these temperatures cotton samples would be essentially dry. Low heat capacities of the dry cottons may account for decreased slopes over the 150-200 C range.

Above 200 C the slopes of the OI vs. chimney temperature curves for the purified cottons increase drastically, the OI for printcloth dropping to 0.103 at 265 C. Temperatures such as these correspond to scorch temperatures for cotton, thus the abrupt decrease in OI values may be due to exothermic oxidation of the fabric surface by oxygen in the environment. Such oxidation would result in heating the fabric nearer its decomposition temperatures and reduction of heat which must be

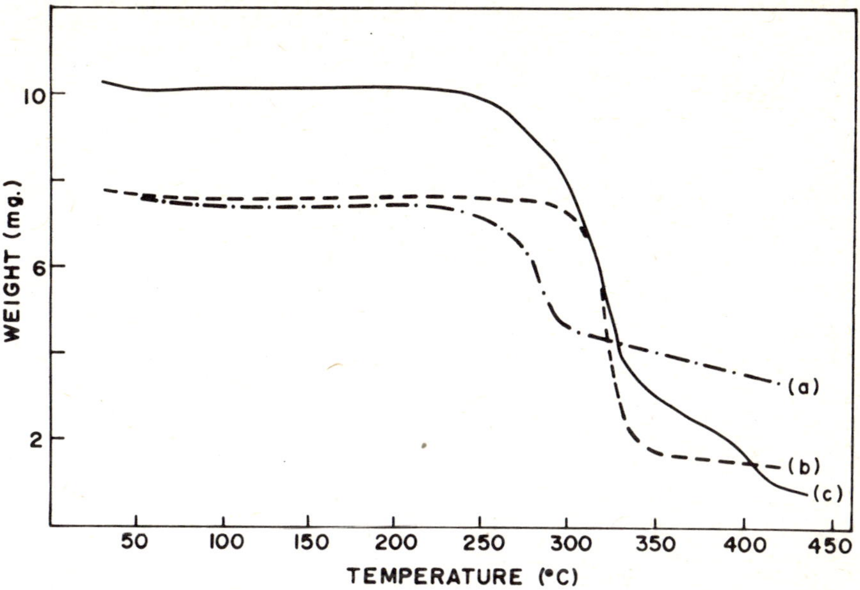

Figure 3. *TGA thermograms for cotton fabrics. (a) Print cloth (3.4 oz/yd^2) treated with THPC-APO (19% add-on); (b) Print cloth (3.4 oz/yd^2); (c) Greige industrial duck (24.3 oz/yd^2).*

supplied by the gas phase. This interpretation is illustrated by DTA and TGA thermograms for cotton print cloth pyrolyzed in air (Figures 2 and 3). Onset of oxidative heating is near 250 C (Figure 2), while weight loss does not occur until the environmental temperature approaches 300 C (Figure 3).

In contrast to the purified cottons, OI values for greige cotton slightly increase above 200 C. It may be also noted that OI values for greige goods are higher than corresponding values for purified cottons over the entire temperature range studied. This has been shown to be attributable to flame-retardant properties of natural growth impurities [1]. Thus, it is reasonable that the OI increases observed (Figure 1) result from action of the natural impurities (inorganic salts, etc.) on the cellulose substrate to give a less flammable sample. This interpretation is borne out in the DTA and TGA thermograms for greige industrial duck in air (Figures 2 and 3). Exothermic activity attributable to interactions between the substrate and its natural impurities is onset near 250 C (Figure 2), and is accompanied by weight loss (Figure 3). This results in a less flammable substrate, but one which is more susceptible to afterglow as illustrated by the decreased onset temperature for the second occurrence of exothermic activity in the DTA thermograms.

Effects of environmental temperature on OI values are greater for flame-resistant cottons than for the corresponding untreated samples. The value for cotton sateen

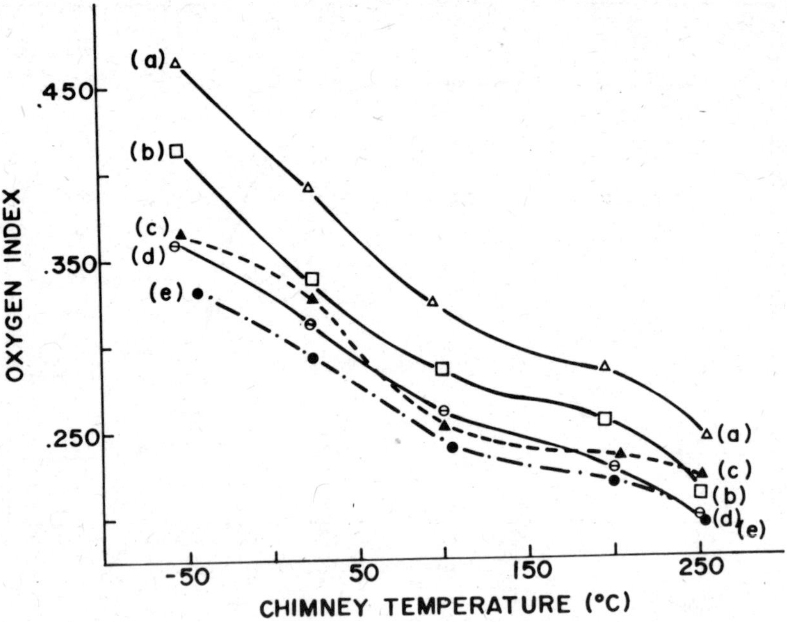

Figure 4. *Temperature-OI relationships for flame retarded cottons (8.6 oz/yd²). Finishes utilized include: (a) Phosphonitrilic ester-bromoform adduct (22% add-on); (b) THPC-Cyanamide (13% add-on); (c) THPOH-NH_3 (30% add-on); (d) THPOH-Amide (25% add-on); and (e) A knit fabric (6.8 oz/yd²) commercially treated with a phosphonate finish.*

treated with a phosphonitrilic ester-bromoform adduct is 0.468 at −54 C, whereas the value at 254 C is reduced to 0.242. This represents approximately 50% reduction in OI values over the temperature range studied. The increased sensitivity of flame-retarded cottons is attributable primarily to the greater heat requirements placed on the gas phase to effect sustained fuel production through substrate decomposition.

All of the flame-retardant cottons studied responded similarly to environmental temperature changes (Figure 4). OI values decrease with increasing temperature until temperatures greater than 100 C are reached, at which point the slopes for the dried samples decrease over the 100 C to 200 C range. Above 200 C, OI values again decrease rapidly with temperature. This latter decrease may be attributable both to exothermic interactions of the substrates with the oxidizing atmosphere in this temperature range (Figure 5) and degradation of the flame retardant in a manner similar to that reported by DiPietro and co-workers [5]. The DTA thermograms of representative flame-retarded cottons (Figure 5) further illustrate the similarities in response of OI values to environmental temperature changes. All of the curves exhibit onset of exothermic activity near 250 C with nadirs near 325 C and 475 C for scorching and afterglow respectively. This would indicate the substrate resulting from the action of heat on each of these flame-retarded cottons to be very similar.

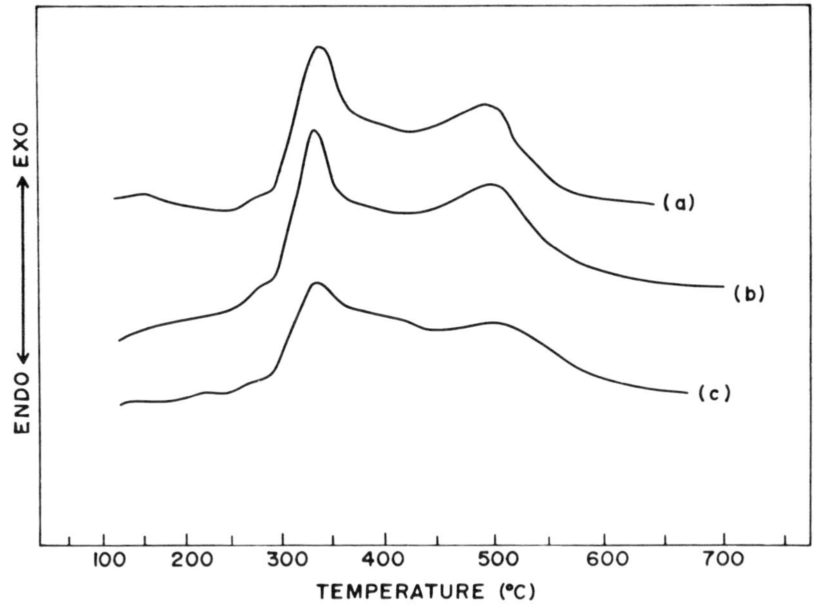

Figure 5. *DTA thermograms for flame retarded cotton fabric (3.4 oz/yd²). (a) THPC-APO (19% add-on); (b) THPC-Amide (23% add-on); (c) THPOH-NH_3 (24% add-on).*

Data presented thus far illustrate that simple linear or curvilinear relationships do not adequately describe responses of OI to environmental temperature changes. It was of interest, therefore, to examine a typical flame-retarded cotton more closely.

The response of OI values to environmental temperature changes for cotton treated with THPC-APO flame-retardant finish (Figure 6) may be characterized by five temperature intervals. The first two intervals, -50 C to 100 C and 100 C to 150 C, are related simply to heating of the sample and its retained water to the point of dryness and heating of the dry sample, respectively. These processes were described earlier. In the 150 C to 200 C region the OI values decrease due to exothermic activity, probably resulting from interaction of the flame retardant with the substrate, which decreases the total quantity of heat which must be supplied by the gaseous phase. In the 200 C to 250 C region the OI values remain fairly constant, as heat produced in the earlier exothermic activity is utilized for by-product volatilizations. Above 250 C the OI values decrease rapidly. These assignments are supported by TGA (Figure 3) and DTA (Figure 5) thermograms of the flame-retarded cottons. The DTA for cotton treated with THPC-APO finish shows exothermic activity onset near 150 C which is overcome by a volatilization endotherm over the 175 C to 250 C region. The TGA thermogram shows weight loss onset near 225 C which, allowing for differences in heating rates, corresponds to the endothermic activity in the DTA curve. The rapid reduction in OI values above 250 C is attributable to exothermic

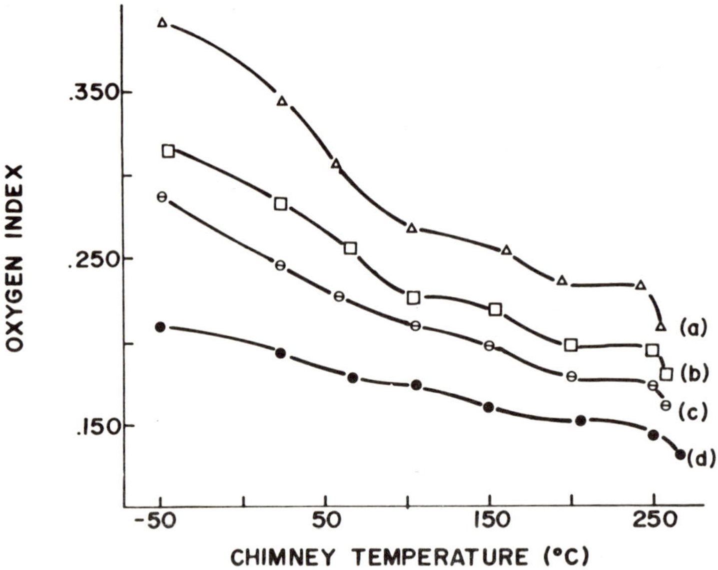

Figure 6. *Temperature-OI relationships for cotton sateen (8.6 oz/yd^2) treated with THPC-APO flame retardant finish: (a) 16.0% add-on; (b) 11.1% add-on; (c) 9.0% add-on; and (d) untreated control.*

interactions of the substrate and flame retardant with the oxidizing atmosphere.

The great decreases in OI values for flame-retarded cottons must receive some further attention at this point. Evaluation of fabric flammability by the OI technique indicates that flame-retarded cottons lose their retardant properties if placed in high temperature environments. Additionally, DiPietro and co-workers [5] demonstrated that flame retardants have diminished effectiveness if subjected to high temperatures, and Olson and co-workers [8] demonstrated that certain flame-retarded cotton fabrics underwent auto-ignition at 400 C in half the time required for the untreated control. They also observed that time required for auto-ignition decreased with increasing add-on of flame retardant. With these data in mind, one must wonder as to the validity of the flame-retardant properties imparted by treatments currently available.

The responses of OI values to environmental temperature changes for the synthetic fibers studied and for wool (Figure 7) is similar to that observed for the cottons, for environmental temperatures below 100 C. The OI decreases in this region can be attributed primarily to heat required to raise the substrates and any retained water toward their respective decomposition temperatures. It must be noted that OI values for nylon 6 and polyester were determined from multilayer sandwiched samples to eliminate the need for stitching with glass or cotton thread, either of which would introduce extraneous variables. The sandwiching technique is effective at low environmental temperatures as the melt is prevented from dripping

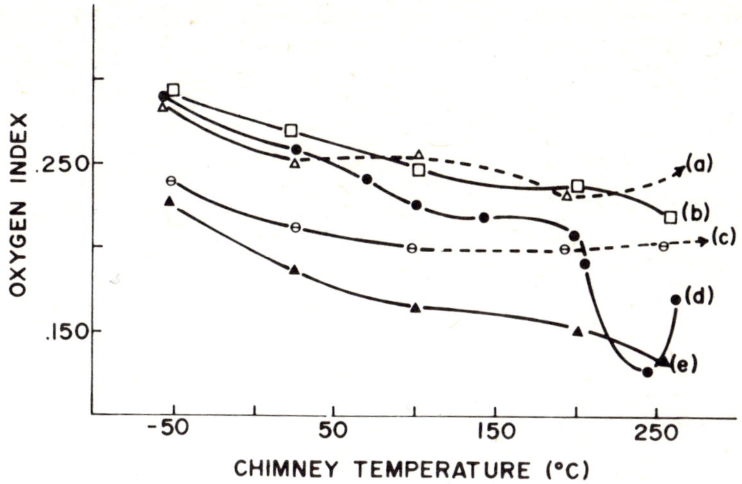

Figure 7. *Temperature-OI relationships for textile fabrics: (a) Nylon 6; (b) Nomex; (c) Polyester; (d) Wool; (e) Cellulose acetate.*

by the rigid substrate below, thus giving an OI value which is more characteristic of the synthetic than of the method of sample preparation. At higher environmental temperatures, however, the sample begins to shrink away from the ignition source thus making ignition more difficult and yielding higher OI values than those actually characteristic for the polymer (represented by dashed lines in Figure 7).

Nomex maintains its high OI values over the entire temperature range studied (Figure 7). This is attributable to its high thermal stability illustrated in its DTA thermogram (Figure 8) and TGA thermogram [8], for which thermal activity is absent until temperatures in excess of 400 C are reached. Once begun, however, the decomposition of nomex is highly exothermic (Figure 8), and the fabric burns with a bright vigorous flame. Thus, at environmental temperatures greater than those represented by this study, Nomex might conceivably become hazardously flammable. It must be borne in mind, however, that Nomex retains approximately 75% residue at 500 C, a property absent in the more conventional polyamides such as wool [8] or nylon.

At environmental temperatures greater than 200 C OI values for wool decrease rapidly, dropping to less than half of the value at room temperature when 250 C is reached (Figure 7). Above 250 C, however, OI values increase sharply. The decrease occurring above 200 C is attributable to exothermic interactions of wool with the oxidizing environment, or scorching, which is reported to occur at these temperatures [9]. The increase in OI values above 250 C is attributable to formation of an insulative foamy char which is produced when ignition is attempted.

Cellulose acetate responded to environmental temperature changes very similarly to cotton cellulose over the entire temperature range studied, thus indicating that the acetate groups have little effect on overall cellulose flammability properties. Nylon 6

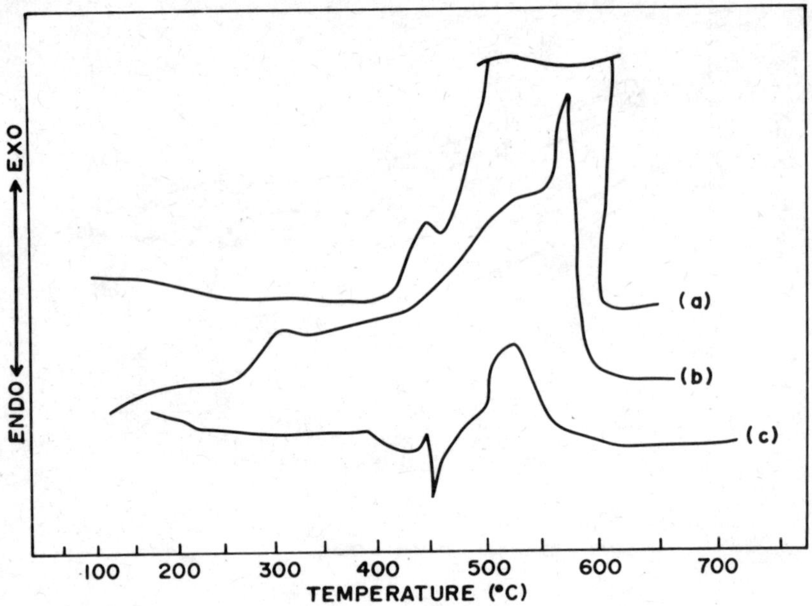

Figure 8. *DTA thermograms for textile fabrics. (a) Nomex; (b) Wool; (c) Nylon 6.*

exhibits OI values near those of Nomex and wool when determined at low temperatures. The tendency of nylon toward viscous flow at moderate temperatures, however, prohibited further investigations. The endothermic activity over the 350 C to 450 C region (Figure 8), attributable to volatilization of decomposition products, suggests that its flammability properties should increase with increasing environmental temperatures.

Temperature-OI relationships for polyester and for polyester-cotton blends with polyester contents at 0, 30, and 50% are illustrated in Figure 9. It is readily apparent that polyester-cotton blends are inherently more flammable than polyester, if the latter is tested without stitching. It is also apparent that the presence of polyester in cotton fabrics affects overall flammability properties very little, as determined by the OI technique. This latter observation may be directly attributable to thermal properties of the component fibers illustrated in the DTA thermograms of the component fibers and their blends (Figure 10). Polyester melts (endotherm near 260 C) well below the decomposition temperature of cotton (exothermic activity in 300 C to 400 C region), but its own decomposition (exothermic activity in 450-580 C region) lies well above that of cotton. Thus cotton, having the lower decomposition temperature, dictates the limits of flammability. The polyester component furnishes additional fuel to the gaseous phase through its decomposition by heat produced in combustion of cotton. The additional fuel increases the vigor of the gas phase oxidation, thus increasing rate of combustion. This is illustrated by comparison of burning times for 2 x 6-inch cotton and cotton-polyester (50/50) print cloth samples. The

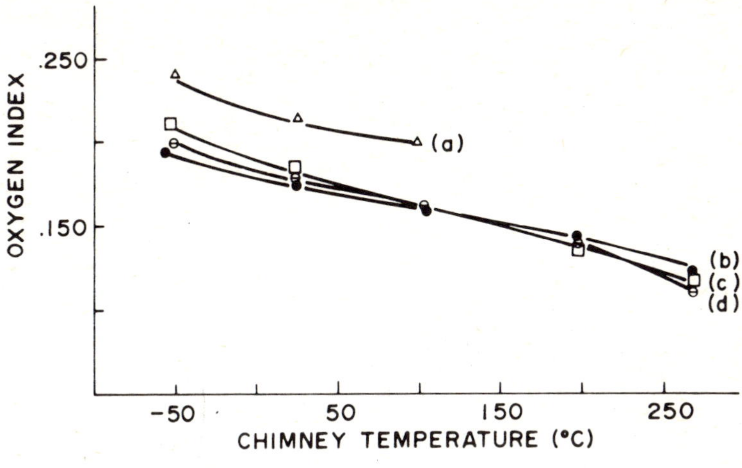

Figure 9. *Temperature-OI relationships for fabrics containing cotton and/or polyester. (a) Polyester; (b) Cotton/polyester (50/50); (c) Cotton; (d) Cotton/polyester (70/30).*

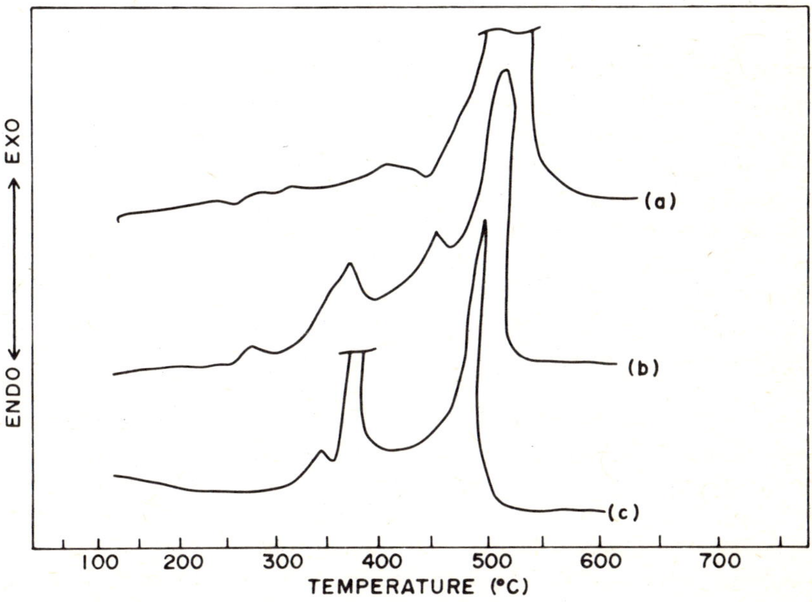

Figure 10. *DTA thermograms for textile fabrics. (a) Polyester; (b) Cotton/polyester (50/50); (c) Cotton.*

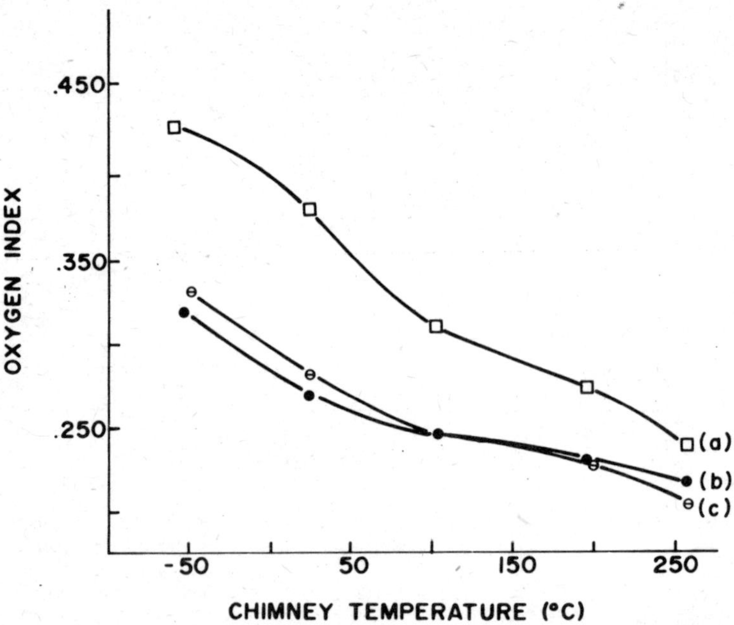

Figure 11. *Temperature-OI relationships for fabrics (3.4 oz/yd^2) treated with THPC-APO. (a) Cotton (19% add-on); (b) Cotton/polyester (50/50, 18% add-on); (c) Cotton/polyester (70/30, 18% add-on).*

cotton sample required 1 min 36 sec to burn from top to bottom and burned with a steady flame. The cotton-polyester (50/50) sample, however, burned with a vigorous, erratic flame and required only 1 min 18 sec to burn from top to bottom.

Principles applied to description of untreated cotton-polyester blend fabrics are also applicable to flame-retarded blends. The OI values over the −50 C to 250 C range for cotton treated with THPC-APO, a flame retardant specifically designed for cotton, lie well above analogous values for treated blend fabrics (Figure 11). The reason for this is that the flame-retarded cotton component is less flammable than the ineffectively flame-retarded polyester, thus it is polyester in this case that dictates the limits of flammability.

The convergence of OI values for flame-retarded cotton with analogous values for cotton-polyester blend fabrics simply reflect the inefficiency of the flame retardant to maintain its effectiveness on cotton at high environmental temperatures. This is further illustrated in OI-chimney temperature relationships for fabrics treated with THPOH-amide (Figure 12) or THPOH-NH_3 (Figure 13) flame-retardant finish. For these systems the OI values for the treated cotton fabric decrease to approximately the corresponding values for the blends.

The similarities in the OI-temperature profiles observed for cotton treated with several representative flame retardants are also observed for analogous profiles for cotton-polyester blends. The DTA thermograms for cotton-polyester blend fabrics

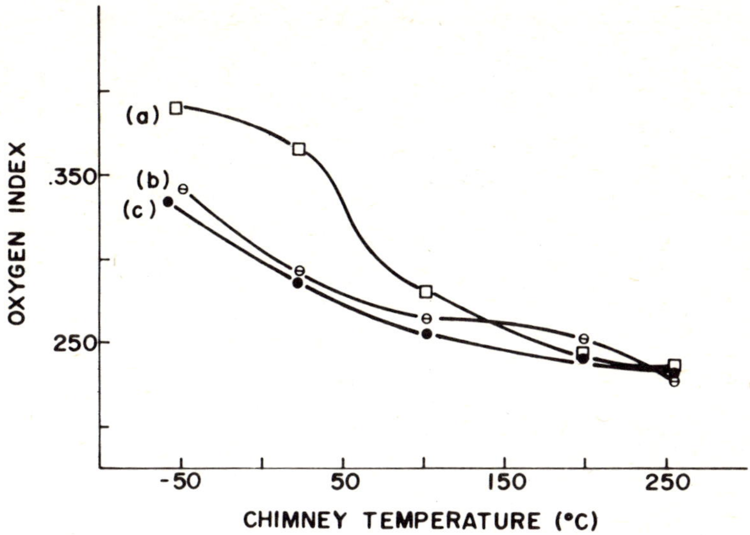

Figure 12. *Temperature-OI relationships for fabrics (3.4 oz/yd^2) treated with THPC-Amide. (a) Cotton (23% add-on); (b) Cotton/polyester (70/30, 22% add-on); (c) Cotton/polyester (50/50, 20% add-on).*

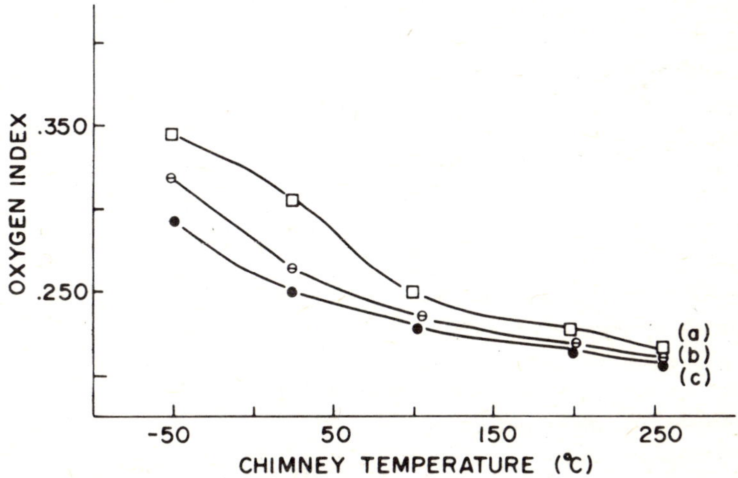

Figure 13. *Temperature-OI relationships for fabrics (3.4 oz/yd^2) treated with THPOH-NH_3. (a) Cotton (24% add-on); (b) Cotton/polyester (70/30, 22% add-on); (c) Cotton/polyester (50/50, 24% add-on).*

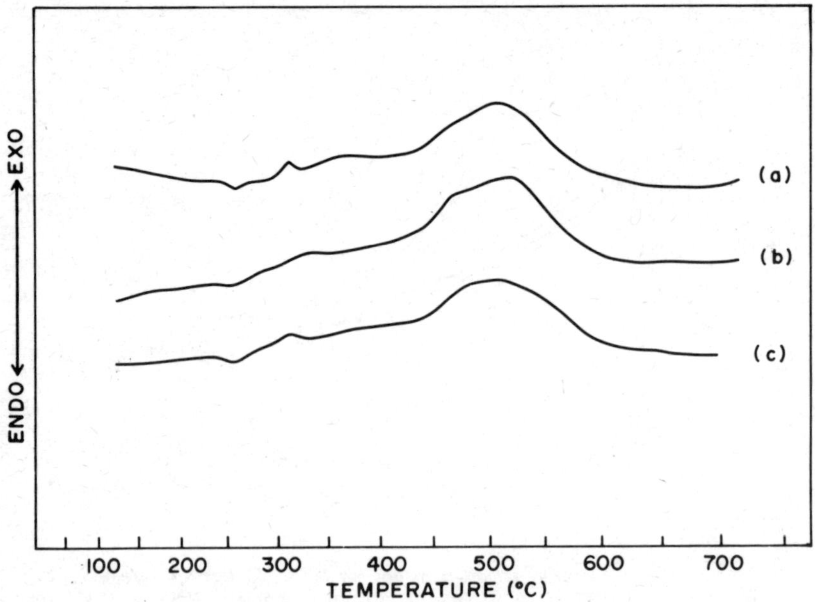

Figure 14. *DTA thermograms for flame retarded cotton/polyester (50/50) blend fabrics (3.4 oz/yd²). (a) THPC-APO (18% add-on); (b) THPC-Amide (20% add-on); (c) THPOH-NH$_3$ (24% add-on).*

treated with representative flame retardants (Figure 14) bear remarkable similarities also. They are, in fact, almost superimposable. This suggests that, as in the case of flame-retarded cotton, the substrate produced in oxidative-thermal degradation of the treated blend fabrics may be the same regardless of which flame retardant represented in this study is used. It is also quite possible that little difference exists in the modes of action of the various flame retardants studied.

SUMMARY AND CONCLUSIONS

The effects of environmental temperature changes on OI values for selected natural and synthetic fibers and their blends were studied for both the untreated and the flame-retarded states. OI values define the minimum level at which gas phase oxidation must occur in order that combustible products be continuously produced by substrate decomposition. OI values obtained, therefore, are greatly affected by external heat sources.

Under conditions of controlled environmental temperatures, thermal activities within the samples or interactions with the oxidizing atmosphere constitute secondary sources of heat which supplement that supplied by the gaseous phase oxidation of volatile pyrolysis products. The net result is a decrease in total heat which must be supplied by the gaseous phase, and consequently lower oxygen demand, reflected by decreased OI values.

REFERENCES

1. J. E. Hendrix, G. L. Drake, Jr. and W. A. Reeves. "Effects of Fabric Weight and Contruction on Oxygen Index (OI) Values for Cotton Cellulose," *J. Fire Flammability.* In press.
2. J. E. Hendrix, J. V. Beninate, G. L. Drake, Jr. and W. A. Reeves. "Effects of Moisture On Oxygen Index (OI) Values for Textiles," *Text. Res. J.* In press.
3. J. E. Hendrix, G. L. Drake, Jr. and W. A. Reeves. Unpublished results.
4. J. E. Hendrix, G. L. Drake, Jr. and W. A. Reeves. "Effects of Temperature on Oxygen Index Values," *Text. Res. J. 41,* 360 (1971).
5. J. DiPietro, H. Barda and H. Stepniczka. "Burning Characteristics of Cotton, Polyester and Nylon Fabrics," *Text. Chem. Color. 3,* 40-48 (1971).
6. American Society for Testing Materials. "Flammability of Plastics Using the Oxygen Index Method." ASTM Designation D-2863-70.
7. W. W. Wendlandt. *Thermal Methods of Analysis,* Interscience Publ, New York, 1964, pp. 152-156.
8. J. E. Hendrix, T. K. Anderson, T. J. Clayton, E. S. Olson and R. H. Barker. "Flammability Measurements and Thermal Decomposition of Textiles." *J. Fire Flammability 1,* 107-140 (1970).
9. W. G. Ashmore. "Surge of New Fiber Activity Reflected In 1959 Man-Made Fiber Table." *Text. World 109* (9), 69-88 (1959).

James E. Hendrix

James E. Hendrix is a National Academy of Science-National Research Council Postdoctoral Research Associate with the Southern Regional Research Laboratory at New Orleans, Louisiana. Prior to accepting the Postdoctoral Research Associateship, he was a cellulose research chemist with FMC Corporation at Princeton, New Jersey. He received his B.S. degree in Textile Science from Auburn University in 1966, his M.S. degree in Textile Chemistry from Clemson University in 1968, and his Ph.D. in Chemistry with a major in Textile Chemistry from Clemson University in 1970.

John V. Beninate

John V. Beninate, a native of New Orleans, Louisiana, received his B.S. (Chemistry) degree from Louisiana State University in 1948. From 1948 until 1957 he was employed by Swift and Company and later by the Celotex Corporation. In 1957 he joined the staff at the Southern Utilization Research and Development Division, U.S.D.A., as a research chemist in the Cotton Finishes Laboratory. He has publications and patents in the textile field including flameproofing, water repellency, soil resistance, and wash-wear finishes. He received a U.S.D.A. Superior Service Award in 1968.

George L. Drake, Jr.

Mr. Drake is a native of Alabama. He received a B.S. degree in Chemistry from the University of Georgia in 1944 and his M.S. degree in Chemistry from the same school in 1949. He has been at Southern Utilization Research and Development Division since 1949. He is the author of numerous publications and patents and is a member of a number of professional and scholastic organizations including, ACS, RESA, and AATCC. In 1961 he became Head of Special Finishes Investigations. He was chosen Federal Civil Servant of the Year by the Federal Business Association of New Orleans in 1962, and was one of the recipients of the U.S.D.A. Distinguished Service Award in 1964. He received the U.S.D.A. Superior Service Award in 1960 and again in 1968. In 1969 he received the Herty Award presented by the Georgia Section of the American Chemical Society.

Wilson A. Reeves

Dr. Reeves is Chief of the Cotton Finishes Laboratory, Southern Utilization Research and Development Division, United States Department of Agriculture, in New Orleans, Louisiana. He received a B.S. degree in Chemistry from the University of Southwestern Louisiana, a M.S. degree from Tulane University, and a Doctor of Science from Clemson University. He has been at the U. S. Government Laboratory doing textile research for nearly twenty years. He is the author of numerous publications and patents and is a member of a number of professional and scholastic organizations. In his current position, Dr. Reeves is leading research in the areas of durable-press finishes, soil-and stain-repellent finishes, mildew-, rot- and light-resistant finishes, and flame-resistant finishes.

Effects of Fabric Weight and Construction on Oxygen Index (OI) Values for Cotton Cellulose

James E. Hendrix, George L. Drake, Jr., and Wilson A. Reeves

*Southern Regional Research Laboratory**
New Orleans, Louisiana 70119

(Received June 14, 1971)

ABSTRACT

The relationship of weight and construction to OI values for untreated cotton cellulose fabrics was investigated. Fabric weight has a substantial effect on OI values. Construction parameters (fabric thickness, thread count, yarn number and weave pattern) have lesser but significant influence. Natural impurities in greige fabrics increase OI values, but also accentuate afterglow. These effects of natural impurities are separate from those of fabric weight and construction.

INTRODUCTION

The flammability characteristics of textile materials represent the composite effects of many factors. Broadly speaking these factors may be classified as fiber type, fabric weight and construction, and flame retardant finish utilized. Effects of various flame retardant finishes on different fiber classes have been a topic of research for many years, particularly during the last decade. Physical factors influencing flame retardancy, however, have received much less attention.

Drake [1] emphasized that weights and constructions of textile fabrics have significant effects on resultant burning characteristics. Lightweight fabrics usually burn much faster than heavier ones, and therefore require higher chemical add-ons to attain desired levels of flame resistance.

Willard and Wondra [2] recently carried out experiments utilizing treated fabrics having different weaves and weights. They observed that chemical requirements necessary to pass the vertical flame test or attain an OI value of 0.26 were substantially less for heavy than for lightweight fabrics. They demonstrated further that a

*One of the laboratories of the Southern Marketing and Nutrition Research Division, Agricultural Research Service, U.S. Department of Agriculture.

direct relationship exists between flame retardant add-on required and fabric air permeability.

Tesoro and Meiser [3] applied the oxygen index technique to the study of flammability properties of single- and multi-component textile fabrics. Values obtained for cellulose ranged from 0.180 to 0.197, a much broader range than the precision (±0.002) of the method would permit. They attributed this variability in OI values to impurities or minor effects of fabric construction. OI values, they explained, are independent of the physical form and dimensions of the sample over a broad range and thus reflect the effects of varying chemical composition of fibrous assemblies essentially independently of geometric parameters of the sample. The oxygen index technique, therefore, was thought to be particularly useful for isolating effects of chemical composition on flammability from complicating variables of fabric construction.

EXPERIMENTAL

All experimental fabrics evaluated were 100% cotton which had been desized, scoured and bleached. Industrial cotton ducks were tested in the greige state, after scouring and again after bleaching. OI values were determined using a commercially-produced oxygen index apparatus with chimney temperatures maintained at 25 ± 1 C.

Test methods employed were:

Oxygen Index	–	ASTM D 2863-70
Fabric Weight	–	ASTM D 1910-64
Thread Count	–	ASTM D 1910-64
Yarn Number	–	ASTM D 1059-69T
Air Permeability	–	ASTM D 737-13T

RESULTS AND DISCUSSION

Experimental fabrics having wide variations in weight, thread count, and thickness were fabricated from yarns of differing numbers. The fabrics equally comprised plain, 3/1 twill, and 5-harness sateen weave constructions. OI values for the experimental fabrics and for selected industrial ducks are presented as functions of fabric weight in Figures 1 and 2. It is clearly evident that OI values increase significantly with fabric weight in both cases, but that other factors are involved also. This is evidenced by the differences in OI values for the experimental fabrics and the industrial ducks.

The experimental fabrics were woven from single yarns and were desized, scoured and bleached. The correlation coefficient for the data population illustrated in Figure 1 is 0.854, which far exceeds the 1% level of significance. The industrial ducks, which had been fabricated from plyed yarns, were scoured and bleached so as to be in a state of purity comparable to the experimental fabrics. The correlation coefficient for the data population illustrated in Figure 2 is 0.979, which is considerably higher than that observed for the experimental fabrics. The ducks, however, lacked the wide

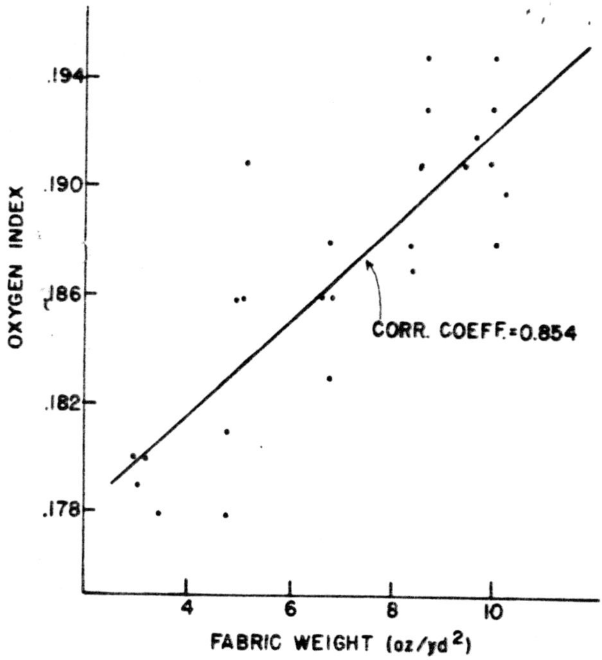

Figure 1. *Effect of fabric weight on OI values for experimental cotton fabrics.*

variations in construction parameters exhibited by the experimental fabrics.

Variables other than fabric weight which might influence OI values are tightness of the weave and sizes of the yarns contained. Generally lower thread counts are required in fabrics made from course yarns as compared to equivalent fabrics made from fine yarns. Thus fabrics made from coarse yarns are thicker and more loosely woven than equivalent fabrics made from fine yarns.

The tightness of the weave may be related to its cover factor, defined by the formula:

$$K_c = K_w + K_f - \frac{K_w K_f}{28}$$

where K_c is fabric cover factor, K_w is warp cover factor and K_f is filling cover factor. Warp and filling cover factors are defined as the ratio of threads per inch to the square root of the yarn count [4, 5]. From the formula it is obvious that low yarn numbers and high thread counts give tightly woven fabrics with large cover factors, compared to equivalent fabrics for which the opposite is true.

The linear relationship between OI values and K_c's is illustrated in Figure 3, the correlation coefficient being 0.888. This higher correlation for these data compared to that for the simple OI vs. fabric weight relationship illustrates the importance of

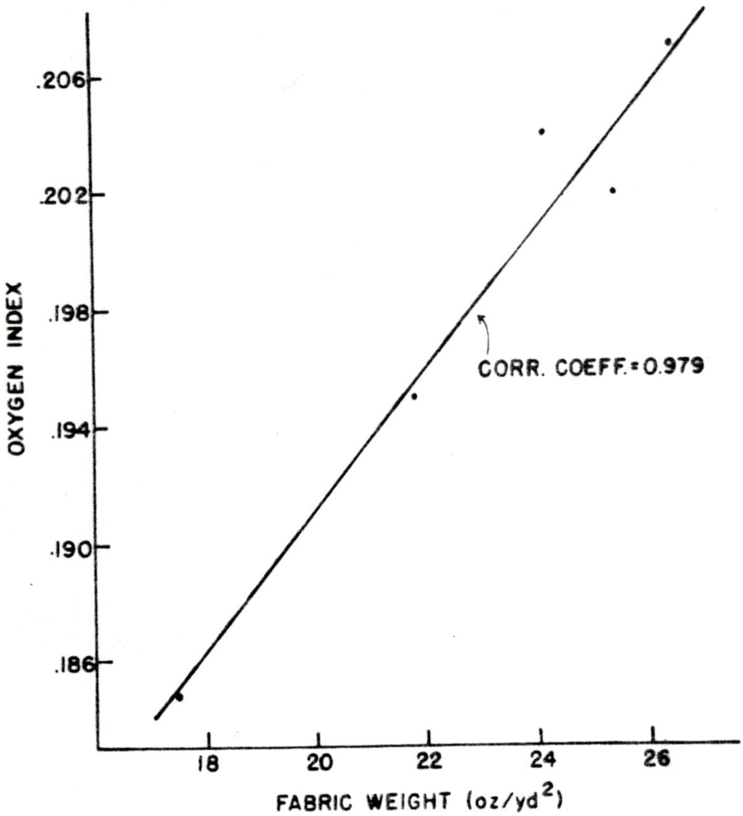

Figure 2. *Effect of fabric weight on OI values for industrial cotton ducks.*

construction in relation to flammability properties. Thus to a first approximation, 79% of the observed variability in OI values can be attributed to variations in K_c, while only 73% is attributable to weight variations. This is because K_c comprises both weight and construction terms.

With the OI vs. K_c relationship thus established as significant, one might consider a more practical method for measuring weave tightness than the rather involved cover factor calculations. Fabric air permeability should vary inversely with weave tightness, and thus might serve as a direct measure of overall construction factors.

Figure 4 illustrates the linear dependency of OI on log [fabric air permeability], the correlation coefficient being – 0.773. This indicates that fabric air permeability will account for only 60% of the variability in OI values, compared to 73% attributable to fabric weight or 79% attributable to weight plus construction. Fabric air permeability, therefore, provides a practical measurement of complex factors affecting OI values, but its significance is less than that of other measurements.

Pictorial representations of OI data presented thus far indicate that OI values are dependent on both fabric weight and construction (looseness or tightness of the

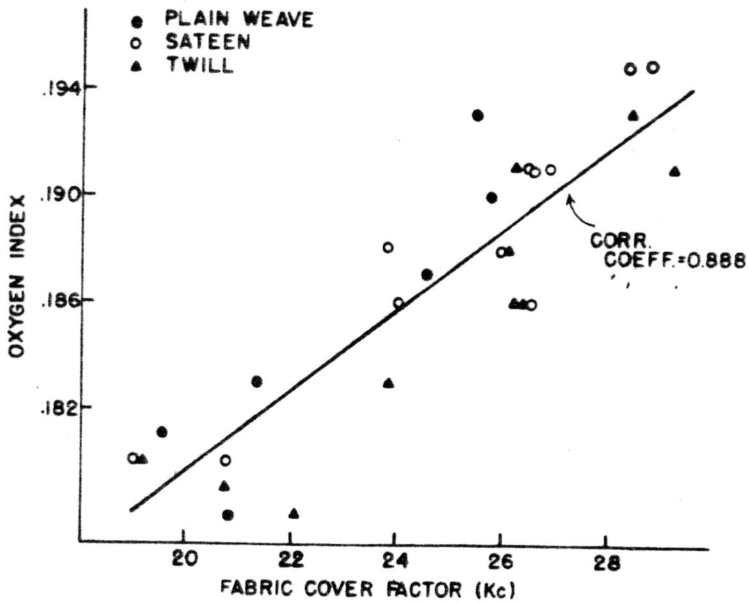

Figure 3. *Relationship between fabric cover factor and OI values.*

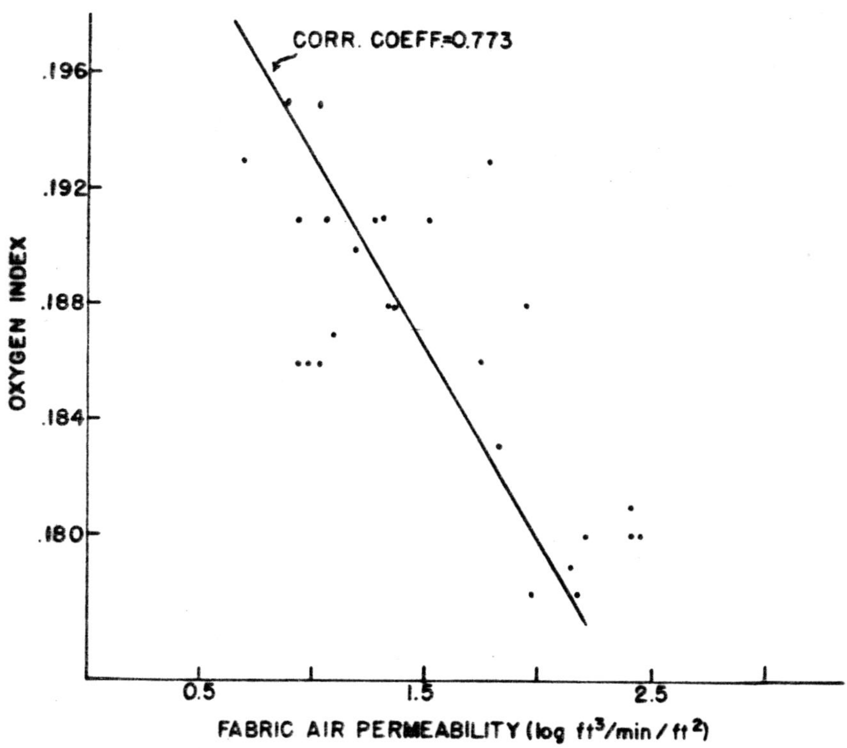

Figure 4. *Relationship between fabric air permeability and OI values.*

weave). It is of interest at this point to determine the extents which OI is dependent on these parameters, and also to examine effects of weave pattern (plain, sateen, and twill). This is easily accomplished by three-way analysis of variance (ANOVA) of OI values with respect to fabric weight, weave type and construction. Summary of the ANOVA is given in Table 1. Fabric weight is shown to be the dominant factor with its significance far exceeding the 1% level. Weave type is next in importance with its significance near the 1% level. Construction was nearly significant at the 5% level; however, this lower significance probably reflects lack of variation in looseness or tightness of the samples tested rather than the actual importance of construction as a source of variance. If construction is examined within weights, the 5 oz. fabrics, which had the widest variations in construction, show an effect which far exceeds the 5% level. The error mean square of 3.5×10^{-6} implies a root mean square of 0.0019 OI units, which is well within the limits of the precision (±0.002) of the test [3].

It has been suggested [3] that variations in OI values for cotton cellulose might be attributable to impurities present in the sample. We have shown by this investigation that large variations in OI values occur which are attributable to factors other than impurities (all samples were desized, scoured, and bleached); however, in view of differences in DTA thermograms reported for greige and purified cotton cellulose [6], and reported flame retardant properties of inorganic salts [7], one would expect OI values for fabrics of varying degrees of purity to be different. Samples of greige industrial cotton ducks were thus selected as a basis for evaluation. These samples were scoured and bleached. OI values were determined at each stage of purification and are presented in Table 2.

The greige samples, all of which contained inorganic salts attained during growth and afterglowed excessively, had substantially higher OI values than the fabrics from either stage of purification. The increased OI values are attributable to catalysis of the cellulose degradation through a path producing more char and less combustible gases [8].

Table 1. Analysis of Variance (ANOVA) of OI Values

Source of Variation	Degrees of Freedom	Sum of Squares	Mean Square	F
Fabric Weight	3	603.5×10^{-6}	201.2×10^{-6}	57.5**
Weave plan	2	60.3	30.2	8.6*
Construction within weight	4	50.5	12.6	3.6†
3.0 oz/yd²	1	0.7	0.7	0.2
4.9 oz/yd²	1	37.5	37.5	10.7*
8.5 oz/yd²	1	8.1	8.1	2.3
10.0 oz/yd²	1	4.1	4.1	1.2
Weave x weight	6	41.7	6.9	2.0
Error	8	28.0	3.5	

**Significant at the 1% level
*Significant at the 5% level
†$F_{5\%(4,8)} = 3.84$; $F_{10\%(4,8)} = 2.81$

Table 2. Effect of Sample Purity on OI Values

Fabric Sample	Description of Purity		
	Greige	Scoured	Bleached
22 oz. Industrial Duck	.231	.196	.203
25 oz. Industrial Duck	.226	.197	.202
27 oz. Industrial Duck	.234	.201	.207

SUMMARY AND CONCLUSIONS

The flammability properties of textile materials are extremely complicated, even in absence of externally applied flame retardants. Determination of these properties for cotton cellulose by use of the OI apparatus revealed that fabric weight has a large effect on flammability. Other factors affecting fabric flammability include those parameters related to fabric construction: thread counts and yarn numbers. An effect was also noted for varying weave patterns.

Aside from fabric weight and construction parameters, sample purity was shown to have a dramatic effect on OI values. Samples still containing normal growth contaminants (inorganic salts, etc.) are inherently less flammable than their purified counterparts, but these samples tend to afterglow excessively.

ACKNOWLEDGMENTS

The authors wish to thank George J. Kyame of this laboratory for supplying the textile fabric samples utilized in this investigation. The authors also thank John D. Tallant for his assistance in statistical analysis of the data.

REFERENCES

1. G. L. Drake, Jr., *Kirk-Othmer, Encyclopedia of Chemical Technology, Vol. 9, 2nd Edition,* New York, Interscience, 1966, pp. 300-315.
2. J. J. Willard and R. E. Wondra, *Text. Res. J. 40,* 203-210 (1970).
3. G. C. Tesoro and C. H. Meiser, Jr., *Text. Res. J. 40,* 430-436 (1970).
4. F. T. Pierce, *J. Text. Inst. 28,* T45-T96 (1937).
5. J. B. Dickson, Unpublished research.
6. R. M. Perkins, G. L. Drake, Jr., and W. A. Reeves, *J. Appl. Polymer Sci. 10,* 1041-1066 (1966).
7. W. K. Tang and W. K. Neill, *J. Polymer Sci.: Pt. C,* 65-81 (1964).
8. C. H. Mack and D. J. Donaldson, *Text. Res. J. 37,* 1063-1071 (1967).

Presented at 162nd National ACS Meeting, Washington, D.C., Sept. 12-17, 1971.

James E. Hendrix

James E. Hendrix is a National Academy of Science-National Research Council Postdoctoral Research Associate with the Southern Regional Research Laboratory at New Orleans, Louisiana. Prior to accepting the Postdoctoral Research Associateship, he was a cellulose research chemist with FMC Corporation at Princeton, New Jersey. He received his B.S. degree in Textile Science from Auburn University in 1966, his M.S. degree in Textile Chemistry from Clemson University in 1968, and his Ph.D. in Chemistry with a major in Textile Chemistry from Clemson University in 1970.

George L. Drake, Jr.

Mr. Drake is a native of Alabama. He received a B.S. degree in Chemistry from the University of Georgia in 1944 and his M.S. degree in Chemistry from the same school in 1949. He has been at Southern Utilization Research and Development Division since 1949. He is the author of numerous publications and patents and is a member of a number of professional and scholastic organizations including, ACS, RESA, and AATCC. In 1961 he became Head of Special Finishes Investigations. He was chosen Federal Civil Servant of the Year by the Federal Business Association of New Orleans in 1962, and was one of the recipients of the U.S.D.A. Distinguished Service Award in 1964. He received the U.S.D.A. Superior Service Award in 1960 and again in 1968. In 1969 he received the Herty Award presented by the Georgia Section of the American Chemical Society.

Wilson A. Reeves

Dr. Reeves is Chief of the Cotton Finishes Laboratory, Southern Utilization Research and Development Division, United States Department of Agriculture, in New Orleans, Louisiana. He received a B.S. degree in Chemistry from the University of Southwestern Louisiana, a M.S. degree from Tulane University, and a Doctor of Science from Clemson University. He has been at the U. S. Government Laboratory doing textile research for nearly twenty years. He is the author of numerous publications and patents and is a member of a number of professional and scholastic organizations. In his current position, Dr. Reeves is leading research in the areas of durable-press finishes, soil-and stain-repellent finishes, mildew-, rot- and light-resistant finishes, and flame-resistant finishes.

Alice M. Stoll and Maria A. Chianta

Naval Air Development Center
Crew Systems Department
Warminster, PA 18974

THERMAL ANALYSIS OF COMBUSTION OF FABRIC IN OXYGEN-ENRICHED ATMOSPHERES*

(Received March 24, 1973)

ABSTRACT: In search of an optimal two-gas space-capsule environment offering minimal hazard from clothing fire, fabric combustion was studied in oxygen-enriched atmospheres of argon, nitrogen, and helium at standard and at hypobaric pressures. From the thermal properties of the gas mixtures used the convective heat transfer coefficient was calculated for each level of pressure and each oxygen concentration. The experimental data showed that at relatively low heating rates, where the effect of the diluent gases could be differentiated from one another, the combustion rate varied inversely with the heat transfer coefficient. At higher heating rates, subtle differences due to specific effects of the diluent gases were over-ridden and destruction rates at each pressure level depended solely upon the mass flow of oxygen, the inert gas serving only to maintain the oxygen concentration. In all instances, the experimental data may be represented by equations of the form $Y = Ce^{nX}$ where Y is the mass flow of oxygen, X is the total heat flow and C and n are constants dependent upon the experimental conditions.

INTRODUCTION

In studies of the effect of oxygen-enriched atmospheres on the burning rate of fabrics [1], it was found that destruction rate could be correlated with the mass flow of both oxygen and diluent. It was observed also that pressure level made little difference in destruction rate when the diluent gas was helium while a measurable effect occurred when the denser gases, nitrogen and argon, were the diluents. This too indicated that the retardant effect of the diluent is a function of mass flow, for the absolute difference in mass of helium from one pressure level to another is negligible while the difference in mass of the heavier gases over the same pressure range is considerable. Consequently, no pressure effect was seen with the light gas while a definite effect was seen with the heavy ones.

The correlation of destruction rate with mass flow suggests that both constants, thermal conductivity and specific heat, which depend on mass, may be significant factors in the damping process. Furthermore, it is well known that the thermal conductivity and the time rate of heat capacity (i.e., the product of mass flow and

*Presented in part before the Heat Transfer Division of the American Society of Mechanical Engineers at the Winter Annual Meeting, New York, New York, November 28, 1972.

specific heat) of helium are much greater than that of nitrogen or argon. Both of these circumstances contribute to more rapid removal of heat from the burning site by helium than by argon or nitrogen and therefore should contribute to the observed superior retardation of burning by helium. Thus, the data suggested that investigation of the role of the heat transport properties of the gas media might assist in explaining the mechanism by which the inert gases retard combustion. The present analysis was made for this purpose.

EXPERIMENTAL MATERIALS AND METHOD

The apparatus and experimental method are described in detail in Reference 1. In brief, the apparatus consists of a pressure chamber which encompasses an ignitor coil of nichrome, a stand for holding the fabric specimen, a thermocouple for measuring (approximately) the temperature of the coil and specimen, a flowmeter for measuring precisely the flow of gas into the chamber and appropriate recording and control instruments. Figure 1 shows schematically the chamber (1 cu. ft. in volume); gas tank with regulator valve and flowmeter; altimeter and vacuum adjust valves for setting pressures; specimen mount; heating coil and power supply; and thermocouple for monitoring the ignitor and specimen temperatures.

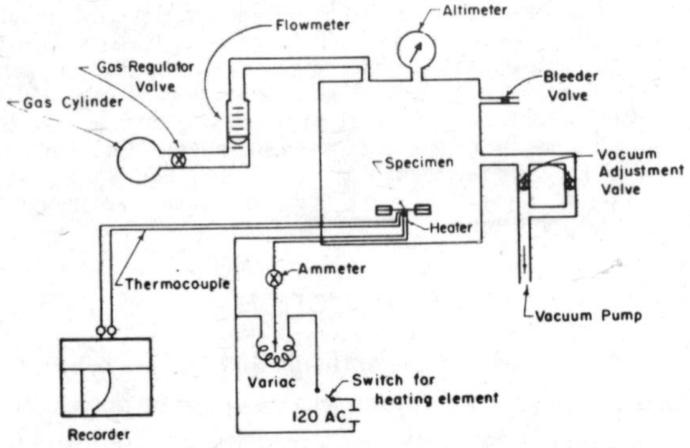

Figure 1. *Pressure chamber with specimen mount.*

The fabric sample was mounted between two square transite plates, 4″ on edge, with a circular central hole 2 5/8″ in diameter. The hole was centered over the ignitor so that the fabric specimen was positioned directly over the ignitor and about 2 mm above it in the early experiments. Later, similar experiments were conducted with the fabric in contact with the ignitor. The fabric used in all experiments was a fire-resistant polyamide, Nomex[1], in a filament fiber, 2/2 twill weave, 3 oz/yd^2 in weight. It was chosen because its flame-resistant quality so inhibited pyrolysis throughout the entire range of oxygen concentration that distinctive

[1] Trade name of E. I. duPont de Nemours textile fiber.

phases in the burning process could be identified more easily in this fabric than in others.

Each specimen was heated to ignition and the heat maintained in the coil until the end of the episode, i.e., throughout burning whether to completion or to self-extinguishment. Three gas mixtures were used, viz., O_2/N_2, O_2/He and O_2/A. Three levels of pressure were used, viz., 14.7, 10.9 and 7.3 psia (equivalent to sea level, 8,000 ft. and 18,000 ft. altitude respectively). At each oxygen concentration the actual flow of oxygen was adjusted for each diluent gas and pressure level to provide the same mass flow of oxygen to the specimen as that delivered at a flow of 15 standard liters/min (SLPM) [1] so that the effects due to each diluent could be separated from those due to oxygen availability alone.

THERMAL PROBLEM, PROCEDURES AND DATA

The objective of the analysis is to determine the heat transport properties of the various gas mixtures as they pertain to the combustion of the fabric.

The total heat input to the system, illustrated in the greatly simplified sketch in Figure 2, is composed of the heat from the ignitor coil, H_h, and the heat generated

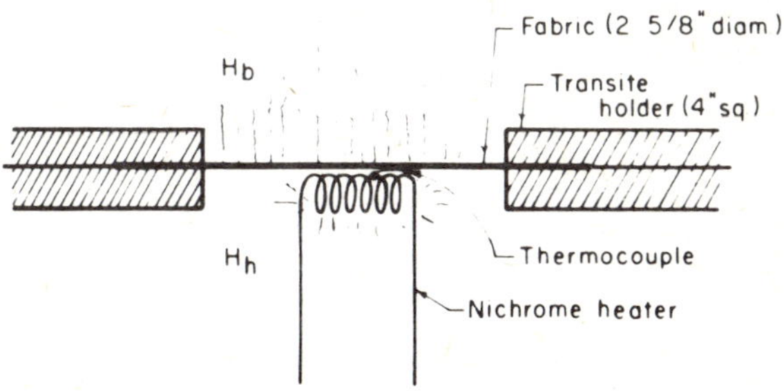

$$H = H_h t + H_b t_b$$

H = Total heat generated

t = Total time from onset of heat to end of burning

t_b = Time duration of fabric burning

Figure 2. *Total heat input.*

by combustion of the fabric, H_b. Since the fabric is not in contact with the coil, it receives heat by radiation and convection as well as hot gas conduction. That part which is supplied by convection is subject to variation as the thermal transport properties of the gas mixtures vary. However, since the gap is small in comparison to the mass of heater coil and fabric involved, these differences may be fairly small and most of the heat may be supplied by radiation. Such considerations do not pertain to the instances in which the fabric was heated by contact with the coil.

In all instances, heat loss proceeds by radiation and convection until the combustion front reaches the holder whereupon conduction heat loss begins to contribute. Since conduction enters only at the end of the burning episode when the burning front of the fabric contacts the holder it may be neglected. Radiation loss, while not insignificant, should be of the same order of magnitude regardless of the gaseous medium since it depends only on the temperature of the fabric as compared with that of the wall. The oxygen content influences the heat flow by the variation it causes in the magnitude and intensity of the combustion reaction, largely a convective process. Thus, of the total heat loss, the greatest variation occurs in the convective portion.

For the analysis of this loss the equation chosen as most nearly approximating the experimental situation is that for the coefficient of heat loss by natural convection from a heated horizontal plate facing upward [2]. This configuration would apply until the fabric burned away whereupon the horizontal plate in that area would be imaginary but the heat generated by the flame front would still be lost as a function of the heat capacity of the gaseous medium. The condition of natural convection was selected as more appropriate than forced convection because the gas flow of 15 SLPM in the relatively large chamber used provided a flow velocity of only 0.088 ft/sec across the specimen. The actual flow required to maintain 15 SLPM as the pressure was reduced, increased to about 20 L/min at 10.9 psia and 30 L/min at 7.3 psia, a velocity of 0.132 and 0.176 ft/sec respectively, or essentially quiet air as determined in conditioned chambers maintained at fixed temperatures [3]. The analysis, however, is complicated by the fact that after the fabric ignites, the gas flow changes from laminar to turbulent as burning progresses. Also, this change occurs at different times after ignition depending on the burning rate. However, the difference effected in the equation (Equation 1 below) is confined to the constant, C, and the exponent, x, which depend upon the particular situation and which are adjusted to adapt to the appropriate flow type:

$$\frac{h_c L}{k_f} = C \left[\frac{L^3 \rho_f^2 \, g \, \beta_f \, \Delta t}{\mu_f^2} \left(\frac{c_p u}{k} \right)_f \right]^x \tag{1}$$

where

- h_c = heat loss by convection, coef., Btu/hr ft^2 °F
- L = length in feet
- k_f = thermal conductivity at t_f, Btu/hr ft^2 °F/ft
- t_f = "film" temperature = $(t_s + t_a)/2$, t_s for surface, t_a for ambient
- C = constant determined by particular situation
- ρ_f = density at t_f, lb/ft^3
- g = acceleration due to gravity, 4.17×10^8 ft/hr^2
- β_f = coef. of volumetric expansion, reciprocal °F
- Δt = temperature difference (fabric – gas)
- c_p = specific heat, Btu/lb °F

μ = viscosity, lb/hr ft = 2.42x centipoises
x = exponent determined by situation

The data required to insert in this equation were derived from a variety of sources and calculated to yield the information in appropriate units.

The specific heat of each gas mixture is determined from Equation 2 [4]:

where:

$$c_{pm} = w\, c_{p1} + (1-w)c_{p2}$$

c_{pm} = specific heat of mixture

$$w = \text{mass fraction of component 1} = \frac{m_1}{m_1 + m_2}$$

where m_1 = weight of gas #1, etc. (2)

and as cited in [5]:

$$w = \frac{(\%D \times MWD)}{(\% O_2 \times MWO_2) + (\%D \times MWD)} \tag{3}$$

where

D = diluent gas = component #1
MW = molecular weight

The viscosity of the binary mixtures is calculated according to the method of C. R. Wilke [6] cited in [4]:

$$\mu_m = \frac{\mu_1}{1 + (x_2/x_1)(\phi_{12})} + \frac{\mu_2}{1 + (x_1/x_2)(\phi_{21})} \tag{4}$$

where

$$\phi_{ij} = \frac{[1 + (\mu_i/\mu_j)^{1/2} (M_i/M_j)^{1/4}]^2}{(4/\sqrt{2})\ [1 + (M_i/M_j)]^{1/2}}$$

where

$M_i + M_j$ = molecular weights of the pure gases
$x_1 + x_2$ = mole fractions
$\mu_i + \mu_j$ = viscosities of pure components

Thermal conductivity of the binary mixture is found by the method of Lindsay and Bromley [7] cited in [4]:

$$k_m = \frac{k_1}{1 + A_1\,(x_2/x_1)} + \frac{k_2}{1 + A_2\,(x_1/x_2)} \tag{5}$$

where

k_m = thermal conductivity of the mixture, other symbols are defined as in Equation 4 above, and A_1 and A_2 are Wassiljewa constants [8] as cited in [4]. These are:

$$A_1 = 1/4 \left\{ 1 + \left[\frac{\mu_1}{\mu_2} \left(\frac{M_2}{M_1} \right)^{3/4} \frac{1 + (S_1/T)}{1 + (S_2/T)} \right]^{1/2} \right\}^2 \frac{1 + (S_{12}/T)}{1 + (S_1/T)}$$

$$A_2 = 1/4 \left\{ 1 + \left[\frac{\mu_2}{\mu_1} \left(\frac{M_1}{M_2} \right)^{3/4} \frac{1 + (S_2/T)}{1 + (S_1/T)} \right]^{1/2} \right\}^2 \frac{1 + (S_{12}/T)}{1 + (S_2/T)}$$

where

T = absolute temperature, °K
S = Sutherland constant

and

$$S_{12} = \sqrt{S_1 \cdot S_2}$$

The density of the gas mixtures varies in proportion to their percentage composition. Therefore,

$$\rho_{O_2D} = x\,(\rho_{O_2}) + (1-x)\,\rho_D \tag{6}$$

where

ρ_{O_2D} = density of oxygen mixture at given temperature and pressure

x = percentage of O_2 present

ρ_D = density of diluent at given temperature and pressure

and

ρ_{O_2} = density of oxygen at same temperature and pressure

Table 1 presents the properties of the pure gases of interest at 80°F as compiled from various engineering handbooks and the data book of the supplier of the gases used [9]. Units are chosen to provide dimensionless numbers in accordance with McAdams' usage [2, pp. 135, 180].

Table 1. Properties of Pure Gases at 80° F and 14.7 psia

Gas	Density lb/ft^3	Sp. Heat Btu/lb °F	Viscosity(u) lb/hr ft	Conductivity Btu/hr ft°F	Diffusivity ft^2/hr	Prandtl *
Helium	0.1106	1.2420	0.0488	0.08640	6.2900	0.702
Oxygen	0.0812	0.2198	0.0499	0.01546	0.8662	0.709
Nitrogen	0.0713	0.2486	0.0432	0.01514	0.8542	0.713
Argon	0.1016	0.1252	0.0541	0.01030	0.8097	0.658

These data were used in a computer program to solve for h_c, the heat transfer coefficient. The value of 1200°F, calculated from the approximate temperature of the flaming fabric (∿ 1290°F as measured by thermocouple) was used for Δt, the temperature difference between the fabric and the gas. The values for C, the constant, and x, the exponent, were determined within the program on the basis of the form of the curve,

$$Y = CX^n \text{ where}$$

$$Y = \frac{h_c L}{k_f};\ X = \frac{L^3 \rho^2 g \beta_f \Delta t}{\mu^2}\left(\frac{c_p \mu}{k}\right)_f \text{ and } n = x$$

Thus, when turbulent flow was indicated by values of X from 2×10^7 to 3×10^{10}, C became 0.14 and n = 1/3. Similarly, in the laminar range, X from 10^5 to 2×10^7, C = 0.54 and n = 1/4 [2]. The resultant data and the constants used in the computations at 14.7 psia are shown in Table 2 for mixtures ranging from pure oxygen to pure diluent in 10% decrements of O_2.

(It is noteworthy that under these conditions the heat transfer coefficient of pure helium is only 1.7 times that of pure oxygen even though its thermal conductivity is almost 6 times as great. At the 50–50 point O_2/He is only 1.3 times as effective as O_2/N_2 as a heat loss medium while argon and its mixtures exhibit even less heat transport capacity than oxygen alone. Furthermore, the relative magnitudes of these values may be expected to pertain in similar heat loss situations irrespective of the dimensions of the surface because they reflect the characteristics of the gaseous media themselves.)

Additional solutions appropriate to the other two levels of pressure were also carried out. The specific heat, thermal conductivity and viscosity depend upon pressure only in the vicinity of the critical point defined for each gas in Table 3.

*Table 2. Heat Transfer Coefficients of Gas Mixtures * †*

Gas Mixture Mole Fraction O_2	Argon	Conductivity Btu/hr ft°F	Density lb/ft³	Specific Heat Btu/lb °F	Viscosity lb/hr ft	Heat Transfer Coefficient Btu/hr ft² °F
1.00	0	0.01546	0.08120	0.21980	0.04990	4.92066
0.90	0.10	0.01492	0.08324	0.20828	0.05038	4.78337
0.80	0.20	0.01439	0.08528	0.19730	0.05084	4.64640
0.70	0.30	0.01386	0.08732	0.18683	0.05129	4.50740
0.60	0.40	0.01334	0.08936	0.17683	0.05173	4.36859
0.50	0.50	0.01282	0.09140	0.16728	0.05215	4.22821
0.40	0.60	0.01231	0.09344	0.15813	0.05256	4.08806
0.30	0.70	0.01180	0.09548	0.14938	0.05297	3.94597
0.20	0.80	0.01129	0.09752	0.14099	0.05335	3.80260
0.10	0.90	0.01079	0.09956	0.13293	0.05373	3.65940
0	1.00	0.01030	0.10160	0.12520	0.05410	3.51690
O_2	Helium					
1.00	0	0.01546	0.08120	0.21980	0.04990	4.92066
0.90	0.10	0.01842	0.07419	0.23381	0.05047	5.29546
0.80	0.20	0.02179	0.06717	0.25080	0.05106	5.65251
0.70	0.30	0.02567	0.06015	0.27181	0.05167	5.99313
0.60	0.40	0.03018	0.05315	0.29849	0.05227	6.31798
0.50	0.50	0.03548	0.04613	0.33345	0.05783	6.42403
0.40	0.60	0.04180	0.03912	0.38130	0.05238	6.66725
0.30	0.70	0.04945	0.03210	0.45075	0.05350	7.10579
0.20	0.80	0.05890	0.02509	0.56070	0.05326	7.57284
0.10	0.90	0.07073	0.01807	0.76116	0.05206	8.00321
0	1.00	0.08640	0.01106	1.24200	0.04880	8.35657
O_2	Nitrogen					
1.00	0	0.01546	0.08120	0.21980	0.04990	4.92066
0.90	0.10	0.01540	0.08021	0.22235	0.04923	4.90877
0.80	0.20	0.01534	0.07922	0.22497	0.04856	4.89695
0.70	0.30	0.01528	0.07823	0.22766	0.04790	4.88486
0.60	0.40	0.01523	0.07724	0.23041	0.04723	4.87410
0.50	0.50	0.01517	0.07625	0.23324	0.04656	4.86352
0.40	0.60	0.01511	0.07526	0.23615	0.04589	4.85191
0.30	0.70	0.01505	0.07427	0.23914	0.04522	4.84036
0.20	0.80	0.01499	0.07328	0.24220	0.04454	4.82914
0.10	0.90	0.01493	0.07229	0.24536	0.04387	4.81772
0	1.00	0.01488	0.07130	0.24860	0.04320	4.80520

*Figures obtained from handbooks and computer; significance of heat transfer coefficient no better than third decimal place.

† $L = 0.219$ ft $P = 14.7$ psia $t_a = 80°F$

Table 3. Critical Point, Pressure and Density of Pure Gases

Gas	Temp (°C)	Pressure (atm)	Density (gm/cm^3)
Argon	−1.22	48.0	0.531
Helium	−267.9	2.3	0.0693
Nitrogen	−147.1	33.5	0.3110
Oxygen	−118.8	49.7	0.430

The conditions of the present study are far from the critical point of any of the gases, therefore no changes were needed in these values. Density is directly proportional to the pressure so that ρ is easily found at each of the new pressure levels as $\rho = p^1/p_{atm} \times \rho_{atm}$ where p^1 is the new pressure, p_{atm} is atmospheric pressure, and ρ_{atm}, the density at atmospheric pressure. These data were used in the same computer program and the heat transfer coefficient found for the levels of 10.9 and 7.3 psia. All values for h_c at the three levels of pressure are shown in Figure 3,

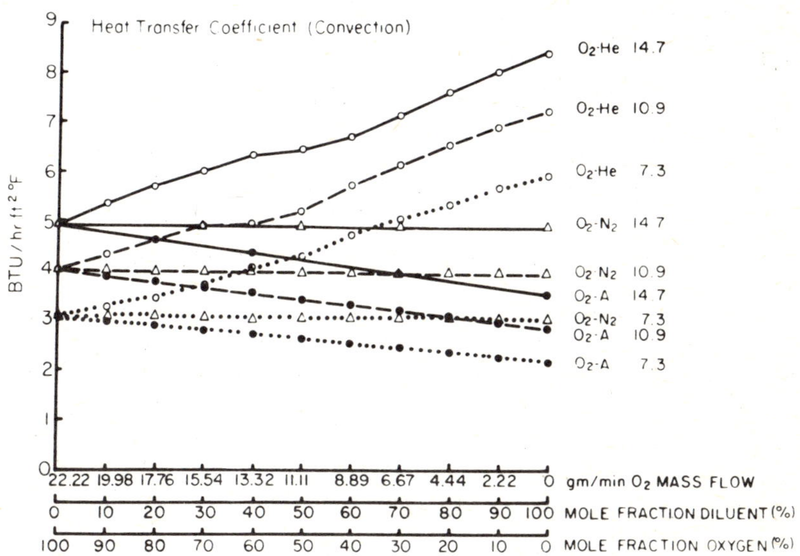

Figure 3. *Convective heat transfer coefficient (h_c) versus mass flow of oxygen.*

correlated with the mass flow of oxygen in gm/min at the mixture flow rate of 15 SLPM. A change in slope occurs in each O_2/He curve at the zone of transition from laminar flow (at the lower O_2 levels) to turbulent; in all other instances the curves are constant in slope and the flow is turbulent.

The total heat generated in the system (H, Figure 2) during any burning episode is the sum of the heat supplied by the heater coil (H_h) and the heat resulting from combustion of the fabric (H_b) multiplied by the burning time. The first term, H_h may be found from the resistance of the nichrome coil, the measured current and voltage and the burning time:

$$H_h = R\,i^2\,t \text{ or } E\,i\,t$$

where

R	=	resistance of the coil	=	2.06Ω
i	=	current	=	9.6 amp
E	=	voltage	=	20 volts

and

t = time.

The second term, H_b, may be determined by DuLong's Rule or Mahler's Rule [10], both of which are based on the caloric value of each chemical element in the material being burned in proportion to its presence in the molecule:

DuLong's Formula:

$$H_b = 14{,}650C + 62{,}100H - 5{,}400(O + N)$$

Mahler's Formula:

$$H_b = 14{,}650C + 62{,}100H(H - 1/8\ O)$$

(present material completely composed of C,H,O and N).

In the present experiments H_h was 192 watts or 45.9 cal/sec and H_b, 11,735 Btu/lb or 6.54 cal/mg. If the heat of combustion greatly exceeds the coil heat, then the latter could be neglected. However, at 6.54 cal/mg a burning rate of about 7 mg/sec produces enough heat to equal that supplied by the coil. Therefore, in assessing the influence of heat transport on burning rate at given oxygen flows, it is necessary to consider the entire heat produced rather than only that due to combustion. The total heat generated, H, may be determined as:

$$H = H_h\,t + H_b\,t_b \tag{7}$$

where

t = total time from onset of heat to end of burning and t_b = time duration of fabric burning, i.e., from ignition to completion or extinction.

Then the total heat flow may be calculated as the coil heat supply rate, 45.9 cal/sec, plus the observed burning rate in mg/sec times 6.54 cal/mg, the heat of combustion of the fabric.

RESULTS AND DISCUSSION

In th[...]nce of heat supplied without contact between the coil and the fabric, the correl[...]s shown in Figure 4 were obtained. Here it is seen that the log of the mass flow of oxygen (ordinate) is directly related to the total heat flow. At each pressure level the curves representative of the three O_2 diluent mixtures converge on the same point at the maximum O_2 flow rate (100%). At concentrations less than 100% O_2, the curves diverge in a manner commensurate with the differences in heat transfer coefficient (Figure 3), i.e., the higher the coefficient at a given O_2 level, the lower the heat flow and (since the coil input remains constant) the slower the burning rate. At increasingly low pressures the burning rates are proportionately slower. However, the heat transfer coefficients at these pressures are also lower, therefore, this trend is in the direction opposite to that expected if heat transport capacity alone were to account for control of the burning rate at any given O_2 mass flow. Furthermore, if heat transport capacity were the sole controlling factor, at points of equal heat coefficients and equal O_2 mass flow, the same destruction rates would be observed irrespective of gas composition. The data in Table 4 show that the rates at these points (located graphically as the crossover points in Figure 3) are not the same. Instead, there is a difference between rates in different gas mixtures at the same pressure level differences (e.g., O_2/A at 14.7 psia and O_2/He at 10.9 psia = a difference of 1.5 mg/sec; O_2/N_2 at 14.7 and O_2/He at 10.9 = 2.1) and in the same gas mixtures at different pressure level differences (e.g., O_2/A at 10.9 and O_2/He at 7.3 = 1.3; O_2/A at 14.7 and O_2/He at 7.3 = 5.9).

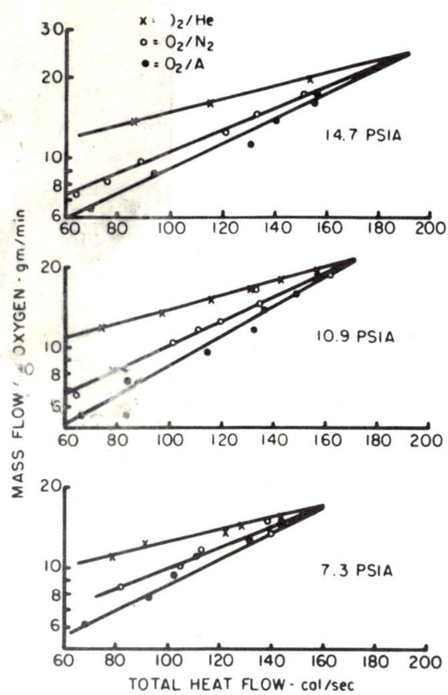

Figure 4. *Oxygen mass flow vs. total heat flow (no contact) at three pressure levels.*

In contrast, the experimental data obtained from heating the specimen by contact with the ignitor, at approximately the same heat output of the coil, reveal no effect due to the nature of the diluent gas. As seen in Figure 5, at each pressure level, all the data points cluster about the same line to yield three parallel curves for the three pressure levels. In this situation, differences in total heat flow attributable to specific effects of the diluent gas are obliterated and the burning rate depends entirely on the oxygen concentration and the pressure level. Tabulating the destruction rate data at equivalent heat coefficient and oxygen flow rates as before now

Table 4. Comparison of Destruction Rates of Fabric in Different Gas Mixtures at Points of Equal Oxygen Mass Flow Rates and Equal Heat Transfer Coefficient Values

O_2 Flow gm/min	Heat Transfer Coefficient (h_c) BTU/hr ft^2 °F	Gas at psia		Dest. Rate mg/sec		Difference mg/sec (#1 - #2)
		#1	#2	#1	#2	
17.76	4.6	O_2/A 14.7	O_2/He 10.9	17.0	15.5	1.5
15.76	4.9	O_2/N_2 14.7	O_2/He 10.9	14.4	12.3	2.1
15.76	3.7	O_2/A 10.9	O_2/He 7.3	16.1	14.8	1.3
13.76	4.0	O_2/N_2 10.9	O_2/He 7.3	12.9	11.1	1.8
11.55	4.2	O_2/A 14.7	O_2/He 7.3	10.9	5.0	5.9
7.87	4.8	O_2/N_2 14.7	O_2/He 7.3	3.6	*	-
6.77	3.9	O_2/A 14.7	O_2/N_2 10.9	4.6	2.8	1.8
3.33	3.3	O_2/A 10.9	O_2/N_2 7.3	Off scale		

*Did not burn.

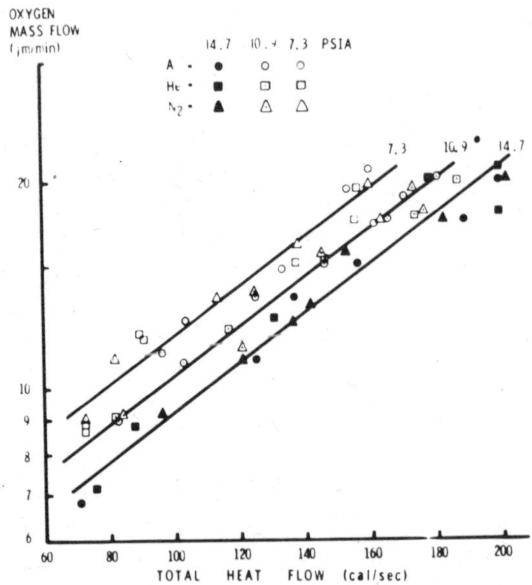

Figure 5. *Oxygen mass flow versus total heat flow (specimen in contact with heating coil) at three pressure levels.*

discloses an orderly difference of about 2 mg/sec between points at 14.7 psia and 10.9 psia and about double this difference between 14.7 and 7.3 psia where the pressure level difference is about double (Table 5). It is possible that the difference due to pressure level may be attributed to chemical kinetics. However, it is also possible that it shows up because the equation used for calculating heat transport capacity applies to natural convection rather than forced convection and the actual flow, however small, does double and triple at the lower pressures. The latter possibility can be explored by using the equation for forced convection but this has not been done yet. It is likely that the differences attributable specifically to the nature of the diluent gases are obscured by the greater heating rate effected by contact of the fabric with the heating coil. This point could be resolved experimentally by removing the coil after ignition of the specimen occurs. However, the whole effect is so weak there is little practical worth in expending the time and resources required to generate these data. Of more importance is the fact that in the face of relatively vigorous heating the efficacy of the inert gas in slowing combustion is based simply on its efficiency as a diluent for the oxygen present. This effect is illustrated in Figure 6 where the destruction rate is correlated with mass flow of diluent. The curves are superimposable so that all data may be represented by the same curves by displacing the destruction rate on the abscissa approximately 2 mg/sec to the right for each decrement of approximately 4 psia in pressure. Now it is seen that the same extent of limitation may be achieved with less than 2 gm/min of He as is obtained with more than 10 gm/min of N_2 or 16 gm/min of A.

COMMENT

The original observations relating destruction rate to the fire-retardant aspects of the gas mixtures are discussed in detail in Reference 11. For the present purpose, the analysis of the thermal aspects of the burning process, it is pertinent to note that the general form of the combustion process may be deduced from the experimental data. As seen in Figure 3, there is a direct and linear relationship (except in the transition zone between laminar and turbulent flow) between the heat transfer coefficient and the mass flow of oxygen. Then, since the total heat flow involved in the combustion process is related directly to the log of the mass flow of oxygen (Figures 4 and 5), the main controlling factor in the burning process, it follows that the combustion process itself may be expressed by equations of general form $Y = CX^n$ which is the form of Equation 1.

Furthermore, the coefficient for mass transfer with respect to the gaseous media is entirely analogous to that for heat transfer [4]. Therefore, the experimental data should bear the same relationship to mass transfer as to heat transfer. In the instance of mass transfer, $Y = h_D L/D$, the dimensionless mass transfer coefficient, analogous to $h_c L/k$, and $X = f\ (Gr_D,\ Sc)$, the Grashof number for mass transfer, and the Schmidt number for molecular diffusivity, respectively, analogous to the Grashof and Prandtl numbers in heat transfer. Of course, these relationships are speculative and far from all-inclusive.

Table 5. Comparison of Destruction Rates of Fabric in Different Gas Mixtures at Points of Equal Heat Transfer Coefficient Values, Fabric in Contact with Ignitor

O_2 Flow gm/min	Heat Transfer Coefficient (h_c) BTU/hr ft^2 °F	Gas at psia		Dest. Rate mg/sec		Difference mg/sec (#1 - #2)
		#1	#2	#1	#2	
17.76	4.6	O_2/A 14.7	O_2/He 10.9	20.2	17.9	2.3
15.76	4.9	O_2/N_2 14.7	O_2/He 10.9	19.2	16.9	2.3
15.76	3.7	O_2/A 10.9	O_2/He 7.3	16.0	13.8	2.2
13.76	4.0	O_2/N_2 10.9	O_2/He 7.3	15.8	13.6	2.2
11.55	4.2	O_2/A 14.7	O_2/He 7.3	12.9	8.8	4.1
7.87	4.8	O_2/N_2 14.7	O_2/He 7.3	6.2	2.1	4.1

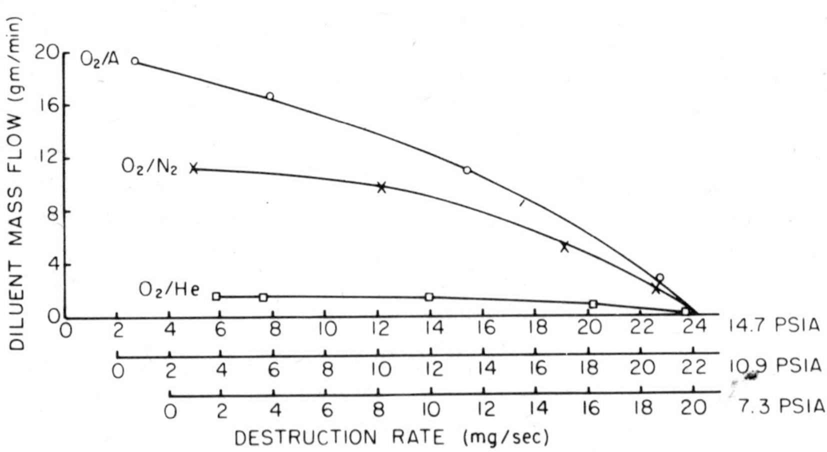

Figure 6. *Mass flow of diluent gases versus destruction rate of fabric.*

Reaction rate theory, too, must have a bearing on the total analysis, for all these mechanisms enter into the combustion process at one stage or another. Indeed, a complete understanding of the broad relationship demonstrated here undoubtedly involves a simultaneous consideration of combinations of all these processes, a formidable task. In the meantime, it is of considerable practical value to be able to express quite accurately for a given combustion situation the effect of variation in composition and pressure of the gaseous medium by use of equations of this familiar general form.

CONCLUSION

It is concluded from the present analysis that flame propagation under the conditions of essentially quiet air flow depends primarily upon the oxygen mass flow available to the specimen. When heating of the specimen is accomplished at a slow rate as in the instance of no contact between ignitor and fabric, differences in damping ability of the three inert gases can be discerned and related to the differences in heat transport capacity of the mixtures. When heating is more intense, as during contact with the ignitor, these differences are obliterated.

It is concluded, further, that whether the heat transport of the gaseous medium or the mass transport on combustion of the material is considered to be the dominant influence, the experimental data fit an equation of the general form $Y = Ce^{nX}$ where Y is the mass flow of oxygen; C and n, constants appropriate to the experimental conditions; and X, the destruction rate or a function thereof (such as heat flow). This observation is of considerable practical importance because it provides a means of predicting from only two or three isolated empirical data points the effect of oxygen dilution by inert gases throughout an entire range of combustion.

REFERENCES

1. M. A. Chianta and A. M. Stoll, "Effect of Inert Gases on Fabric Burning Rate," Aerospace Med. 40: #12, Dec. 1969, pp. 1304–1306.
2. W. H. McAdams, Heat Transmission, 3rd ed., McGraw-Hill Book Co., Inc., New York, Toronto and London, 1954, p. 180.
3. A. P. Gagge, "Standard Operative Temperature," *Temperature: Its Measurement and Control in Science and Industry*, Reinhold Publishing Corp., New York, N.Y., 1941, p. 550.
4. R. G. Eckert and Robert M. Drake, Jr., *Heat and Mass Transfer*, McGraw-Hill Book Co., Inc., New York, 1959, p. 494.
5. Ibid., p. 440.
6. C. R. Wilke, *J. Chem. Physics* 8: 517, 1950.
7. A. L. Lindsay and L. A. Bromley, *Ind. Eng. Chem.* 42: 1, 508, 1950.
8. A. Wassiljewa, *Physik Z.*, 5: 737, 1904.
9. *Matheson Gas Data Book*, 4th Ed., The Matheson Co., Inc., East Rutherford, N.J., 1966.
10. *Scott's Standard Methods of Chemical Analysis*, N. H. Furman, Ed., 5th Ed. Vol. I, D. Van Nostrand Co., Inc., Princeton, N.J., New York, Toronto and London, Mar. 1958, p. 1187.
11. M. A. Chianta and A. M. Stoll, "Fire Retardance of Mixtures of Inert Gases and Oxygen," Aerospace Med. 44: #2, Feb. 1973, pp. 169–173.

A. M. Stoll

Alice M. Stoll, Head of the Biophysics Laboratory, Crew Systems Department, Naval Air Development Center. Background in the biophysics and physiology of burns and biotechnology of heat transfer. Interest in present work derived from accidental burning of experimental subjects in simulated space capsule clothing fire which occasioned a study of O_2-enrichment effects on fabric flammability. Heat transfer analysis was made on experimental data obtained in this investigation.

M. A. Chianta

Maria A. Chianta, Research Chemist in Biophysics Laboratory, Crew Systems Department, Naval Air Development Center, responsible for conduct of study of O_2-enriched atmospheres of inert gases suitable for space capsule environment offering life support with minimal fire hazard. Data in present report was generated during the phase of study dealing with fabric flammability and subjected to heat transfer analysis as part of an effort to determine the damping mechanism of various gas mixtures.

AUTHORS' NOTE

Opinions and conclusions contained in this report are those of the authors. They are not to be construed as necessarily reflecting the views or endorsement of the Department of the Navy.

Burning Rate and Heat Transfer from Vertically Burning Cotton Cloth Panels

Norman J. Alvares,
Stanford Research Institute, Mendo Park, California
and
Perry L. Blackshear
Department of Mechanical Engineering, The University of Minnesota

(Received October 14, 1969)

ABSTRACT

ı apparatus was developed to measure the burning rate and heat transfer rate of vertically hanging cotton cloth panels. Of particular interest were the burning characteristics of the cloth upon approach to an isothermal wall. This information was necessary to describe the heat transfer environment between burning clothing and skin, and between burning drapery and walls, or upholstered objects. The measured quantities were: (1) the mass loss rate, (2) the radiant heat transfer rate, (3) the total heat transfer rate, and (4) the mass of deposited condensate per unit area (at distances between the wall and cloth where flame quenching occurs). It was observed that the rate of mass loss increased to a maximum as the distance between the wall and cloth was reduced to 3 in. Both the total and radiant heat transfer rates attained their maximum levels at a distance of 1.5 in., and convective flux at 0.75 in. Significant quantities of condensate were deposited at a distance between the cloth and wall of 0.5 in.

INTRODUCTION

Every year, scores of people and millions of dollars are lost to the economy because of uncontrolled fires. While laws have been made to provide guidelines for the construction of houses and buildings with a minimum of fire hazards, these laws are negated by the trappings that are used for personal comfort and decoration. A large proportion of the upholstery fabrics, rugs, and drapes used in the modern home are made of flammable materials, and these items determine whether or not a room will become enveloped in an uncontrollable fire. Similarly, the garment industry persists in making clothing (particularly for feminine use) of flammable materials. Cotton dresses and nightgowns, ignited at the hemline, burst into intense flames which result in, at worst, death, and at best, permanent disfigurement to the wearer.

Studies by the IIT Research Institute [1] indicate that before a room could become fully enveloped in fire, a significant item in the room must be burning. This ensures that the heat load in the room is sufficient to distill combustible gases from the rest of the room's contents. These gases collect at the ceiling and ignite. When the thermal energy provided by these processes is of sufficient magnitude, all other combustible materials in the room will ignite. This phenomenon has been named "Flashover." If the drapes, curtains, or wall hangings contain flammable materials, they become very important in the buildup of the heat load to the point of flashover. In addition, drapes are quite often the initial ignition source in room fires. Thus, the heat transfer from the drapes to the walls and nearby furniture is of primary importance in the study of practical fire prevention in the home.

A parallel problem is the heat transfer from burning clothes to skin. This is understandable when we recognize that thermal damage to skin and the ignition of cellulosic materials follow similar response parameters, i.e., there exist critical rates of energy deposition before the skin will exhibit damage, or before cellulosic materials will ignite.

Stoll [2] has shown that the surface of the skin must attain a temperature of 44°C before skin damage occurs, and that irreversible skin destruction is sustained at a surface temperature of 72°C. At all temperatures above 44°C, the skin sustains damage regardless of whether or not the temperature pulse is increasing or decreasing. Thus, the most important parameter for the assessment of injury is the skin temperature history. Buettner [3], Henriques [4], and Hardy [5], as well as Stoll, have studied the relationships between heat transfer and skin injury for idealized thermal energy boundary conditions. Thus, it should now be fruitful to describe some real thermal environments to which the skin may be exposed, so that skin temperature histories may be approximated for similar exposures.

The ignition of cellulosic solids by thermal radiation has been studied extensively in the past [6, 7]. Lee, et al [8] shows that the lowest irradiance (H), at which cellulosic materials (simulated upholstery panels) will ignite to flaming combustion is: $H = 0.25$ cal cm^{-2} sec^{-1}* for a duration of about 100 seconds. In this case, the estimated surface temperature at the time of ignition is 300°C. At lower surface temperatures, irreversible changes will occur in the structure of the cellulose, but flaming ignition will not occur. Very little information is available concerning the heat produced by burning fabrics. Less is known about the potential of burning items in the spread of fire to adjacent items. Webster [9] has measured the total heat transfer from burning fabrics to a nearby surface and has observed a maximum total heat transfer rate (H^*)** of about 0.7 cal cm^{-2} sec^{-1}. This rate is quite capable of producing both thermal damage to the skin and the ignition of cellulosic fuels. These data, however, are limited to one distance between the cloth and the surface and, therefore, do not constitute an adequate study of the problem.

It was the goal of this research to determine the heat transfer parameters for the two parallel situations reviewed above. These measurements indicate

---

* Irradiance is a term particular to radiant heat exchange only. Hereafter, whenever the symbol H appears, it is to be understood that the units are: cal cm^{-2} sec^{-1}.

** H^* will be defined as total heat transfer rate, which includes all modes of heat transfer.

the burning rate ($\dot{m}$)* and the heat transfer rate as a function of distance between the cloth and an isothermal wall.

When flame does not touch the wall, it is clear that the dominant modes of heat transfer to any adjacent surface is radiation and convection. In the case of clothing fires, however, the fabric is very close to the skin, and flame quenching will occur. This could result in a flux of condensing pyrolysis products depositing on the skin. If the enthalpy of condensation of the products is large enough, this process could contribute significantly to the heat transfer and the resulting damage. Techniques were devised to differentiate between radiative, convective, and condensation heat transfer to the wall.

The measurements indicate that heat transferred from normal-sized, cotton-containing drapes and wall hangings is of sufficient magnitude to be a significant contributor to the fire spread in enclosure fires. If an upholstery item is within 12 inches of an actively flaming drape, fire spread in the item would be significantly enhanced, and if the extent of the drapes was large, ignition of the item is possible. Damage to skin will occur at any distance within 12 inches of flaming cloth, with severe damage occurring at the distances normally associated with wearing apparel.

INSTRUMENTATION AND EXPERIMENTAL TECHNIQUE

Burning rate and heat transfer measurements were not made simultaneously because of the complexity of maintaining the various instruments required for each task. The burning rate measurements were made first because the experience gained in observing the burning characteristics of the samples at various wall distances would complement the heat transfer measurements. Due to this order, we shall designate the burning rate measurements as Phase I, and the heat transfer measurements as Phase II, and discuss them independently.

The first task of the effort was to choose a representative material that would adequately simulate both clothing and drapery fabrics. Since cotton is the most common fabric used in this country, a moderately heavy, pure cotton fabric (Osnaburg muslin, wt = 22 mg, cm^{-2}) was selected as the experimental material.

Phase I—Burning Rate:

The burning rate is defined as the rate of mass loss due to combustion, and mass loss is measured by weighing the specimens during combustion. A strain gauge force transducer was employed for this purpose. It was installed in the weighing gallows shown schematically in Fig. 1. Vertically orientated cloth specimens six feet long by one foot wide were connected by a system of pulleys to the calibrated strain gauge. The mass loss of the specimens during combustion resulted in a change of voltage from the gauge which was continuously monitored by the strip chart recorder.

The first set of measurements was made using free-hanging cloth samples, which hereafter will be referred to as "free-burns." The technique included

---

* $\dot{m}$ = the amount of mass loss to the material, per unit time, due to combustion.

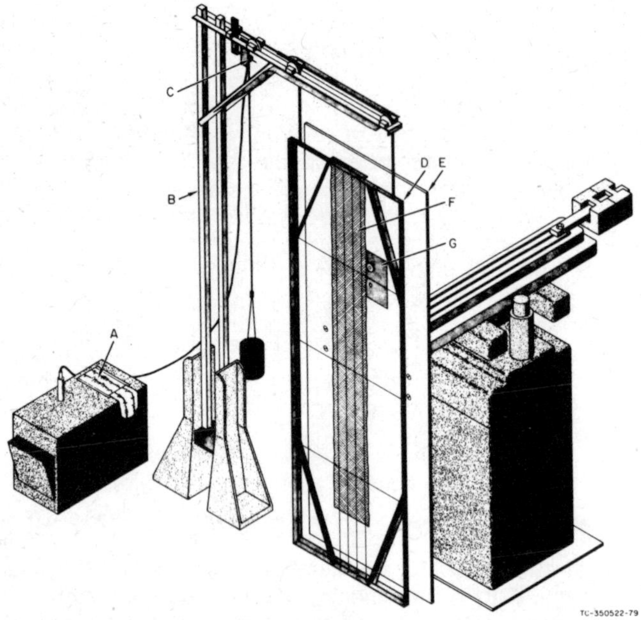

Figure 1. *Schematic detail of the experimental apparatus, where the letter designators show, respectively, a, mass loss recorder; b, weighing gallows; C, load cell; d, retaining screen; e, isothermal wall; f, cloth sample position; and g, thermal instrument mount on the isothermal wall.*

spraying a small amount of alcohol across the bottom of the cloth, igniting it with a match, and recording the mass loss. It was noted that the rate of mass loss was significantly altered by changes in atmospheric humidity, with the response that the mass loss was inversely proportional to relative humidity. Since environmental control was not practical, quantitative data gathering was deferred until the fall, when atmospheric variations would not cause extreme changes in the humidity.

The free-burning cloth panels were quite stable during the flaming combustion period, and did not appear to lose cohesive strength until after the subsequent glowing phase had begun to die out. However, if the burn occurred near a vertical surface, the draft created by the flame caused the cloth to move toward the wall. This behavior could not be tolerated since close approach to the wall was a necessary prerequisite for performance of the research. Therefore, a constraining device was constructed to control the movement of the cloth. It consisted of a grid of thin steel wires that was attached to an aluminum frame. Measurements of $\dot{m}$ for free-burns and constrained burns were found to be the same (in the absence of near walls), so that no correction factor had to be applied to the frame burns.

The mean flame-standoff-distance (Y*) is defined as "The separation distance between the luminous flame front and surface to the solid fuel" [10]. This parameter is important to flame-spread measurements, since flame-quenching by an external surface may commence at the boundary defined by the flame edge. Measurements were made of Y* by placing a scribed metal rod normal to the surface of the cloth and photographing the rod during several burns. For both free- and flame-burns, these measurements showed that Y* was about 3.0 inches.

An isothermal wall was constructed of aluminum plate (2 feet by 6 feet and 1/4-inch thick) reenforced with angle iron along each long edge. Three-eights-inch copper tubing was intuitively distributed on the back, fastened down by clips, and covered by heat "Thermon" transfer cement. The plate was vertically oriented on the horizontal traveling table of a surface grinder. A thermometer was inserted in the back of the plate for temperature control.

Phase II—Heat Transfer:

After the burning rate measurements were finished, a square opening was cut in the top section of the isothermal plate to receive a mounting containing heat sensing elements. The elements were salvaged from thermal radiation calorimeters used initially for measuring the thermal energy emitted from atomic bombs. Their calibration was accomplished by use of standard instruments obtained from the Naval Radiological Defense Laboratory.

It was obvious from the burning rate experiments that the wall calorimeters would require a black that was sturdy enough to stand up to periodic cleaning. Thus, the sensing elements were blackened with black paint (Nextel Velvet Coating Black) which, when dry, forms a strong, impervious surface that retains its properties to temperatures as high as 300°C. Similar black paints have reported absorptivities of 0.97 [12]. Two elements were installed in the mounting plate. A silver chloride infrared filter was located in front of the topmost element so that it would absorb only the radiative flux from the burning cloth, while the unfiltered instrument would sense the total energy. The filtered element and the total element are from now on abbreviated as Cal-F and Cal-T, respectively.

Two water-cooled radiometers were constructed and calibrated by the same techniques as the radiometric instruments described in Reference 11. However, each instrument contained two separate sensing elements [13] made with 1-mil constantan foils. The foils were coated with the 3M black, and on each instrument one foil was covered by a sapphire (Al_2O_3) infrared filter. The radiometers were used to discriminate between the radiative and convective flux on the free side of the burning cloth. The total, and filter sensing elements of the radiometer, will hereafter be abbreviated as Rad-T and Rad-F, respectively.

The quantity of condensate collected on the wall was measured by placing a 1-mil foil, with dimensions of 7.3 and 9.5 cm, in intimate contact with the wall, just above the instrument station, and weighing the foil before and after experiments where flame quenching occurs.

A flame detector probe with 400 d-c volts impressed across the gap was located next to the sensors in the wall mount. The electrons and ions generated in the flame create a conduction path in the gap between the elements of the probe, thus closing the circuit between the probe and recorder. In

Figure 2. *Photographic detail of the experimental apparatus.*

this way, the detector senses the flames as soon as they reach the level of the instruments.

The chronology and techniques of the heat transfer and burning rate measurements were similar. The first measurements were at the maximum distance (δ) between the wall and cloth (essentially $\delta = \infty$). The last measurements were made at $\delta = 0.5$ inch, where flames were partially quenched between the wall and cloth. Only the calibration details were different. For $\dot{m}$, a weight calibration was the practice before each run, while a millivolt calibration was required for each circuit before heat transfer measurements.

Figure 2 is a photograph of the overall apparatus for heat transfer and mass loss measurements.

RESULTS

The format used for the presentation of the results will follow that used for the experimental section, i.e., the burning rate and the heat transfer data will be isolated into Phases I and II, respectively. However, first will be discussed a parameter that is common to both the burning rate and the heat

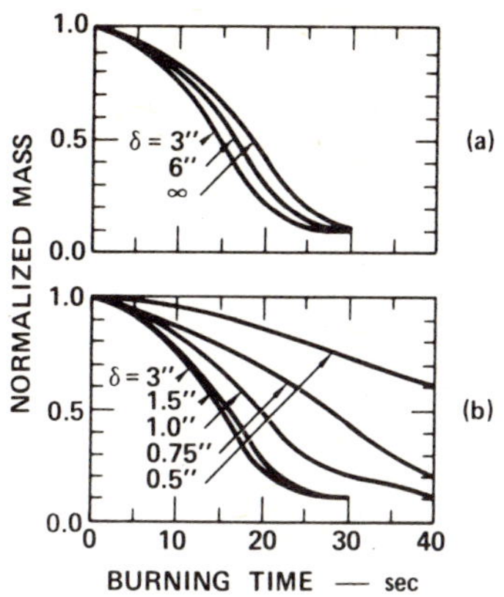

Figure 3. *Composite of mass loss data for all δ's.*

transfer measurements, the temporal appearance of the cloth during combustion.

As previously stated, the specimen is ignited by spraying alcohol across the bottom of the cloth and lighting it with a match. After two to three seconds, the alcohol burns out and the cloth becomes the major source of flame. The flame length then grows to a turbulent height of about five feet. By the time this state is reached, the bottom of the cloth apparently can no longer support flaming combustion, and the entire mass of flames moves upward along the cloth. The flames then attain the top of the frame and the nonflaming bottom boundary progresses up the cloth at a decelerating rate until it reaches the top, where the flame goes out. The length of the turbulent flame mass is roughly constant for a time period of 10-15 seconds, starting at about ten seconds past ignition. Following the bottom boundary of the flame is a pulse of glowing combustion, which appears to pass through a maximum intensity at a finite time after the flame has passed. As the glowing combustion dies out, the remaining hot ash loses mechanical strength, and falls away from the wire grid. As δ is reduced to Y^* (~3 in.), the rate of flaming combustion appears faster, and the glowing phase apparently lasts longer. Also, the height of the flame between the cloth and the wall is much longer than the height of the flame burning up the outside of the specimen. When δ is less than one inch, the flame starts uniformly at the bottom of the cloth, but then begins to move up the edges from which it slowly migrates toward the center, producing copious quantities of smoke between the cloth and wall. When approximately two-thirds of the specimen have been consumed, active flaming sometimes resumes.

Phase I—Burning Rate:

Figures 3a and 3b are composite reproductions of weight loss at the various δ's of measurement. Figure 3a shows smooth curves of the data taken from the measurements at $\delta = \infty$ to $\delta = 3$ in., while Fig. 3b shows the data from $\delta = 3$ in. to $\delta = 0.5$ in. Clearly indicated is the increase of m from $\delta = \infty$ to $\delta = 3$ in., and the decrease as δ becomes less than Y^*. The shape of the mass loss data is similar until $\delta = 0.5$ in., i.e., the slope is steep during the flaming phase, and abruptly decreases when active flaming stops and glowing begins. Notice that the mass loss due to glowing appears to be minimum at $\delta = 3$ in. Figure 4 is an isotemporal plot of the normalized mass loss versus δ, over the flaming phase of the measurements. These data display more dramatically the mass loss efficiency at $\delta = 3$ in. The mass loss measurements were terminated soon after the active flaming phase, since at least 80 percent of the mass was by then consumed.

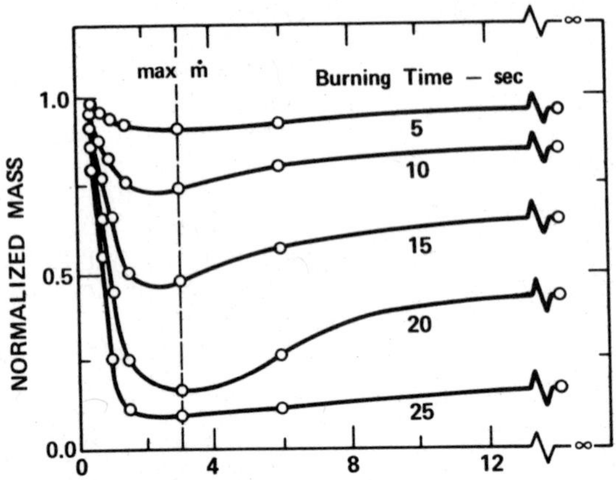

Figure 4. *Isotemporal plots of normalized mass loss versus δ.*

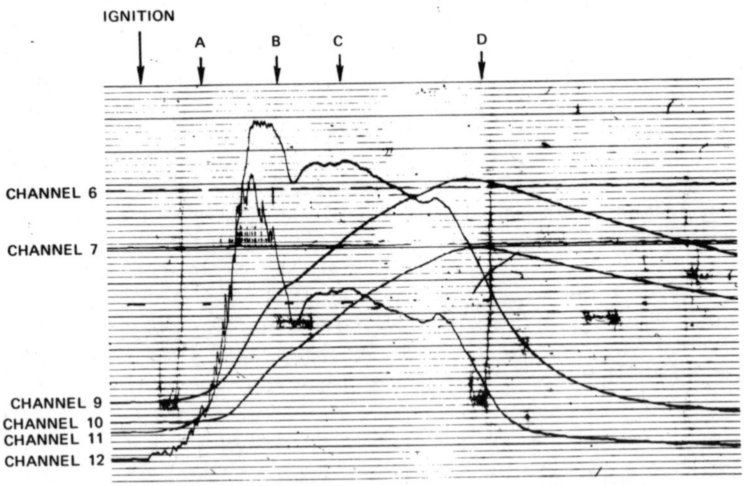

Figure 5. *A photoreproduction of strip chart recorder heat transfer data where: ch 7 through ch 12 are respectively, the flame detector, an event marker, Cal-T, Cal-F, Rad-F, and Rad-T, and where the letter designators are: a, flame detected at thermal instrument station: b, flame leaves thermal instrument station: c, flame-out: and d, breakup.*

Phase II—Heat Transfer:

Figure 5 is a photoreproduction of raw data for a measurement at $\delta = 3$ in. The galvanometer traces have been identified by channel number. Channels 7 through 12 are, respectively, the flame detector, a remote pulser, Cal-T, Cal-F, Rad-F, and Rad-T. Note that the peak intensity recorded by the radiometers corresponds to the steepest slope of the calorimeters, and that the flame detector indicates active flaming in this region. The sensitivity of the Rad-T had been reduced by more than a factor of 2, resulting in the relatively small

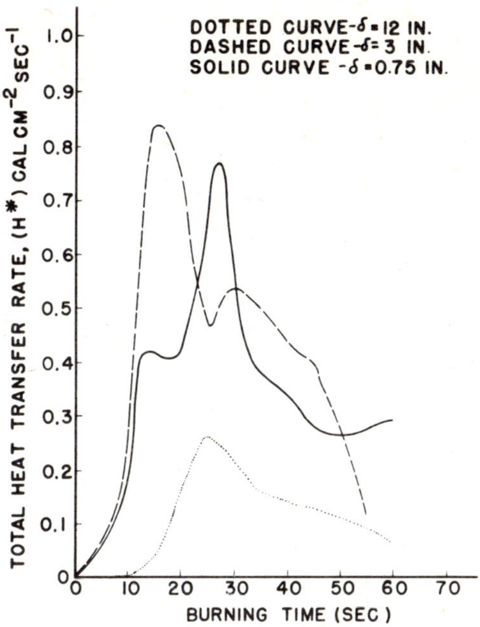

Figure 6. *Cal-T data showing the total heat transfer rates at δ's of 12, 3, and 0.75 in., respectively.*

signal magnitude indicated in the figure. The apparent noise of Rad-T is an indication of the degree of turbulence produced by the active flaming from the cloth. In order to realize information from the radiometer traces, deflection data were taken from smooth curves drawn through the apparent noise. The letter designators, at the top of the figure, indicate thermal milestones of a test, and are identified as follows: A, flame detected at the instrument station; B, flame bottom lifts from station area; C, flame-out, and D, breakup.

Figure 6 includes plots of differentiated Cal-T data at $\delta = 12$, 3, and 0.75 in., respectively. These curves show definite characteristics particular to measurements at each distance, and are quite similar to the response of Rad-T. At $\delta = 12$ in., the view factor of the detector includes the entire specimen, so that the local characteristics of the burn are washed out, and the energy received by the detector is integrated over the total burn time. At $\delta = 3$ in., the local heat transfer characteristics become apparent. We see that the peak intensity occurs when the flame is present at the instrument station, followed by an obvious minimum which can be attributed to the junction between flaming and glowing combustion, after which the glowing combustion peak occurs. Then the intensity of thermal energy decreases, in a variable way, until breakup. At $\delta = 0.75$ in., the cloth is within Y^*, resulting in flame-quenching between the cloth and the wall. This behavior is apparent

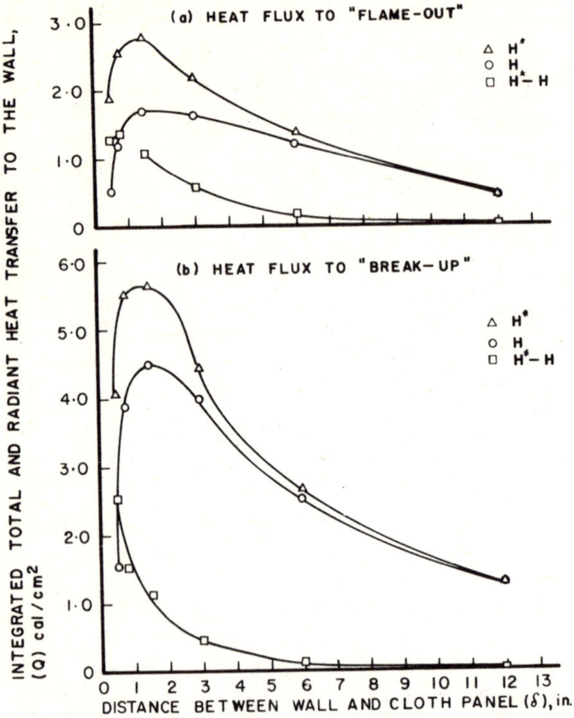

Figure 7. *Integrated total and radiant heat transfer as a function of δ, where (a) shows data to "flame-out," and (b) shows data over entire building time to "breakup".*

upon perusal of the output of the sensors. Cal-T shows a rapid rise in heat transfer, following closely the trend of the $\delta = 3$ in. data. This rise then levels off for almost 10 seconds. Then the major pulse of energy occurs, peaks, and then dies out to a lower, almost constant level. The flame detector identifies this structure as a flame-nonflame-flame sequence which indicates that the 10-second period between the two flame pulses is the result of flame quenching. This observation is supported by the presence of condensate on the wall after the flaming phase of the burn.

Integrated heat transfer data are described in Figs. 7a and 7b. The top portion (a) of the figure contains curves of H* and H for the time period between ignition and flame-out. The bottom portion (b) contains curves of H* and H for the time period between ignition and breakup. Included in both sections of the figure is a plot of the delta, H*-H, which is an indication of the convection flux for $\delta \geqq 1.5$ in. and the combination of convection and condensation flux for $\delta < 1.5$ in.

The outer limit of condensation deposition was observed to be at $\delta = 1$ in. No weight measurements were made until $\delta = 0.75$ in., where mass accumulation at the wall averaged 3.6×10^{-4} g cm^{-2}. At $\delta = 0.5$ in., the mass accumulation was found to be 1.1×10^{-3} g cm^{-2}, which is a factor of 3 increase between the two measurements.

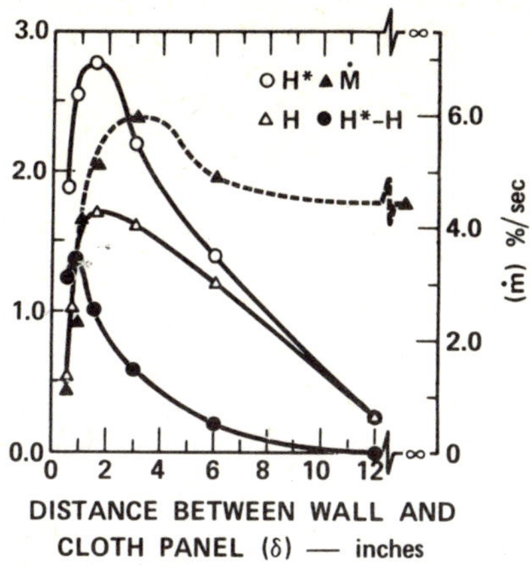

Figure 8. *Comparison of burning rate with integrated heat transfer data as a function of δ.*

DISCUSSION

The goals of this research were to ascertain and to quantize the heat transfer processes that take place in the region between a flaming cloth panel and an isothermal wall. The measurements have brought to light several parameters of interest which help to describe the burning and heat transfer characteristics of the vertical cloth panels at various diminishing distances between the cloth and the wall. These characteristics can be summarized as follows:

$\delta \geqq 12$ in.

$\dot{m}$ = constant

H* = H = constant (A decreasing value as δ increases)

3 in. $\geqq \delta \geqq$ 1.5 in.

$\dot{m}$, tends to maximum
H*, tends to maximum
H, tends to maximum

1.5 in. $\geqq \delta \geqq$ 0.5 in.

$\dot{m}$, decreases
H*, decreases
H, decreases
(H*-H is maximum at δ = 0.75 in.)

Figure 8 compares $\dot{m}$ with the integrated values of H*, H, and H*-H, as a function of distance from the wall, for the burning period which is included by the time of ignition, and flame-out. Note that the $\dot{m}$ peak occurs at 3 in., the H and H* peaks at 1.5 in., and the H*-H peak at 0.75 in. These observations are now discussed independently.

1. *δ = 3 in., Peak H* = 0.82* (values are derived from rate measurements similar to the data in Fig. 6). At δ = 3 in., visual inspection of the burns has resulted in the observation that the flame length between the cloth and the wall is half again as long as the flame length on the free side of the cloth. Since δ = 3 in. ≅ Y*, we may conjecture that during the active flaming phase of the burn, the region between the cloth and the wall is entirely filled with hot combustion gases. It is possible that the setting of δ = 3 in. provides an air entrainment rate which produces the optimum fuel-air ratio for this combustion geometry, thus, maximizing $\dot{m}$.

2. *δ = 1.5 in., Peak H* = 0.97* (values are derived from rate measurements similar to the data in Fig. 6). At δ = 1.5 in., the wall is well within Y* and $\dot{m}$ has decreased. Here, the flame height on the wall side of the cloth is twice as long as it is on the free side, indicating a considerable buoyancy in the gas mixture between the cloth and the wall. Both the integrated H and H* reach their maximum value at this distance. The increase in H* is quite large and is undoubtedly due to the enhanced convective interaction that must

occur within the flame boundary. The maximum in integrated H is probably due more to the increase in burning time than to the radiant intensity of the hot cloth and flame gas emissions. The air entrainment rate is large, as indicated by the flame height, but the amount of air available for combustion is limited by the magnitude of δ, which is probably the reason why $\dot{m}$ is decreased from the $\delta = 3$ in. value.

3. δ = *0.75 in., Peak H** = *0.77* (values are derived from rate measurements similar to the data in Fig. 6). At $\delta = 0.75$ in., H^*-H (an indication of the convective heat transfer component) attains its maximum value, while both H and H^* suffer abrupt decreases. $\dot{m}$ has been reduced to almost one-third of its maximum value, and very close to one-half of its free-burn value. The reduction in H is the most dramatic heat transfer change. It may be caused by the reduction of flame-spread rate. Apparently, the initial velocity of the combustion gases between the cloth and wall are not sufficient to produce air entrainment rates that penetrate to the central regions of the panel. In this interior region, the cotton cloth is subject to the hot products of combustion from below and the heat flux from the free side of the flames, resulting in pyrolysis of the cloth. However, since there is no air available, combustion will not occur, and the pyrolytic gases convect up along the interface between the wall and the cloth. During this convection process, some of these gases diffuse to the wall and condense. Eventually, the flames from the edge and bottom join at about three-fourths of the distance to the top of the cloth, and active flaming ensues. Until this coalescence, convection is the major heat transfer contributor to the wall. Figure 6 shows these temporal heat transfer effects quite well.

Figure 7 shows that the heat transfer characteristics of the total burn, including the glowing region, are similar in distribution to the heat transfer in the flaming region alone. The only difference is that radiative transfer appears to play a more important role for the total burn. The total energy deposited on the wall after flame-out is equal to the total energy deposited during active flaming. In relative terms, approximately one-half of the total energy is released in the first third, and the other half, during the remaining two-thirds of the total burning time. This relationship is apparently valid for the heat transfer measurements at any distance between the wall and the cloth.

Isothermal pyrolysis measurements indicate that 75 percent of the products of pyrolysis of cellulosics are noncondensable gases and condensable vapors [14]. Blackshear and Murty [15] have reported results from the pyrolysis of cellulosic dowels, which show temperature plateaus around 100 and 370°C. This indicates that at low to moderate heating rates, water vapor is the first product of the breakdown of the cellulosic molecule. Isothermal pyrolysis data [16], however, suggest that a homogeneous pyrolysis gas mixture, consisting of water vapor and the other components, is produced at different rates throughout molecule breakdown.

The condensate collected on the gold foil was also deposited on a silver chloride window for intrared analysis. The spectrum of this substance was compared to spectrum of the tar-fraction, derived from the pyrolysis of pure cotton [17], and was found to agree quite well. The main constituent of the tar-fraction is levoglucosan, which is quite soluble in water. The condensate deposited on the wall was observed to be quite liquid soon after a burn had been made. If the wall was not cleaned quickly, however, the deposit became "gummy" and hard to remove; if left for a period of a day or so, it changed to a plastic-like solid. Infrared analysis of the initial, liquid-like condensate shows a strong absorption peak at 3.0 microns, which is not present in the spectrum of the solid-like condensate. Absorptions in this spectral region are characteristically ascribed to OH bonds. Thus, it appears that water is lost from the condensate upon aging. The temporal properties of the wall deposit can, therefore, be explained in two ways. First, reference 15 would indicate that water vapor (the first product of pyrolysis) is deposited on the wall prior to the tar-fraction, which is released later at about 300°C. Upon contact with the wall, it forms a solution with the water layer already present there. Or, secondly, using the conclusions of the isothermal pyrolysis studies of reference 16, the tar-fraction is either in solution with the water vapor, because of the pyrolysis mechanism, or is an intimate mixture of water vapor, tar-fraction, and other things.

Since there exists no thermodynamic information concerning the unknown constituents of the wall deposits, and since water appears to be a major component of these deposits, estimates regarding the upper limit of heat deposited by condensation will be made, using the properties of water vapor. It will be assumed that the vapors diffuse to the wall in the saturated state, condense at the wall temperature, and deposit their condensation enthalpy to the wall. For steam to water, at 20°C, the condensation enthalpy is 585 cal g. Using the measured quantity of condensate deposited at the wall at $\delta = 0.5$ in. (1.06×10^{-3} g cm^{-2}), the total heat transferred to the wall by saturated condensing steam is 0.62 cal cm^{-2}. That this is a rough estimate is obvious, but recall that some unknown constituent of the condensate has been observed to evaporate at low temperatures. Imagine the rate of evaporation when exposed to the glowing cloth after the active flaming phase of the burn is over. Thus, the amount of condensate deposition, measured by the techniques described herein, may not be nearly as large as would be obtained if the sample were to be extracted immediately after flaming had ceased. The uncertainty of these results and deductions require caution in interpretation. Therefore, it will be assumed that the total heat transfer to the wall, by the condensation of volatile pyrolysis products, is of the order of tenths of cal cm^{-2} or less. The relative importance of the condensation flux in transferring heat to the wall at $\delta = 0.5$ in. is of the same order as the radiative flux. It appears then that the convection process is most efficient in this region and is responsible for at least half, if not more, of the total heat transfer. The amount of condensate deposited at $\delta = 0.75$ in. is about one-third of the amount deposited at $\delta = 0.5$ in.; thus, the deposited energy is also one-third of whatever it is at $\delta = 0.5$ in.

SUMMARY

The heat transfer parameters, for the case of vertically oriented burning cloth panels near an isothermal wall, have been measured as a function of distance from the wall. The motivation for these measurements was to describe the thermal environments to which the wall was exposed, and apply these data to the real situations of burning clothing next to the skin, or to flaming drapes next to walls and upholstery items. The results were to allow prediction as to the response of skin and flammable room items to the measured thermal exposures.

Skin Burn Parameters:

Stoll reports that skin exposed to irradiance levels as low as $H = 0.1$ produces skin temperatures that exceed 44°C, which is the defined threshold temperature for skin damage. Perusal of the thermal energy curves contained in Fig. 6 shows that H^* magnitudes at all the measurement distances exceed $H^* = 0.1$ by significant margins over relatively long periods of time. Similarly, at $\delta \leqq 6$ in., H^* exceeds the maximum exposure rate to produce irreversible tissue damage. Thus, for the experimental conditions described herein, thermal damage of extreme severity will result to the skin at all $\delta \leqq 6$ in. This is not a startling or unexpected result. Intuition would be qualitatively as informative. However, these data do show the potential danger of rather common-place clothing articles, such as cotton nightgowns. Also illustrated is the duration during which a garment can produce significant amounts of thermal energy. The flaming combustion phase is only responsible for approximately 50 percent of the total thermal energy, while the glowing phase yields the rest. Thus, to reduce skin damage by clothing fires, it is not only important to put out the flames, but also to remove the still glowing ashes as quickly as possible.

Ignition and Flashover Parameters:

The ignition response of cotton cloth and simulated upholstery panels exposed to thermal radiation has been explored elsewhere (i.e., references 6, 7, 8, and 18). In order for upholstery panels to be ignited by thermal radiation, they must be exposed to an irradiance of $H = 0.5$ for a duration of 20 to 30 seconds [18]. The burning cloth specimens of this research that produced average H^*'s (high enough and for long enough periods of time for the ignition of upholstery panels) were located in the region of 3 in. $\geqq \delta \geqq 0.75$ in. from the wall.

It is obvious that increasing the size of the panel will result in an increase in the irradiance to a unit area of the surface at the wall. However, angle factor calculations [19] show that this irradiance enhancement approaches an asymptote of $2H$ at $\delta = 12$ in. as the width of the panel is increased to 4 ft. Also, larger panels may take longer to burn. Thus, it is quite possible that the enhancement of H^* and of burning duration due to an increase in panel size may add up to a significant ignition threat to upholstery items exposed to flaming drapes.

It was also shown, in reference 18, that irradiances as low as $H = 0.1$ would increase the burning rate of upholstery panels by a factor of 3. Thus, the enhancement of the burning rate of previously ignited items by exposure to H^* from the burning drapes would significantly affect the chance for a room to shift from a mode of individual item burns, to the point of flashover. This could be a synergistic progression, i.e., ignition and/or intensification of the combustion of an upholstered object by the energy derived from the drapes could result in the continuation of the fire to another adjacent item, which in turn could be intensified by the energy from the first burning item until the whole room is ripe for flashover. Conversely, if the drapes have not, as yet, been ignited in a room where other items are already on fire, their ignition and the resultant energy production may add just the right amount of energy to the room to produce flashover.

At the present time, techniques of fire-proofing cotton materials have been developed that would reduce the hazards of fire initiation and propagation to practically zero [20]. These methods are rather complicated and costly, but they are still in the research stage, and if the techniques were optimized for industrial production, the costs would probably decrease to an economical level. For temporary fire protection, mixtures of common salts, in low solution concentrations, have proved to be very effective for ignition inhibition and combustion retardation [21]. Further research is now in progress, with the goal of modifying the cellulose molecule in such a way as to completely change its pyrolytic characteristics without radically modifying any of its more desirable qualities.

NOMENCLATURE

H	=	Irradiance ($cal \cdot cm^{-2} sec^{-1}$)
H*	=	Total heat transfer rate ($cal \cdot cm^{-2} \cdot sec^{-1}$)
δ	=	Distance between cloth panel and isothermal wall
$\dot{m}$	=	Mass loss per unit time
Y*	=	Mean flame-standoff-distance
Cal-F	=	Thermal radiation detector at the wall
Cal-T	=	Total heat transfer detector at the wall
Rad-F	=	Free side thermal radiation detector
Rad-T	=	Free side total heat transfer detector

REFERENCES

1. F. J. Vodvarka and T. E. Waterman, "Fire Behavior, Ignition to Flashover." Final Report, Task 2-C, Multitask Contract No. OCD-PS-64-50, OCD Work Unit 2536-C, IITRI Project N6059, IIT Research Institute, Technology Center, Chicago, Illinois 60616.
2. Alice M. Stoll, "Heat Transfer in Biotechnology, Vol. 4, J. P. Hartnett and T. F. Irvine, *Advances in Heat Transfer,* Vol. 4, 1967, Academic Press.
3. Konrad Buettner, "Effects of Extreme Heat and Cold on Human Skin. I. Analysis of Temperature Changes Caused by Different Kinds of Heat Applications," *J. Appl. Physiology,* Vol. 3, No. 12, 691-702 (June 1951).
4. F. C. Henriques, Jr. and A. R. Moritz, "The Conduction of Heat to and Through Skin and the Temperatures Attained Therein. A Theoretical and an Experimental Investigation." *The Am. J. of Pathology,* Vol. 23, 531-719 (1947).
5. James D. Hardy, "Effects of Thermal Radiation on Man," A Decade of Basic and Applied Science in the Navy. U. S. Government Printing Office, Washington, D. C. 1957.
6. R. H. Renner, S. B. Martin, and R. E. Jones, "Parameters Governing Urban Vulnerability to Fire from Nuclear Bursts (Phase I)." USNRDL-TR-1040, U. S. Naval Radiological Defense Laboratory, San Francisco, California, 30 June 1966.
7. D. L. Simms, "Damage to Cellulosic Solids by Thermal Radiation," *Combustion and Flame,* Vol. 6, 303-318 (Dec. 1962).
8. B. T. Lee and N. J. Alvares, "The Effect of Irradiance on the Ignition and the Rate of Fire Spread Along Composite Cellulosic Materials," 1967 Spring meeting, Western States Section, The Combustion Institute, La Jolla, California.

9. C. T. Webster, H. G. H. Wraight, and P. H. Thomas, "Heat Transfer from Burning Fabrics," *J. of the Textile Inst.,* Vol. 53(1), T-29-T37 (1962).
10. Franklin J. Kosdon, "Combustion of Vertical Cellulosic Cylinders in Air," 1968 Spring Meeting, Western States Section, The Combustion Inst., Univ. of Southern California.
11. R. J. Jenkins, "Calorimeters and Radiometers for the Measurement of Large Thermal Irradiances," USNRDL-TR-839, 6 April 1965.
12. G. G. Gubareff, J. E. Janssen, and R. H. Torborg, "Thermal Radiation Properties Survey. 2nd Edition", Honeywell Research Center, Minneapolis, Minnesota, 220-221, 1960.
13. T. E. Waterman, "A Calorimeter for Separating Radiative and Convective Heat". *Fire Technology,* 109-114, May, 1968.
14. J. J. Brenden, "Effect of Fire-Retardant and Other Inorganic Salts on Pyrolysis Products of Ponderosa Pine," *Forest Products J.,* 69-72 (February 1965).
15. P. L. Blackshear, Jr., and K. A. Murty, "Heat and Mass Transfer To, From, and Within Cellulosic Solids Burning in Air," Tenth Symposium (International) on Combustion, 911-23, The Combustion Institute, 1965.
16. A. E. Lipska and F. A. Wodley, "Isothermal Pyrolysis of Cellulose," 1968 Spring Meeting, Western States Section, The Combustion Institute, University of Southern California.
17. S. L. Madorsky, V. E. Hart, and S. Straus, "Pyrolysis of Cellulose in a Vacuum," *J. of Res. of NBS 56(6),* 343-54 (June 1956).
18. N. J. Alvares, and L. L. Wiltshire, "Ignition and Fire Spread in a Thermal Radiation Field," USNRDL-TR-68-56, 7 May 1968
19. E. M. Sparrow, and R. D. Cess, "Radiation Heat Transfer," Appendix A, Brooks/Cole Publishing Company, Belmont, California, 1967
20. R. M. Perkins, G. L. Drake, Jr., and W. A. Reeves, "DTA and TGA Studies of Flame-Resistant Fabrics," *J. of Appl. Pol. Sci. 10* (1966)
21. G. H. Trylon, "Fire Protection Handbook," 12th Ed., National Fire Protection Association, 60 Battery March Street, Boston, Mass., 1962

Norman J. Alvares

Norman J. Alvares is a Mechanical Engineer in fire research at Stanford Research Institute at Menlo Park, California. He received his A.B. degree in Physics from San Francisco State College in 1956 and his M.S. degree in Mechanical Engineering from University of Minnesota in 1969. He worked at the Naval Radiological Defense Laboratory before joining Stanford Research Institute. His experience includes thermal radiation measurements at nuclear tests, the optical properties of high temperature metals and coatings, ignition and fire spread, and the effects of combustion products on fire extinguishing foams.

Perry L. Blackshear, Jr.

Perry L. Blackshear, Jr., is a Professor at the University of Minnesota. He received his B.S. and M.S. degrees from Georgia Institute of Technology and his Ph.D. degree from Case Institute of Technology, all in Mechanical Engineering. He worked at the Lewis Research Center of NASA before coming to the University of Minnesota. His experience includes combustion research, combustion instrumentation, combustion instability, heat and mass transfer processes, and factors affecting red blood cell destruction.

NARASIMHAN RANGAPRASAD, C. M. SLIEPCEVICH, AND J. REED WELKER

The University of Oklahoma
Flame Dynamics Laboratory
Norman, Oklahoma 73069

THE PILOTED IGNITION OF COTTON FABRICS

(Received September 8, 1973)

ABSTRACT: Piloted ignition times were measured for several cotton fabrics. The results are generalized into a single curve based on a simple lumped-parameter heat transfer equation and compared with the ignition times of other thin cellulosic materials from the literature.

INTRODUCTION

THE LITERATURE CONTAINS considerable data on the piloted ignition of wood [7, 9, 11, 13, for example] and some data on piloted ignition of polymers [4, 11], but little data on the piloted ignition of fabrics [12]. This work was undertaken to measure and correlate the piloted ignition times of cotton fabrics subjected to thermal radiation. Piloted ignition times are generally considered to be the shortest time in which ignition can occur. They therefore represent the minimum time required for a fabric to begin burning when exposed to a heat source. Piloted ignition occurs faster than spontaneous (non-piloted) ignition because the pyrolysis products do not have to be heated to the ignition point by the energy source. Piloted ignition is more realistic than spontaneous ignition in ordinary fires involving clothing because the clothing must be quite near a flame or other heat source in order to ignite unless the source is quite large.

Two general types of fabrics were used, one obtained especially for these studies and one obtained locally. The special materials were furnished by the Southern Regional Research Center, U.S. Department of Agriculture. The fabrics were unbleached, ecru in color, and woven using the same basic fiber and construction, the only major difference being the fabric weights, which were 0.019, 0.020, 0.022, 0.024, 0.026, 0.029, and 0.035 g/cm^2. Two fabrics were purchased locally as examples of commercially available fabrics. One of these commercial fabrics was bleached white and the other was black; each weighed 0.013 g/cm^2.

PROCEDURE

The piloted ignition times were measured using 10 cm square samples in the ignition cabinet described by Koohyar, *et al*, [5] and used previously to study the ignition of wood [12] and polymers [4]. The cabinet is designed to provide a known level of radiant heat flux incident on the sample surface. The radiant energy source may be either a hydrocarbon flame or high temperature tungsten filament lamps. Flame radiation is used because it simulates the radiant heat from an actual fire much better than any artificial source, including the gas-fired radiant panel. Tungsten lamp radiation is used to provide a comparison with other ignition data from the literature.

The sample is positioned vertically in a sample holder that keeps the exposed surface smooth and facing the radiant energy source. A water-cooled shield is placed between the sample and the energy source until the source has reached steady state. The radiant flux is monitored continuously during this warmup period. When the source has reached steady state the shield is rapidly withdrawn and the sample is exposed to the radiant flux. The pyrolysis products rise from the sample past a three-dimensional gas flame near the top of the sample. This gas pilot flame is designed so that the ignition time is independent of its location. Ignition is recorded by a photosensitive cell that views across the front of the sample.

The incident radiant flux (or irradiance) is varied by changing the distance between the sample and the source or by injecting oxygen at the base of the flame. In the case of the tungsten lamps, the lamp filament temperature is maintained constant because of the effect of radiant energy distribution on ignition time [4, 12]. Incident irradiances up to 3 cal/cm^2-sec can be provided for either flame radiation or tungsten lamp radiation. Complete details of the procedure are given by Rangaprasad [8].

RESULTS AND DISCUSSION

Figure 1 shows the typical result of a series of ignition experiments for the lightest and heaviest of the ecru fabrics. The ignition time is dependent on three parameters, the incident irradiance, the fabric weight and the wavelength of the incident radiation. The dependency on all three can be predicted from a simplified heat balance.

Consider the fabrics to represent a portion of an infinite slab and assume the slab is thin enough that no thermal gradient exists within its boundaries. If the slab is inert and opaque, and if it is assumed that no heat is lost from the slab by convection or reradiation, the heat balance gives

$$c\, w_o\, \Delta T_i = \bar{\alpha} H_i\, t_i \tag{1}$$

c	=	specific heat
w_o	=	weight of material per unit area
ΔT_i	=	temperature rise at ignition
$\bar{\alpha}$	=	average absorptance
H_i	=	incident irradiance
t_i	=	ignition time

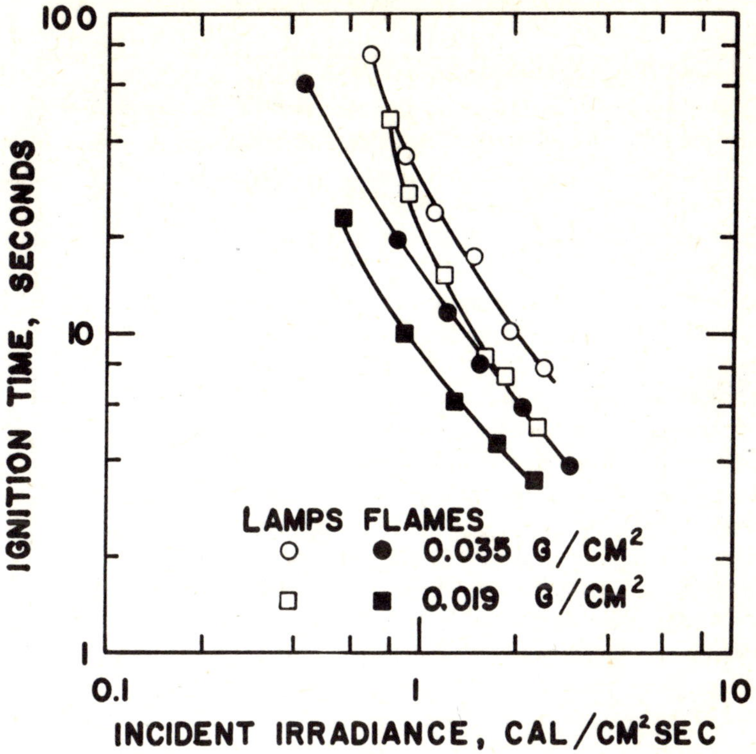

Figure 1. Piloted ignition time for cotton fabrics.

Equation 1 obviously does not account for all the factors that influence ignition because it is so simplified, but it does indicate how the four major factors, incident irradiance, fabric weight, absorptance, and ignition time are related. Figure 1 includes the tacit assumption that the primary variables are the ignition time and the incident irradiance. That assumption is valid because those two factors vary over the widest range during an ignition experiment. However, the fabric weight and average absorptance also affect the ignition time, the fabric weight directly and the average absorptance inversely, as indicated by Equation 1.

Consider first the fabric weight. If the fabric is thin enough to preclude thermal gradients, the heat required to raise the fabric to the ignition temperature is directly related to the fabric weight (or the product of density and thickness). Figure 2 shows the result when the ignition times are corrected for changes in fabric weight. The ratio t_i/w_o is shown to be dependent only on the incident irradiance and the source of radiation. Only the data for fabrics having unit weights of 0.019 and 0.035 g/cm^2 are shown in Figure 2.

The source of radiation causes differences in ignition times because of differences in the spectral absorptance of the fabric. These differences occur not only for fabrics [12], but for wood [13] and synthetic polymers [4] as well. In fact, it is a good generalization that spectral absorptances are important in ignition of

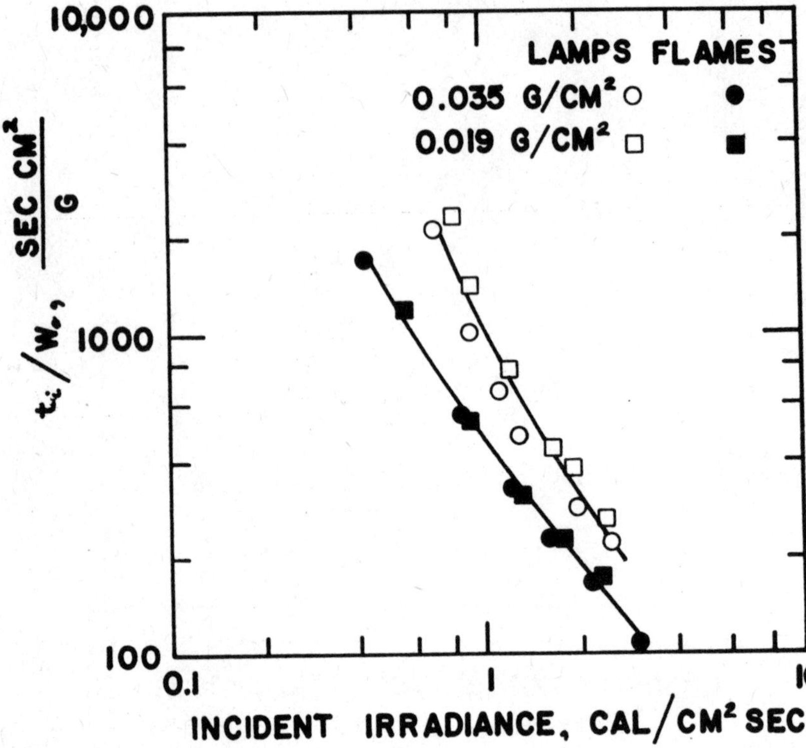

Figure 2. Piloted ignition time corrected for fabric weight.

nearly all light-colored combustible materials. The average absorptance varies with exposure time because of charring of the exposed surface, so for convenience, the average absorptance of the unheated material is used to show the effect. The average absorptance is calculated as described in an earlier paper [12], and because spectral data are not available for the fabrics used in this work, the average absorptances from that paper are used. Their values are given in Table 1 for reference.

It is interesting to note that the average absorptance for flame radiation is about the same regardless of the fabric color and that the average absorptance for tungsten lamp radiation is slightly higher than that for flame radiation for a black fabric while the average absorptance for natural fabrics is significantly lower for lamp radiation than it is for flame radiation. Similar behavior was found for natural and black painted wood and for light and dark synthetic polymers [4, 13].

Figure 3 shows a plot of the fabric ignition data in which the ignition times are considered a function of the initially absorbed irradiance, $\bar{\alpha}H_i$. Data for all fabric weights between 0.013 and 0.035 g/cm^2 are shown. In order to utilize the grouping of parameters suggested by Equation 1, the ignition time is divided by w_o c ΔT.

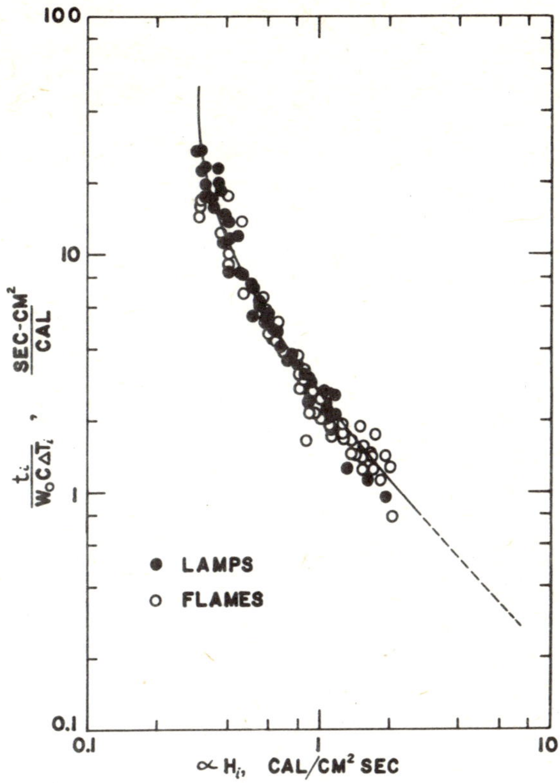

Figure 3. Generalized correlation of piloted ignition data for cotton fabrics.

The temperature rise, ΔT, is taken as the temperature rise at which the maximum weight loss rate is recorded when the material is subjected to thermogravimetric analysis [8]. The result is a fairly tight grouping of the ignition data around a single line, with the degree of scatter of the data approximately within the accuracy that ignition data can be measured. At low irradiances, where the ignition times are longest, the data scatter over a larger range and tend to curve upward. In practice, an asymptote is reached for average absorbed irradiances below about 0.3 cal/cm^2-sec; although the material will char and perhaps glow at lower heating rates, it does not give a flaming ignition. The slope of the curve at higher irradiances is approximately -1, a result that would be expected according to Equation 1, which predicts the inverse relationship between ignition time and heating rate. At higher heating rates the heat losses, which are approximately constant regardless of heating rate, are of lesser importance. At low irradiances, heat losses are important and in fact limit the temperature rise of the samples.

The literature contains several examples where the "critical irradiance" is determined by plotting irradiance versus the reciprocal of ignition time and extrapolating the curve to "infinite" time. That same procedure can be followed with the data on Figure 3, giving the result shown in Figure 4, where $\bar{\alpha}H_i$ is plotted versus $w_o\, c\, \Delta T/t_i$. The line in Figure 4 represents the same data as the line in Figure 3; the individual data points are omitted for clarity. The resulting line in Figure 4 is straight and intercepts $\bar{\alpha}H_i$ at the critical *absorbed* irradiance of about 0.3 cal/cm^2-sec. The total steady state energy loss of 0.3 cal/cm^2-sec required to balance that critical absorbed irradiance would require a fabric temperature of about 260°C if (1) both sides of the fabric lose heat by convention and radiation, (2) the convection coefficient is 2.5 (10^{-4}) cal/cm^2-sec-°C as suggested by Alvares, *et al*, [3], and (3) the average emittance of the fabric is assumed to be 0.8, a value consistent with measurements [12]. The calculated fabric temperature attained at steady state for critical irradiance thus approaches the temperature at which degradation begins to occur for slowly heated cotton fabrics [8].

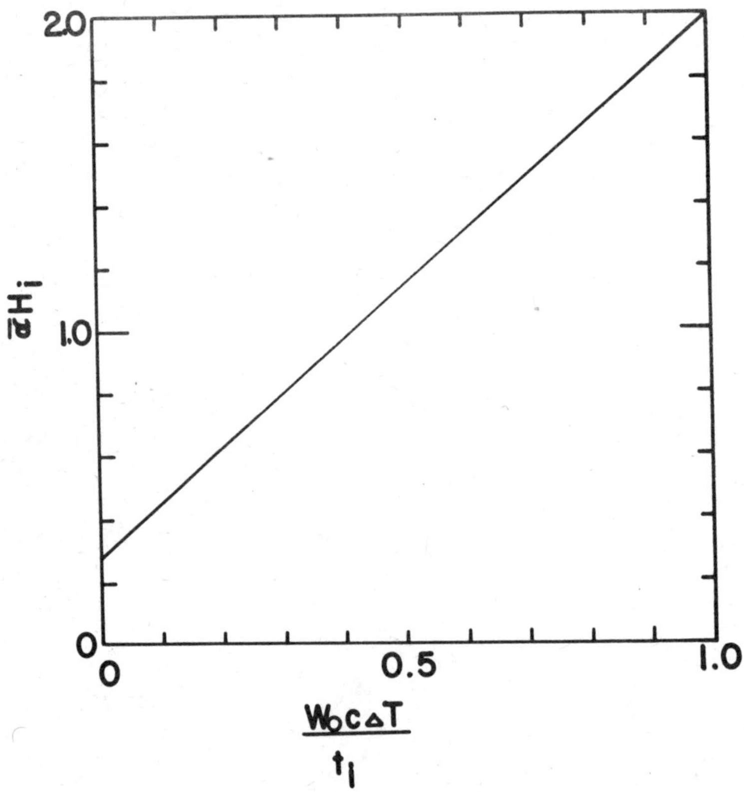

Figure 4. Extrapolation of ignition data to determine critical irradiance.

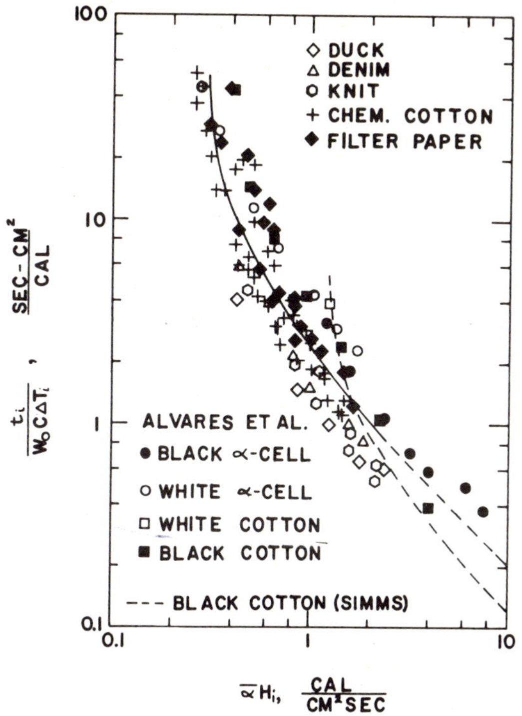

Figure 5. Comparison of ignition data for thin cellulosic materials.

The literature contains data that can be compared to those presented here. Simms [10] presented unpiloted ignition data for fabrics, and his results might be expected to be similar to the data of Figure 3 at high irradiances. Figure 5 compares the curve Simms drew through his data with the curve from Figure 3. The comparison is fair at higher irradiances, but the asymptote is reached at higher irradiances by the unpiloted data, as would be expected. In addition, some data from the work of Alvares [1], Lee and Alvares [6], and Alvares and Anderson [2] are shown; some of these points are for α-cellulose and some are for cotton fabrics. They are all for non-piloted ignition and the ignition times should be higher than for piloted ignition at low irradiances. In fact, it is quite unusual to obtain unpiloted flaming ignition of cellulosic materials at such low irradiances. Most ignition tests, such as those of Simms [10], have found the extrapolated "critical irradiance" to be higher than those at which Lee and Alvares [6] were able to ignite cotton fabrics.

Figure 5 also compares data on piloted ignition of commercially obtained cotton duck, denim, and knit fabrics from earlier work at the University of Oklahoma [11]. Finally, some piloted ignition data for chemical cotton and filter paper taken by Rangaprasad [8] are shown. The commercial fabrics (duck, denim, and knit) ignited a little faster than the fabrics tested in this work, the chemical cotton in about the same time, and the filter paper required longer exposure, especially at low irradiances. Chemicals added to these samples during processing may have influenced their ignition times. Flame-retarded or treated fabrics will not have the same ignition times as the fabrics in these tests.

CONCLUSIONS

Figure 3 shows that the primary factors in the time required to ignite cotton fabrics are the incident irradiance, the average absorptance, and the fabric weight. The "critical absorbed irradiance," below which piloted ignition will not occur, is about 0.3 cal/cm^2-sec. The correlation of Figure 3 can be used to predict piloted ignition times not only for cotton fabrics, but to a first approximation, most thin cellulosic materials as well. For cellulosic materials more than about 0.1-cm thick, the correlation of Reference 12 can be used to estimate piloted ignition time.

Table 1. Average Absorptances for Cotton Fabrics (12)

Fabric Color	Radiation Source	
	Flames	Tungsten Lamps
Black	0.75	0.82
Natural	0.72	0.46

REFERENCES

1. N. J. Alvares, Private communication (May 1967).
2. N. J. Alvares, and T. H. Anderson, "Ignition Hardening of Cellulosic Materials," USNRDL-TR-67-125, U.S. Naval Radiological Defense Laboratory, San Francisco (10 August 1967).
3. N. J. Alvares, P. L. Blackshear, and K. A. Murty, "The Influence of Free Convection on the Ignition of Vertical Cellulosic Panels by Thermal Radiation," Central States Section, Combustion Institute Meeting, University of Minnesota (March 1969).
4. J. Hallman, J. R. Welker, and C. M. Sliepcevich, "Ignition of Polymers," *SPE Journal, 28,* 43 (1972).
5. A. N. Koohyar, J. R. Welker, and C. M. Sliepcevich, "An Experimental Technique for the Ignition of Solids by Flame Irradiation," *Fire Technology, 4,* 221 (1968).
6. B. T. Lee, and N. J. Alvares, "The Effect of Irradiance on the Ignition and the Rate of Fire Spread along Composite Cellulosic Materials," 1967 Spring Meeting, Western States Section, The Combustion Institute, La Jolla, California (April 1967).
7. E. B. Moysey, and W. E. Muir, "Pilot Ignition of Building Materials by Radiation," *Fire Technology, 4,* 46 (1968).
8. Narasimhan Rangaprasad, "The Piloted Ignition of Thin Materials," Ph.D. dissertation, University of Oklahoma, Norman, Oklahoma (1972).
9. D. L. Simms, "On the Pilot Ignition of Wood by Radiation," *Combustion and Flame, 7,* 253 (1963).
10. D. L. Simms, "Ignition of Cellulosic Materials by Radiation," *Combustion and Flame, 4,* 293 (1960).

11. J. R. Welker, and C. M. Sliepcevich, "Susceptibility of Potential Target Components to Defeat by Thermal Action," University of Oklahoma Research Institute Report No. OURI-1578-FR, Norman, Oklahoma (1970).
12. J. R. Welker, H. R. Wesson, and C. M. Sliepcevich, "Ignition of Alpha-Cellulose and Cotton Fabric by Flame Radiation," *Fire Technology, 5,* 59 (1969).
13. H. R. Wesson, J. R. Welker, and C. M. Sliepcevich, "The Piloted Ignition of Wood by Radiation," *Combustion and Flame, 16,* 303 (1971).

Narasimhan Rangaprasad

Narasimhan Rangaprasad received his B.E. at the University of Madras in 1967, his M.E. at the Indian Institute of Science in 1969, and his Ph.D. at the University of Oklahoma in 1972. He is presently employed in India.

C. M. Sliepcevich

C. M. Sliepcevich is a Research Professor of Engineering at the University of 'klahoma at Norman, Oklahoma. He received his B.S. degree in Chemical Engineer-g from the University of Michigan in 1941, his M.S. degree in Chemical Engineer-g from the University of Michigan in 1942, and his Ph.D. in Chemical Engineering m the University of Michigan in 1948. Before joining the faculty of the iversity of Oklahoma in 1955, he was a faculty member at the University of chigan and a senior chemical engineer at Monsanto Chemical Company. His perience includes thermodynamics, process control, reaction kinetics and talysis, high pressure design, energy scattering, and cryogenics.

Reed Welker

J. Reed Welker is a Research Engineer at the University of Oklahoma Research Institute at Norman, Oklahoma. He received his B.S. degree in Chemical Engineering from the University of Idaho in 1959, his M.S. degree in Chemical Engineering from the University of Idaho in 1961, and his Ph.D. degree in Chemical Engineering from the University of Oklahoma in 1965. His experience includes flame dynamics, mass transfer thermodynamics, and atmospheric dispersion of pollutants.

CARLOS J. HILADO

Research and Development Department
Union Carbide Corporation
South Charleston, West Virginia 25303

FIRE STUDIES OF BEDDING MATERIALS

(Received July 26, 1973)

ABSTRACT: Increased concern with the flammability of bedding materials prompted a limited investigation of the fire behavior of readily available materials, to obtain guidance regarding the proper use of existing materials and the development of new and improved materials. A variety of tests, ranging in scale from small specimens to full mattresses with sheets and pillows, were performed.

Currently available materials and technology have been shown capable of giving innerspring and urethane foam core type mattresses adequate resistance to small ignition sources. The presence and involvement of sheets, pillows, and pillowcases increases the severity of exposure and introduces many variables affecting the resistance of the mattress to ignition and involvement.

It is the opinion of the author that a bedding system cannot be considered to have a given level of resistance to ignition unless that level of resistance is possessed by all the components of the system: mattresses, sheets, pillows, pillowcases, blankets, and sleepwear.

Fire experience indicates that population habits such as smoking in bed may be the major factor in bedding fires. The extent to which bedding materials must be improved to attempt to offset human carelessness is an issue which has not been completely resolved.

INTRODUCTION

INCREASING CONCERN FOR public safety has resulted in greater attention to the flammability behavior of a variety of consumer products. The amendment of the Flammable Fabrics Act has enlarged the scope of this scrutiny from clothing to other products such as blankets and floor coverings.

Bedding materials have been involved in a significant number of fires year after year, primarily because of the widespread habit of smoking in bed. A study of 3,145 single-fatality fires reported to the National Fire Protection Association from February 1966 through January 1968 [1] showed that 525 were clothing fires and 2,620 were nonclothing fires, and of the nonclothing fires 307 were caused by smoking in bed and 237 were caused by smoking on upholstered furniture.

Cigarettes and matches are among the principal sources of ignition in building fires. Of the 996,600 building fires in the United States in 1971, 118,400, or 12 percent, were attributed to smoking and matches, and 70,400, or 7 percent, were attributed to children and matches [2]. Of the 992,000 building fires in the United States in 1970, 107,200, or 11 percent, were attributed to smoking and matches, and 63,800, or 6 percent, were attributed to children and matches [3]. Of the 973,000 building fires in the United States in 1969, 111,800, or 11 percent, were attributed to smoking and matches, and 79,400, or 8 percent, were attributed to children and matches [4].

Bedding materials can be exposed to two types of fire exposure: exposure to a small ignition source such as a lighted match or cigarette, as in cases of occupants smoking in bed and children playing with matches; and exposure to a developing or fully developed room fire involving other articles in the room or the contents of adjacent rooms in the structure, with initial ignition coming from any of a variety of causes.

In the first type of exposure, the behavior of the bedding materials has an important bearing on life safety in that resistance to ignition can prevent a fire, resistance to flame spread can hinder its development, and limitation of combustion products can reduce physical injury and problems of escape. In the second type of exposure, the behavior of the bedding materials is much less significant to the safety of the occupant of the bed, because the temperatures required for ignition tend to be well above lethal levels. The combustion products would be the major concern of occupants of adjacent rooms and of persons combatting the fire.

Bedding materials have traditionally been combustible materials, and requirements of comfort, convenience, and cost will continue to call for substantial use of combustible materials. The flammability characteristics of these materials have generally been considered to present no hazard than could be considered an acceptable risk by the majority of consumers. Increases in population density, especially in urban areas, and changes in population habits have resulted in both increased probability of fire exposure in terms of frequency, and increased probability of fire injury in terms of number of persons affected. The person who smokes in bed probably has in a given radius not only more neighbors but also more neighbors with the same habit. The resultant need to reduce the probability and consequences of fires has been reflected in growing government concern.

Fire tests of bedding and upholstery materials have been performed by the Southwest Research Institute under a contract awarded by the National Bureau of Standards [5, 6]. As a result of these tests and investigations of fire records, the Department of Commerce has issued a flammability standard for mattresses [7] and found a need for a flammability standard for upholstered furniture [8]. In these standards and in earlier standards for carpets and rugs and children's sleepwear [9, 10, 11], the government approach has been to reduce fire hazard by reducing the probability of ignition by small ignition sources.

Pursuing the philosophy that manufacturers should, as a matter of public responsibility, explore methods of improving the safety of their products without the need for forcible government action, Union Carbide Corporation has undertaken a limited investigation of the flammability characteristics of readily available bedding materials, to obtain guidance as to the proper use of existing materials and the development of new and improved materials. This paper presents the results of this work.

PLAN OF INVESTIGATION

The work was divided into three phases: first, a study of the behavior of various types of mattresses and sofa-bed cushions when subjected to small ignition sources, to determine their relative susceptibility to ignition or progressive smoldering; second, a study of the behavior of various types of mattresses furnished with sheets and pillows of different materials, when exposed to a small ignition source, to determine the relative fire resistance of complete bedding systems; and third, a study of the smoke producing characteristics of various mattress constructions, to determine the relative sight obscuration which results. Mattress behavior was emphasized because mattresses are the major product of the bedding industry and its most identifiable one.

The investigation had several limitations which were recognized at the outset. The lack of room-scale fire test facilities prevented the acquisition of information on the effect of placement, ventilation, and interaction with other materials, and limited thermal and optical measurements to the vicinity of the materials involved. The almost complete absence of published work on bedding fire tests resulted in the dependence of the procedures of each test on the results of previous tests, to at least some extent. Information was almost entirely qualitative at the start of the test program, and became more quantitative as experience was accumulated.

The articles tested were limited to those which were readily available, or which could be constructed by existing methods from readily available materials. The fire-retardant samples of cotton cover and urethane foam were commercially available to the bedding industry, and are not necessarily representative of the best technology available at the time this paper is read. Modacrylic and aromatic polyamide sheets and pillowcases were tested, but there were no fire-retardant cotton analogs which had the permanent fire retardance and esthetics of these two types of materials.

TESTS ON MATTRESS SECTIONS

The purpose of this series of tests was to determine the relative susceptibility of representative types of mattress constructions to ignition or smoldering when subjected to small ignition sources. Three types of small ignition sources were selected: a cigarette, a match, and a methenamine pill; the first two are prominent causes of bedding fires, and the third is a standard ignition source specified in government standards for carpets [9, 10].

Mattresses were tested without sheets and pillows in order to isolate the effect of mattress construction and materials from interaction with the behavior of sheets and pillows. The samples were tested in the bone-dry condition to provide the most unfavorable conditions and to avoid the effects of humidity variations. These conditions were not unrealistic, since they can be approached by humidity levels in the winter months.

Test Procedure

Sections one foot wide, each the full width of the mattress, were cut from each mattress sample. The cutting of sections from the innerspring constructions presented special problems, because the special saw required to cut through the metal innerspring generated enough heat and sparks to scorch sisal padding in the vicinity of the cut. The use of water on the saw blade provided sufficient cooling to prevent ignition in most cases, but one smoldering fire had to be extinguished with water. As a result of this experience, spark and hot-surface ignition methods were considered for inclusion in the study. These types of ignition are widely used in fire research on liquids and vapors, but were judged not to be representative of enough actual fire experiences.

The test sections were conditioned for one hour at 105°C in a circulating-air oven and then sealed in polyethylene containers until the time of test. A sheet of asbestos-cement board, 1/4-inch thick, supported the test section in a large laboratory hood. Three ignition sources were placed on top of each sample: a lighted non-filter cigarette (regular length), a lighted methenamine pill (Eli Lilly, No. 1588), and a lighted wooden match. Figure 1 shows a typical arrangement.

Figure 1. *Typical arrangement for tests on mattress sections. Ignition sources, left to right: cigarette, methenamine pill, match. Test No. 7, time 15 seconds.*

Test Results

Fifteen tests were performed on mattress sections. These are summarized in

Table 1. Four tests involved innerspring constructions, two of these with fire-retardant covers; one test involved individually wrapped coil springs with padding and quilting; four tests involved latex foam rubber core constructions, two of these with fire-retardant covers; and six tests involved urethane foam core constructions, one of these with fire-retardant cover and one with both fire-retardant cover and fire-retardant core.

The lighted cigarette failed to ignite any mattresses of the innerspring and urethane foam core types, and damage was limited to a small charred area with a little darkening or melting out of the underlying material. The lighted cigarette produced progressive internal smoldering in all the latex foam rubber core mattresses, and the affected area attained a 12-inch diameter after 22 minutes when the cover was not fire-retardant, and after 35 to 44 minutes when the cover was fire-retardant cotton. The use of cut sections to show internal construction was particularly illustrative in this case since it permitted detection of smoldering behavior, but burning or smoldering rate increased when the area affected by flame or smoldering reached the cut edge with its greater access of oxygen. The presence of cut edges may provide no more access of air than the "breathing" action caused by normal movement of an occupant sleeping on the mattress, but this point would be difficult to resolve. For the purposes of this study, the valid portion of the test was considered to end when the affected area reached the cut edge, which corresponded to a diameter of 12 inches.

In several tests, the lighted methenamine pill and lighted match started spreading fires which reached the cigarette before the cigarette could show any serious effects. In such cases, the test was repeated with only a lighted cigarette as the ignition source.

The internal smoldering of the latex foam rubber produced dense and irritating smoke, grayish white at first, changing to yellowish brown as smoldering progressed and turning to black when the smoldering changed to burning. Two samples of latex foam rubber mattresses ignited upon contact with air after being carried out of the laboratory to alleviate smoke conditions; the ignition delay time was about 30 seconds, and the smoldering rubber was deliberately ignited with a match in subsequent tests as a safer method of disposal.

The lighted wooden match was quickly extinguished when in contact with the fire-retardant cotton cover used in six of the tests, and with the plastic-coated cover used in one test. Spreading fires resulted in all other tests, or progressive smoldering in latex foam rubber.

Damage from a lighted methenamine pill was limited to small charred areas in the four tests employing fire-retardant cotton covers on innerspring and urethane foam core type mattresses. Progressive smoldering occurred in latex foam rubber despite the use of a fire-retardant cover, which increased the time required to involve a 12-inch diameter area from 20 minutes to 31–32 minutes.

Some of the differences in behavior described are shown by Figures 2 to 16.

Table 1. Tests on Mattress Sections

Test No.	Mattress Construction: Core	Padding	Cover	Ignition Source	Result	Time, minutes for Affected Area to Reach 12 inches Diameter
1	Innerspring unit	Latex sisal pad, felt batt	Spun rayon, quilted 3/4 inch urethane foam	Cigarette	Small charred area	
				Methenamine pill	Spreading fire	3.5
				Wooden match	Spreading fire	3.5
3	Innerspring unit	Latex sisal pad, 1 inch urethane foam	Spun rayon, quilted 1/2 inch urethane foam	Cigarette	Small charred area	
				Methenamine pill	Spreading fire	3.5
				Wooden match	Spreading fire	3.5
8	Innerspring unit	1/4 inch latex sisal pad, 3/4 inch urethane foam	Cotton FR	Cigarette	Small charred area	
				Methenamine pill	Small charred area	
				Wooden match	Small charred area	
7	Innerspring unit	1/4 inch tufflex FR, 3/4 inch urethane foam	Cotton FR	Cigarette	Small charred area	
				Methenamine pill	Small charred area	
				Wooden match	Small charred area	
9	Coil springs, individually wrapped	1 inch urethane foam	Rayon, quilted two 1/4 inch layers urethane foam	Cigarette	Small charred area	
				Methenamine pill	Spreading fire	4
				Wooden match	Spreading fire	4+
2	Latex foam rubber, 4 inch	None	Spun rayon, quilted felt	Cigarette	Progressive smoldering	22
				Methenamine pill	Progressive smoldering	20
				Wooden match	Progressive smoldering	15
4	Latex foam rubber, 4 inch	None	Spun rayon, quilted 1/2 inch urethane foam	Cigarette	Progressive smoldering	22
				Methenamine pill	Spreading fire	3
				Wooden match	Spreading fire	3
11	Latex foam rubber, 4 inch	None	Cotton FR	Cigarette	Progressive smoldering	44
				Methenamine pill	Progressive smoldering	32
				Wooden match	Small charred area	

Table 1. Tests on Mattress Sections (continued)

Test No.	Mattress Construction: Core	Padding	Cover	Ignition Source	Result	Time, minutes for Affected Area to Reach 12 inches Diameter
15	Latex foam rubber, 4 inch	None	Cotton FR	Cigarette	Progressive smoldering	35
				Methenamine pill	Progressive smoldering	31
				Wooden match	Small charred area	
5	Urethane foam, 4 inch	None	Rayon, quilted 1/4 inch urethane foam	Cigarette	Small charred area	
				Methenamine pill	Spreading fire	5
				Wooden match	Spreading fire	5
6	Urethane foam, 5.5 inch	None	Rayon	Cigarette	Small charred area	
				Methenamine pill	Spreading fire	6
				Wooden match	Spreading fire	6
14	Urethane foam, 1.5 pcf, 4 inch	None	Cotton, plastic coated	Cigarette	Small charred area	
				Methenamine pill	Small charred area	
				Wooden match	Small charred area	
13	Urethane foam, 1.5 pcf, 4 inch	None	Cotton	Cigarette	Small charred area	
				Methenamine pill	Spreading fire	6
				Wooden match	Small charred area	
12	Urethane foam, 1.5 pcf, 4 inch	None	Cotton FR	Cigarette	Small charred area	
				Methenamine pill	Small charred area	
				Wooden match	Small charred area	
10	Urethane foam FR, 2.0 pcf, 4 inch	None	Cotton FR	Cigarette	Small charred area	
				Methenamine pill	Small charred area	
				Wooden match	Small charred area	

Figure 2. *Test No. 1, time 3 minutes. Regular cover, padded innerspring core.*

Figure 3. *Test No. 3, time 3 minutes. Regular cover, padded innerspring core.*

Figure 4. *Test No. 4, time 3 minutes. Regular cover, latex foam rubber core.*

Figure 5. *Test No. 5, time 5 minutes. Regular cover, urethane foam core.*

Figure 6. *Test No. 6, time 5 minutes. Regular cover, urethane foam core.*

Figure 7. *Test No. 9, time 5 minutes. Regular cover, individually wrapped coil springs.*

Figure 8. *Test No. 13, time 5 minutes. Regular cover, urethane foam core.*

Figure 9. *Test No. 2, time 15 minutes. Regular cover, latex foam rubber core.*

Figure 10. *Test N 40 minutes. Fire retardant cover, latex foam rubber core.*

Figure 11. *Test No. 15, time 35 minutes. Fire retardant cover, latex foam rubber core.*

Figure 12. *Test No. 7, cut-away view after test. Fire retardant cover, padded innerspring core.*

Figure 13. *Test No. 8, cut-away view after test. Fire retardant cover, padded innerspring core.*

Figure 14. *Test No. 10, cut-away view after test. Fire retardant cover, fire retardant urethane foam core.*

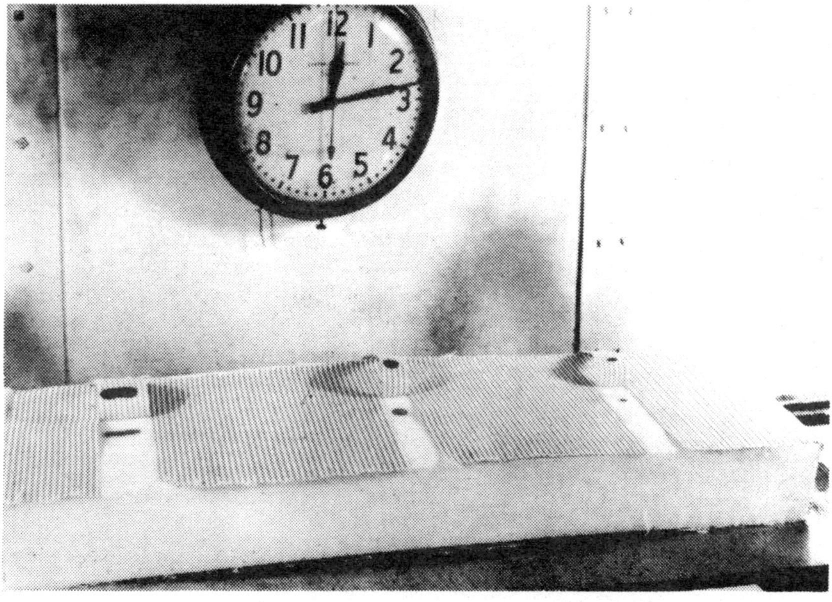

Figure 15. *Test No. 12, cut-away view after test. Fire retardant cover, urethane foam core.*

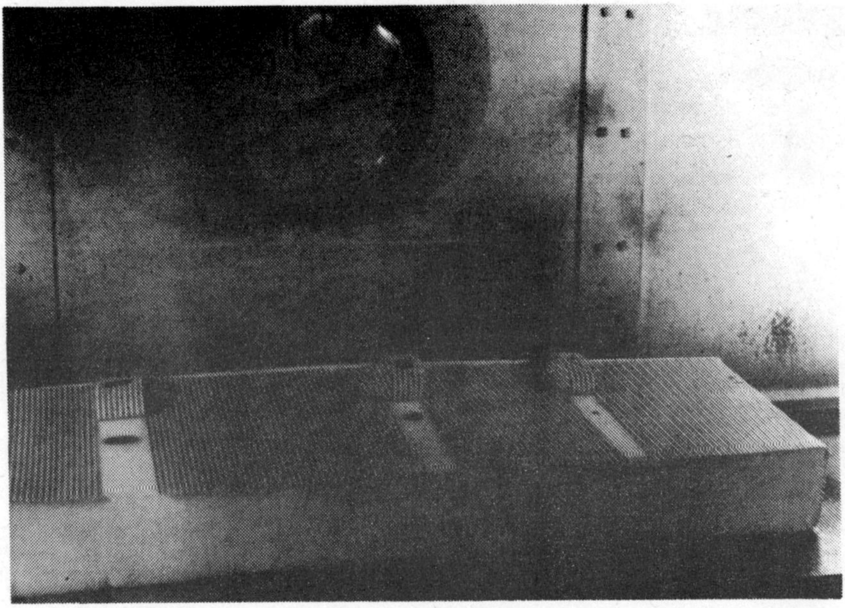

Figure 16. *Test No. 14, cut-away view after test. Plastic coated cover, urethane foam core.*

Type of Ignition

Spreading fires or progressive smoldering were produced by a lighted methenamine pill in 10 of the 15 tests, by a lighted match in 7 of the 15 tests, and by a lighted cigarette in 4 of the 15 tests. The methenamine pill averaged out as the most severe ignition source, and the lighted cigarette the least severe. If only spreading fires are considered, the methenamine pill produced these results in 7 of the 15 tests, and the match in 6 of the 15 tests. These two ignition sources tend to be similar in severity of exposure. For test purposes the methenamine pill is preferred because it is on the average more severe and seems more reproducible. The reproducibility of this type of exposure is shown in Table 2.

Table 2. Reproducibility of Methenamine Pill Exposure

Test No.	Burning Time, sec.
7	90
8	100
10	90
11	95
12	85
13	100
14	94
15	90

Ease of Extinguishment

Fires in rayon covering and in urethane foam were relatively easy to extinguish with water or carbon dioxide, but fires in felt batt and in latex foam rubber were difficult to extinguish with water and almost impossible to extinguish with carbon dioxide. Smoldering latex foam rubber produced large quantities of smoke when hit by water.

TESTS ON SOFA-BED CUSHIONS

The purpose of this brief series of tests was to determine the relative susceptibility of some types of sofa-bed cushions to ignition when exposed to small ignition sources. The smaller size of the cushions relative to mattresses made it impossible to use more than two ignition sources simultaneously. On the basis of experience with mattress sections, only the cigarette and methenamine pill were used.

Test Procedure

Each sample cushion was cut in half to expose the entire cross section, and a half-section was used as the test specimen. Conditioning for one hour at 105°C in a circulating-air oven was followed by storage in a polyethylene container until the time of test. The specimen was supported by 1/4-inch thick, asbestos-cement board. Two ignition sources were placed on top of each specimen: a lighted non-filter cigarette (regular length) and a lighted methenamine pill (Eli Lilly, No. 1588). Figure 17 shows a typical arrangement.

Test Results

Three tests were performed on sofa-bed cushion sections. These are summarized in Table 3. The lighted cigarette produced no ignition in any of the tests. The open-mesh fabric covered cushions were ignited by the methenamine pill and eventually consumed, but the vinyl fabric covered cushion did not ignite. The presence of polyester fiberfill between the fabric covering and the urethane foam cushion seemed to accelerate flame spread, probably due to greater access of air. The resistance of the vinyl fabric to ignition is attributed to the substantial chlorine content of polyvinyl chloride.

Some of the described differences in behavior are shown in Figures 18 to 20.

TESTS ON MATTRESSES WITH SHEETS AND PILLOWS

The purpose of this series of tests was to determine the effect of fire involvement of sheets and pillows on the behavior of various types of mattress constructions. The choice of sheets and pillows is the prerogative of the mattress owner, and beyond the control of the mattress manufacturer. The exposure severity of burning

Figure 17. *Typical arrangement for tests on sofa-bed cushion sections. Ignition sources: cigarette, methenamine pill. Test No. 1, time 15 seconds.*

Table 3. Tests on Sofa-Bed Cushion Sections

Test No.	Cushion	Cover	Ignition Source	Result
1	Urethane foam, 5-inch	Vinyl fabric	Methenamine pill Cigarette	Small charred area Small charred area
2	Urethane foam, 4-inch Polyester fiberfill	Open-mesh fabric	Methenamine pill Cigarette	Spreading fire Small charred area
3	Urethane foam, 5-inch	Open-mesh fabric	Methenamine pill Cigarette	Spreading fire Small charred area

sheets and pillows is greater than that of the small ignition sources employed in the preceding tests, but the response of the mattress under this exposure had not been previously studied.

Because of the tendency exhibited by latex foam rubber core mattresses to undergo progressive smoldering and produce large quantities of dense smoke, this type of mattress was not used in these tests. The majority of tests in this series involved mattress constructions which exhibited satisfactory resistance to small ignition sources.

Figure 18. *Test No. 1, time 10 minutes. Vinyl fabric cover.*

Figure 19. *Test No. 2, time 5 minutes. Open mesh fabric cover.*

Figure 20. *Test No. 3, time 5 minutes. Open mesh fabric cover.*

Three types of pillows were used: one with 100 percent latex foam rubber filling, one with 100 percent polyester fiberfill filling, and one filled with 50 percent down and 50 percent crushed duck feathers. Down-and-feather pillows were available with 25/75, 50/50, and 75/25 ratios of down to feathers, and the 50/50 ratio was selected as representative. All pillows had 100 percent cotton covering. These three types of pillows together are believed to constitute the large majority of the pillow market.

In the majority of tests in this series, the sheets and pillowcases were 100 percent cotton, fine comb percale. None of the cotton sheets, pillowcases, or pillow covers were fire retardant. The fire-retardant cotton materials generally available to the bedding industry at the time did not have the permanent fire retardance and esthetics of the modacrylic and polyamide materials which were used in several tests.

Test Procedure

The size of the hood available for the full-scale tests limited sample size to twin bed size mattresses (38 by 75 inches). In most of these tests, the mattress was covered with two cotton sheets, the upper sheet turned down at the end where a single pillow, covered with a pillowcase, was placed. Only single sheets of modacrylic and polyamide were used due to material limitations. The entire assembly was supported by a 1/4-inch thick sheet of asbestos-cement board, with a similar sheet of asbestos-cement board along the back wall of the hood to prevent damage to the structure.

To obtain some measure of incident radiation from the burning pillow and pillowcase initially, and from the burning mattress and sheets subsequently, a radiometer purchased from Heat Technology Laboratories was positioned at various distances along the top surface of the sheets covering the mattress. The radiometer was moved from location to location along the centerline, with the receiving surface at right angles to the top surface of the mattress and 1.5 inches above it.

To obtain a measure of the temperatures reached in the vicinity of the pillow during the test, thermocouples were positioned on the top surface as shown in Figure 21. In cases where burning was expected in the interior of innerspring type mattresses, thermocouples were positioned in the interior of the mattress as shown in Figure 22.

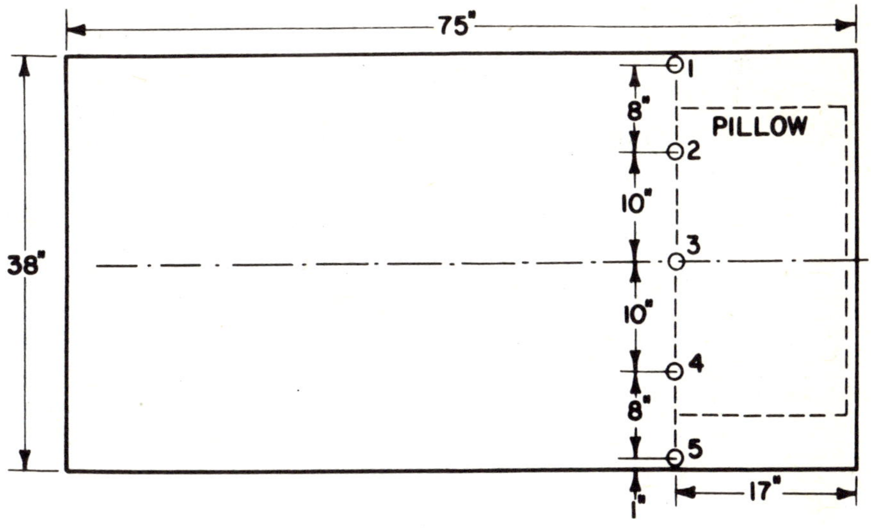

Figure 21. *Locations of thermocouples on top surface of mattress.*

Because the test procedure gradually developed in sophistication as the series progressed, heat flux and temperature measurements were employed only in the later tests.

The mattresses were not given special conditioning because of the lack of available conditioning facilities for articles of this size. Review of the environmental conditions prevailing during storage, transfer, and testing indicated that relative humidity was 30 ± 20 percent; this was less severe than the bone-dry conditioning used in the preceding phase but more severe than typical laboratory conditioning to 50 or 65 percent relative humidity.

Ignition was accomplished by placing a lighted methenamine pill (Eli Lilly, No.

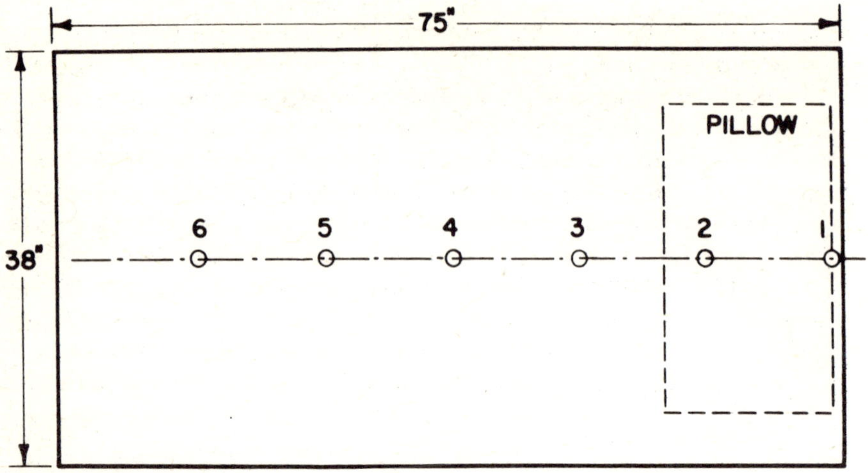

Figure 22. *Locations of thermocouples in core of innerspring mattress.*

1588) on top of the cased pillow. Flame generally progressed from pillowcase to pillow covering to pillow filling, and from pillowcase to sheets to mattress.

Typical test arrangements are shown by Figures 23 to 25.

Figure 23. *Typical arrangement for tests on mattresses with sheets and pillows. Test No. 4.*

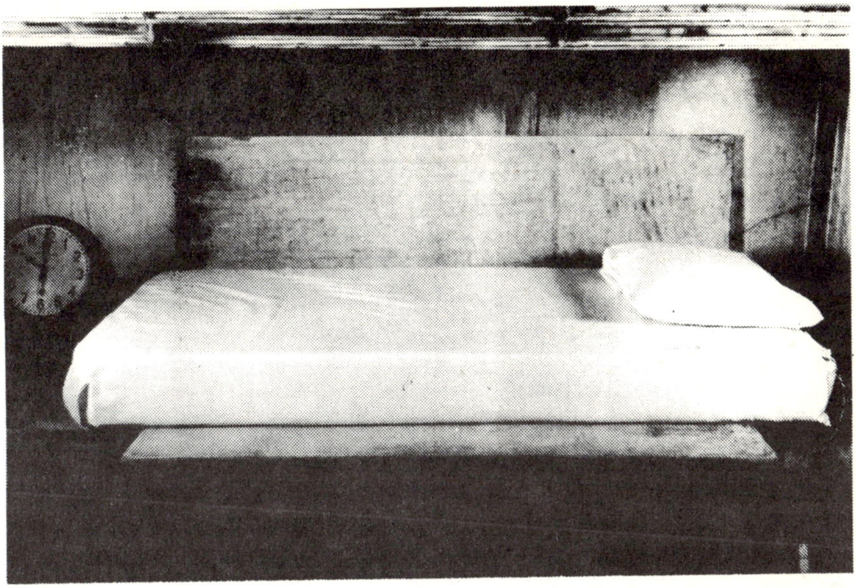

Figure 24. *Typical arrangement for tests with thermocouples on top surface. Test No. 10.*

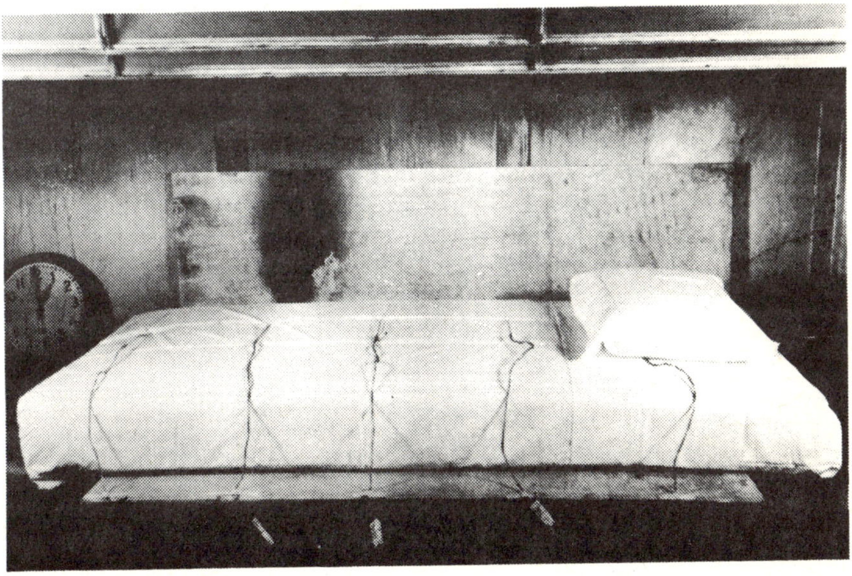

Figure 25. *Typical arrangement for tests with thermocouples in core. Test No. 13.*

Test Results

Thirteen tests were performed on mattresses with sheets and pillows. These are summarized in Table 4. Eight tests involved innerspring type mattresses and five tests involved urethane foam-core type mattresses.

Pillows with latex foam rubber filling were used in most of the tests because they provided the most severe fire exposure. The latex foam rubber filling burned steadily and generated considerable heat. Polyester fiberfill filling melted back gradually and the thick liquid was ignited by the burning pillowcase and pillow covering. Down-and-feather filling was scorched by the burning pillowcase and pillow covering but did not burn, and most of the filling was intact after the test. Because the down-and-feather pillow kept the upper half of the pillowcase and pillow cover away from the mattress while they were burning and tended to smother the lower half, the net fire exposure is believed to be less severe than the burning of an empty pillow cover and pillowcase laid on the mattress.

In all tests with cotton sheets and cotton pillowcases, the mattress was eventually ignited, and the time to ignition was a function of exposure severity. One mattress construction employing fire-retardant urethane foam became involved after 9 minutes exposure to a burning latex foam rubber pillow, 12 minutes exposure to a burning polyester fiberfill pillow, or 23 minutes exposure to a burning down-and-feather pillow. In the last case there was a significant induction period between exposure and ignition, and in that case the mattress ignited in the area next to the far wall where the asbestos-cement boards increased exposure intensity by geometric configuration and increased exposure time by their high heat capacity.

Mattresses with innerspring cores exhibited sudden releases of gases which had accumulated in the essentially hollow interior. These gases seemed to be combustible vapors generated by thermal decomposition of the mattress materials, and internal burning was observed after the gas release. It is not known at this time whether the evolved combustibles mixed with the air already present in the innerspring core to form a flammable composition which was then brought to autoignition temperature; or whether rapid pressure build-up within the mattress, inadequately relieved by diffusion through the padding and cover, resulted in a rupture in the fire-weakened mattress, through which the flame front moved into the interior. The sensing devices used did not have sufficient sensitivity to resolve this point.

Despite the use of fire-retardant materials, the exposure severity of burning sheets and pillows was sufficient to overcome the resistance of all the mattresses tested. Two approaches to reducing severity were to reduce the fraction of incident heat flux absorbed by the mattress by using sheets for fire protection, and to reduce the incident heat flux by minimizing involvement of the pillow.

Using the first approach, the use of a polyamide sheet on top of the mattress prevented ignition of one mattress and significantly delayed the ignition of another;

Table 4. Tests on Mattresses with Sheets and Pillows

Test	Mattress Construction				Pillow			Involvement		Time, minutes to Event							
No.	Core	Padding	Cover	Sheet	Filling	Cover	Pillowcase	Pillow	Mattress	A	B	C	D	E	F	G	H
4	Innerspring unit	68% blended cotton felt, 32% sisal pads	Rayon	Cotton	100% Latex foam rubber	Cotton	Cotton	Yes	Yes	5	7	8	12	14			
13	Innerspring unit	68% blended cotton felt, 32% sisal pads	Rayon	Cotton	100% Latex foam rubber	Cotton	Cotton	Yes	Yes								
6	Innerspring unit	3/4 inch urethane foam FR, 1/4 inch tufflex	Cotton FR	Cotton	100% Latex foam rubber	Cotton	Cotton	Yes	Yes	6	6	8	12	15	20	25	28
7	Innerspring unit	3/4 inch urethane foam FR, 1/4 inch tufflex	Cotton FR	Cotton	100% Latex foam rubber	Cotton	Cotton	Yes	Yes	4	5	6	11	14	20		25
10	Innerspring unit	3/4 inch urethane foam FR, 1/4 inch tufflex	Cotton FR	Modacrylic	100% Latex foam rubber	Cotton	Cotton	Yes	Yes	5		8		25			
11	Innerspring unit	3/4 inch urethane foam FR, 1/4 inch tufflex	Cotton FR	Polyamide	100% Latex foam rubber	Cotton	Cotton	Yes	No	7							
9	Innerspring unit	3/4 inch urethane foam FR, 1/4 inch tufflex	Cotton FR	Modacrylic	100% Latex foam rubber	Cotton	Modacrylic	No	No								
8	Innerspring unit	3/4 inch urethane foam FR, 1/4 inch tufflex	Cotton FR	Polyamide	100% Latex foam rubber	Cotton	Polyamide	No	No								
5	100% urethane foam	None	Rayon	Cotton	100% Latex foam rubber	Cotton	Cotton	Yes	Yes	7	8	9	12	14	16		20
12	100% urethane foam	None	Rayon	Polyamide	100% Latex foam rubber	Cotton	Cotton	Yes	Yes*	5			10				
1	100% urethane foam FR	None	Cotton FR	Cotton	100% Latex foam rubber	Cotton	Cotton	Yes	Yes	4	5	7	12	15	20		30
2	100% urethane foam FR	None	Cotton FR	Cotton	100% Polyester fiberfill	Cotton	Cotton	Yes	Yes	4	5	8	10	15	20		
3	100% urethane foam FR	None	Cotton FR	Cotton	50% down, 50% feathers	Cotton	Cotton	Yes	Yes*	6	8	12		20	36		40

Event A — fire spreads over edge of pillow
B — fire spreads to sheets

Event C — fire spreads to top edge of mattress
D — pillow ceases burning
E — fire reaches 50 percent of mattress length

Event F — fire reaches 60 percent of mattress length
G — fire reaches 70 percent of mattress length
H — fire reaches 80 percent of mattress length

*Significant induction period before involvement of mattress.

the first mattress contained fire-retardant materials, and the second did not. The performance of the polyamide sheet is ascribed to its high strain release temperature of about 750°F [12] which enabled it to maintain integrity of the fabric structure even in the process of charring. Modacrylic tended to shrink under fire exposure, leaving areas of the substrate material exposed. Any fire resistant fabric which chars in place without distortion would probably provide the protection desired.

Using the second approach, modacrylic and polyamide pillowcases prevented ignition of the latex foam rubber pillow by a lighted methenamine pill. The latex foam rubber pillow did not appear susceptible to progressive internal smoldering when covered with a fire-retardant pillowcase.

Figures 26 to 38 show some of these results in photographs.

Figure 26. *Test No. 3, time 20 minutes. Down-and-feather pillow, fire retardant mattress.*

Reproducibility of Fire Exposure

The selection of a latex foam rubber pillow with cotton pillow cover and pillowcase, ignited by a methenamine pill, as a standard fire source in full-scale tests gave good reproducibility, considering the variations possible in the manner in which a pillow is stuffed into a pillowcase and laid on a mattress. This source of variation was minimized by having the same technician make the bed for each test. The precision with which the various events in the burning process were repeated is shown by Table 5.

Figure 27. *Test No. 2, time 20 minutes. Polyester fiberfill pillow, fire retardant mattress.*

Figure 28. *Test No. 1, time 20 minutes. Latex foam rubber pillow, fire retardant mattress.*

Figure 29. *Test No. 5, time 20 minutes. Cotton sheet, cotton pillowcase, regular mattress.*

Figure 30. *Test No. 12, time 20 minutes. Polyamide sheet, cotton pillowcase, regular mattress.*

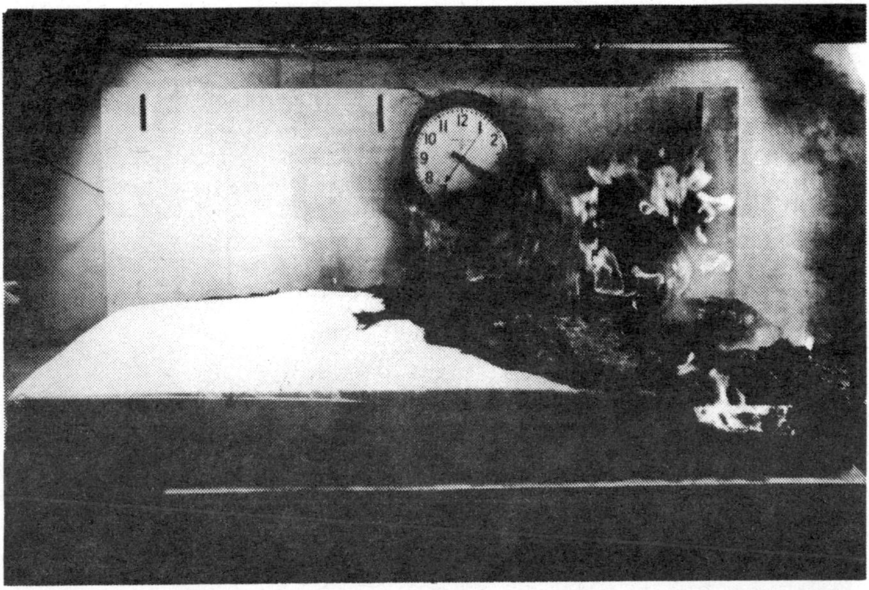

Figure 31. *Test No. 4, time 20 minutes. Cotton sheet, cotton pillowcase, regular mattress.*

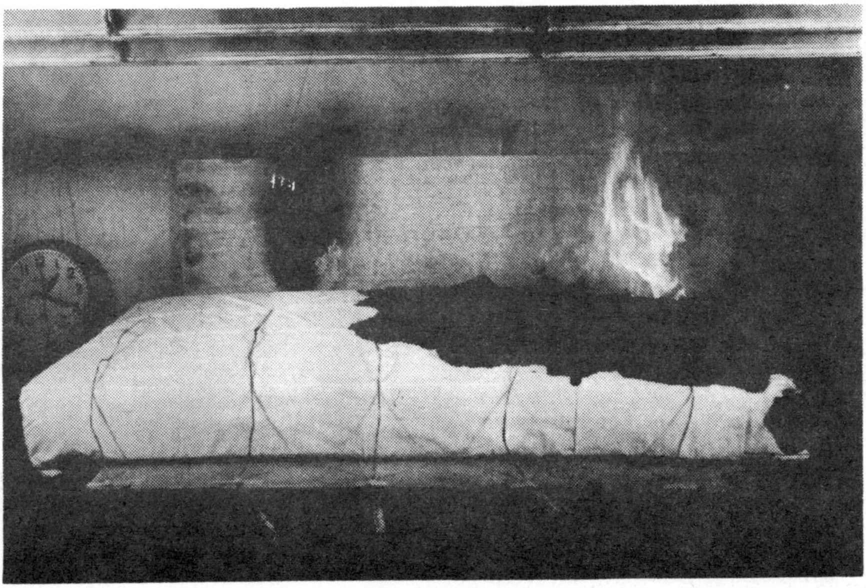

Figure 32. *Test No. 13, time 20 minutes. Cotton sheet, cotton pillowcase, regular mattress.*

Figure 33. *Test No. 6, time 20 minutes. Cotton sheet, cotton pillowcase, fire retardant mattress.*

Figure 34. *Test No. 7, time 20 minutes. Cotton sheet, cotton pillowcase, fire retardant mattress.*

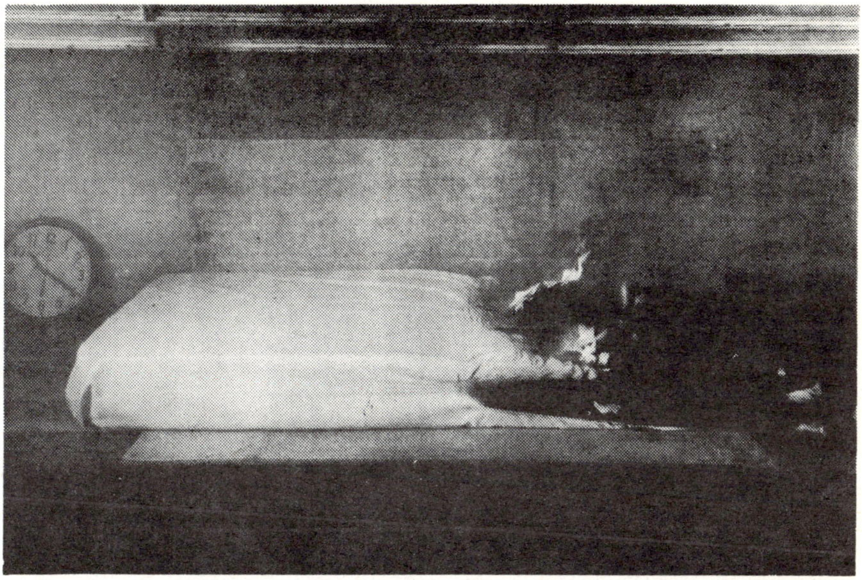

Figure 35. *Test No. 10, time 20 minutes. Modacrylic sheet, cotton pillowcase, fire retardant mattress.*

Figure 36. *Test No. 11, time 20 minutes. Polyamide sheet, cotton pillowcase, fire retardant mattress.*

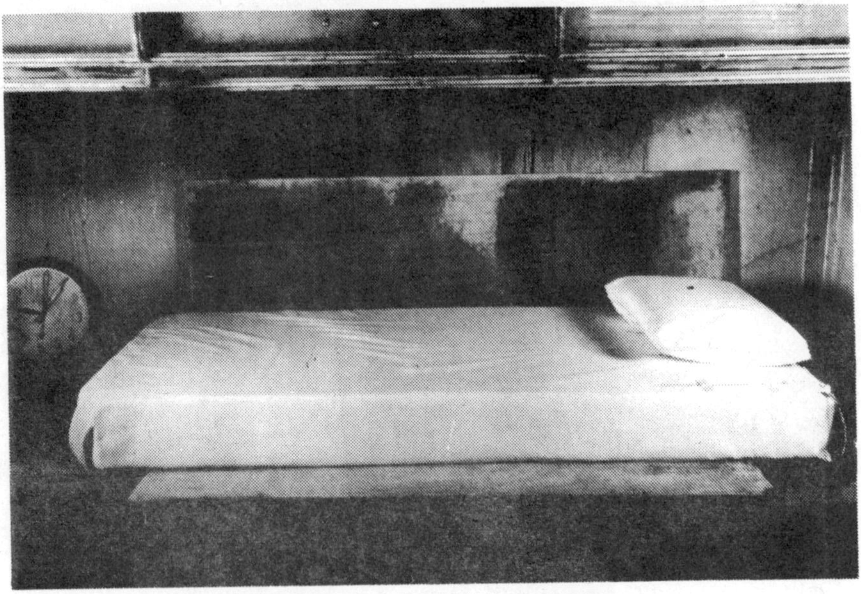

Figure 37. *Test No. 9, time 5 minutes. Modacrylic sheet, modacrylic pillowcase, fire retardant mattress.*

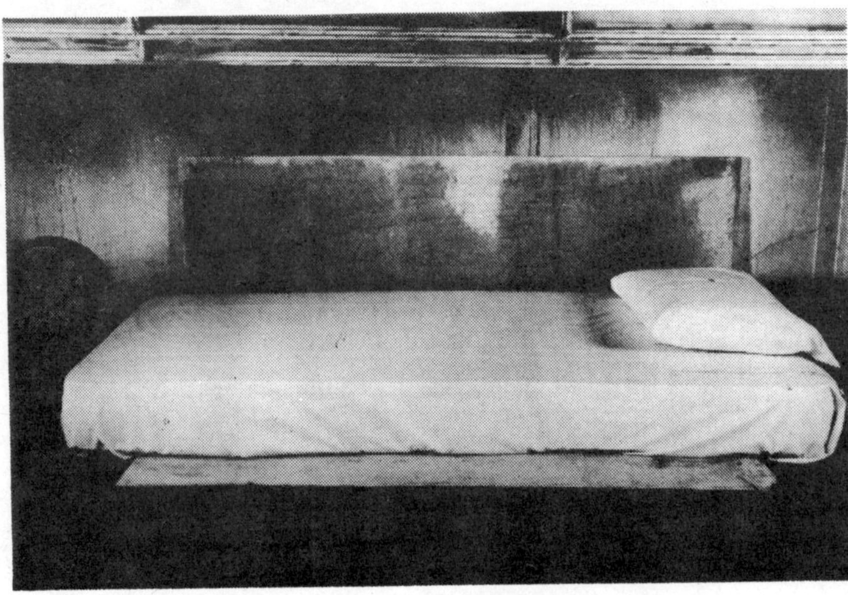

Figure 38. *Test No. 8, time 4 minutes. Polyamide sheet, polyamide pillowcase, fire retardant mattress.*

Table 5. Reproducibility of Fire Exposure Burning Latex Foam Rubber Pillow in Cotton Pillowcase Ignited by Methenamine Pill

	Time, minutes, for			
Test No.	Fire Spread Over Edge of Pillow	Fire Spread to Sheets	Fire Spread to Top Edge of Mattress	Pillow to Cease Burning
1	4	5	7	12
4	5	7	8	12
5	7	8	9	12
6	6	6	8	12
7	4	5	6	11

The temperatures recorded in the vicinity of the burning pillow are shown in Figures 39 to 42. The differences between the tests indicate some interaction with the mattress, with some mattresses contributing more combustible volatiles or conducting away more heat. The use of fire retardant materials in or on top of the mattress appears to reduce exposure severity.

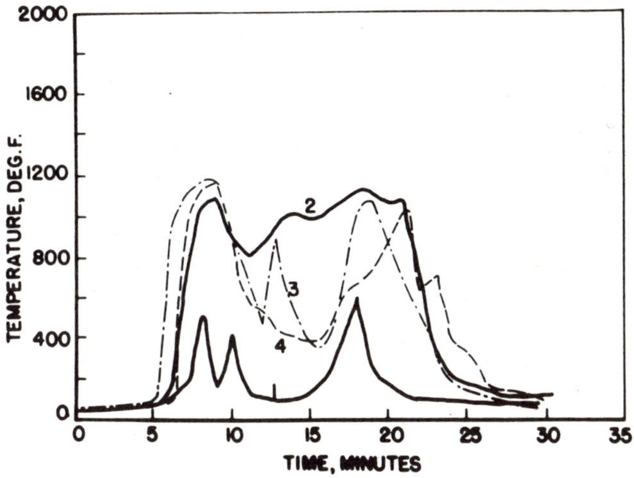

Figure 39. *Temperatures at surface near pillow, Test No. 6. Latex foam rubber pillow, cotton pillowcase, cotton sheet, fire retardant mattress. Thermocouple locations shown in Figure 21.*

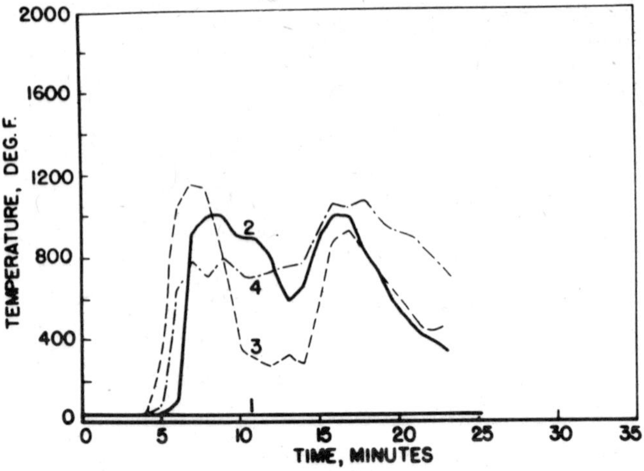

Figure 40. *Temperatures at surface near pillow, Test No. 10. Latex foam rubber pillow, cotton pillowcase, modacrylic sheet, fire retardant mattress. Thermocouple locations shown in Figure 21.*

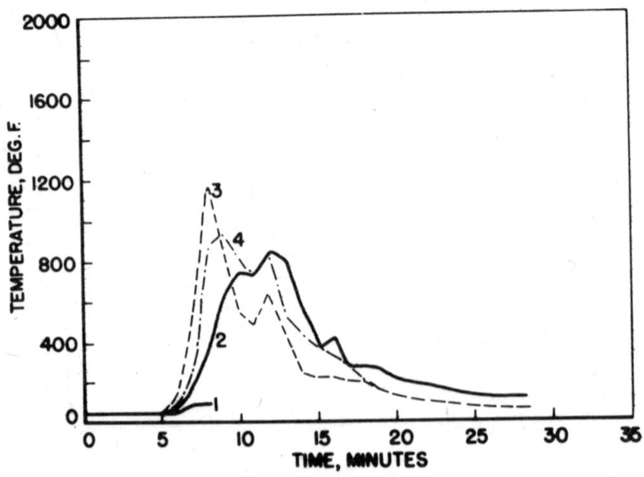

Figure 41. *Temperatures at surface near pillow, Test No. 11. Latex foam rubber pillow, cotton pillowcase, polyamide sheet, fire retardant mattress. Thermocouple locations shown in Figure 21.*

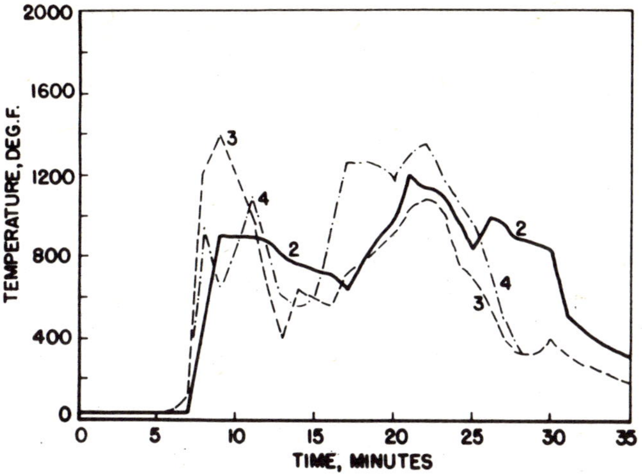

Figure 42. *Temperatures at surface near pillow, Test No. 12. Latex foam rubber pillow, cotton pillowcase, polyamide sheet, standard mattress. Thermocouple locations shown in Figure 21.*

Surface Temperatures

The temperatures reached at the mattress surface once a pillow becomes fully involved are in the vicinity of 1000°F, as shown by Figures 39 to 42. These temperature levels explain why supposedly fire-retardant mattress materials exhibit insufficient resistance to ignition. The ignition temperatures of most available padding and core materials are below 1000°F. Table 6 lists some ignition temperatures obtained in a Setchkin apparatus[13].

Table 6. Flash-Ignition Temperatures of Some Cushioning Materials

Material		Flash-Ignition Temperature, °C
Flexible urethane foam		280
Flexible urethane foam, fire-retardant, Sample	1	405
	2	364
	3	394
	4	407
	5	397
	6	390
	7	416
Styrene-butadiene rubber foam		270
Neoprene rubber foam, fire-retardant		395

The value of protective sheets may lie in keeping the flame front from penetrating to the mattress itself, so that any flammable gas mixtures produced would have to be heated to the autoignition temperature, rather than ignited by a flame front at the lower flash-ignition temperature.

Core Temperatures

Once burning occurs in the interior of an innerspring type mattress, temperatures in the vicinity of 1200°F can be developed by combustion in this relatively well-insulated enclosure. Figures 43 to 44 present the temperatures recorded inside two mattresses, one with and one without fire-retardant components. The fire-retardant mattress exhibited internal combustion earlier in the test; this may be

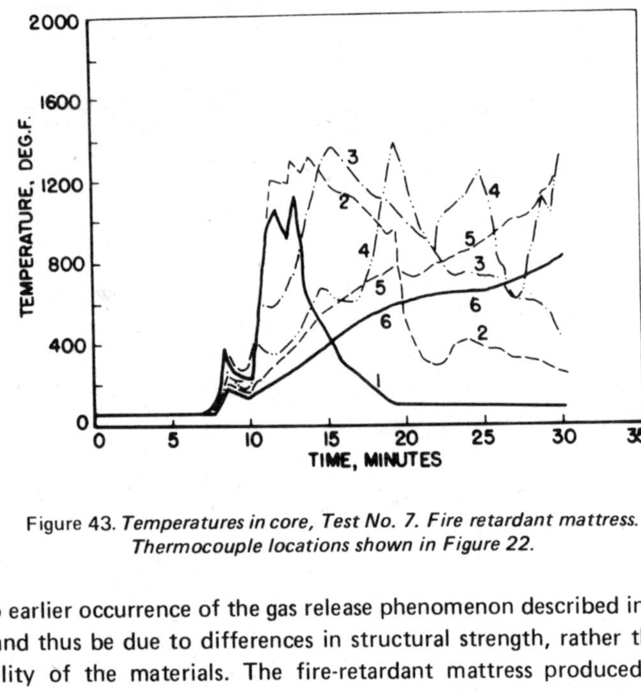

Figure 43. *Temperatures in core, Test No. 7. Fire retardant mattress. Thermocouple locations shown in Figure 22.*

related to earlier occurrence of the gas release phenomenon described in a preceding section, and thus be due to differences in structural strength, rather than inherent flammability of the materials. The fire-retardant mattress produced lower peak temperatures and smaller areas under the temperature-time plots, indicating substantially less heat produced during burning.

Heat Flux

Heat flux data were difficult to obtain with continuity because the radiometer had to be moved frequently to escape the advancing flame front. Most measurements were less than 1.0 Btu/ft^{2}sec, and reached 2.0 Btu/ft^{2}sec only when the flames were almost in contact with the radiometer. From the data obtained, it

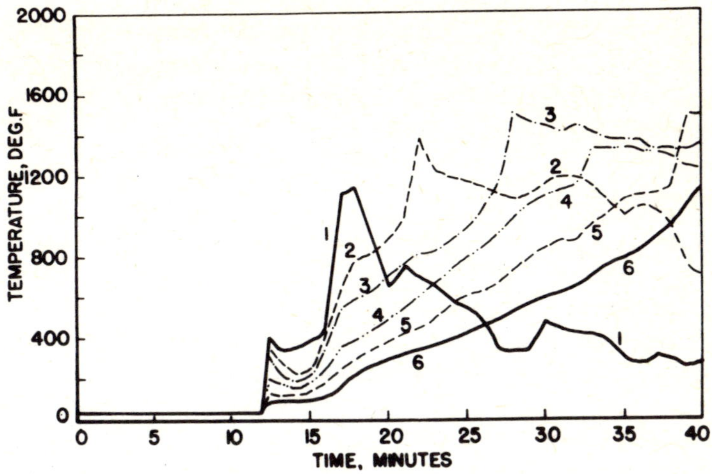

Figure 44. *Temperatures in core, Test No. 13. Standard mattress. Thermocouple locations shown in Figure 22.*

would appear that to attain the irradiance level of 2.2 Btu/ft^{2}sec used in the National Bureau of Standards smoke density test [14], a material would have to be positioned 6 to 12 inches from a burning latex foam rubber pillow at the stage of full involvement. The period of full involvement of the pillow and maximum heat flux is less than five minutes.

The particular radiometer used for these measurements had previously been checked in the NBS apparatus and found to agree within 2 percent with the radiometer used to calibrate the apparatus.

SMOKE DENSITY TESTS

The lack of room-scale test facilities prevented the measurement of smoke produced by burning complete mattresses under controlled conditions, and limited smoke density tests to those which could be performed in laboratory test chambers. Two test chambers were available: the Rohm and Haas XP2 smoke density chamber [15] and the National Bureau of Standards smoke density chamber [14].

It was recognized that the data obtained in small-scale laboratory tests do not necessarily predict behavior in actual fire situations. This series of tests was intended to provide guidance as to relative smoke-producing characteristics, using the means at hand, and to indicate the relative usefulness of different laboratory techniques in simulating the smoke conditions observed in the larger-scale tests described in earlier sections.

Apparatus

The Rohm and Haas XP2 smoke chamber [15] is a cabinet measuring 30 inches high, 12 inches wide, and 12 inches deep, completely enclosed except for 1-inch high ventilating openings around the bottom. For standard tests, the heat source is a propane-air flame from a pencil-tip burner, applied at a 45-degree angle to the bottom of a horizontal specimen. A horizontal photometer path, 20 inches above the bottom of the chamber, is used for measuring light absorption. The instrument purchased from George Eysenbach of Bristol, Pennsylvania, was equipped with a photometer providing an overall sensitivity range of 100 to 1.

The National Bureau of Standards smoke chamber [14] is a completely enclosed cabinet, measuring 3 feet wide, 3 feet high, and 2 feet deep, in which a specimen 3 inches square is supported vertically in a frame such that 2-9/16 inches square is exposed to heat under either flaming or nonflaming (smoldering) conditions. The heat source is an electric furnace, adjusted with the help of a circular foil radiometer to give a heat flux of 2.2 $Btu/ft^2 sec$ (2.5 $watts/cm^2$) at the specimen surface. A vertical photometer path for measuring light absorption is employed to minimize measurement differences due to smoke stratification which could occur with a horizontal photometer path at a fixed height; the full three-foot height of the chamber is used to provide an overall average for the entire chamber. The instrument purchased from American Instrument Company of Silver Spring, Maryland, was equipped with a photomultiplier providing an overall sensitivity range of 1000 to 1.

Test Procedure

In an effort to establish procedures most relevant to the burning of complete bedding systems, three groups of tests were conducted on mattress materials. In the first series of tests, one-inch cubes of core materials, and one-inch squares of cover and pad materials, were subjected to the standard Rohm and Haas XP2 test procedure involving a four-minute exposure to a burner flame fueled by propane at 40 psi pressure.

In the second series of tests, sections of mattress constructions, each 2 inches square and the depth of the constructions, were exposed to a lighted methenamine pill (Eli Lilly, No. 1588) in the XP2 smoke chamber with the burner turned off and no other ignition source present. For innerspring constructions, the innerspring was replaced by the existing specimen support screen, because the metal springs would not burn and would merely complicate specimen preparation. For solid foam core constructions, the thickness of core material was reduced to 2 inches, so that the total thickness of material on the specimen support screen would not exceed 3 inches in any case. By this approach, total specimen height was restricted to 2 ± 1 inches, and the variation in distance from the photometer path was correspondingly reduced.

In the third series of tests, sections of mattress constructions, each 3 inches square and with the same depth as in the second series of tests, were subjected to the standard NBS test, involving 20 minutes exposure to an electric furnace delivering a heat flux of 2.2 Btu/ft^{2}sec, under both flaming and nonflaming conditions.

Test Results

Smoke density data obtained in these three series of tests are presented in Tables 7 to 9. In the first series of tests, dealing with individual materials, the cover materials, with the exception of the plastic coated cover, produced little smoke. Among the pad materials, two urethane foam samples had moderate smoke and the rest little or none. Latex foam rubber produced very dense smoke. There was moderate smoke from fire-retardant urethane foam and little smoke from non-fire-retardant urethane foam.

Table 7. Smoke Produced from Mattress Components Rohm and Haas XP2 Test

Material		Light Absorption, percent	
		Maximum	Final
Core Materials (Specimens 1 x 1 x 1 inch)			
Latex foam rubber, Sample	A	96.3	50.0
	B	99.7	64.3
	C	97.7	59.7
Flexible urethane foam, Sample	A	5.0	0
	B	4.3	1.0
	C (1.5 pcf)	8.7	0.3
	D (1.5 pcf)	7.3	1.0
Flexible urethane foam, fire-retardant, 2.0 pcf		27.0	4.3
Pad Materials (Specimens 1 x 1 inch)			
Latex sisal pad, 1/4 inch, Sample	A	0.7	0
	B	6.0	0.3
	C	1.3	0
Felt batt, 1 inch		0	0
Tufflex, 1/4 inch		0	0
Flexible urethane foam, Sample	A, 1 inch	6.7	0.3
	B, 3/4 inch	20	4.5
	C, 3/4 inch	24.3	3
Cover Materials (Specimens 1 x 1 inch)			
Rayon, quilted with	3/4 inch urethane foam	4.3	0
	1/2 inch urethane foam	1.7	0.3
	1/2 inch urethane foam	1.3	0
	1/4 inch urethane foam	1.3	0
	felt	0	0
Cotton, fire-retardant, Samples A, B, C		0	0
Cotton		0	0
Cotton, plastic coated		24.3	3

Table 8. Smoke Produced from Mattress Constructions Modified Rohm and Haas Test

Material	Ignition	Absorption, percent		Specific Optical Density		Time to Ds = 16 min.
		Max.	Res.	Maximum, Dm	Final, Df	
Spun rayon cover, urethane foam 3/4" (a), felt batt, latex sisal pad	Yes	81	2	64.9	0.8	3.17
Spun rayon cover, felt (a), latex foam rubber 2"	No	86	0	76.8	0	2.58
Spun rayon cover, urethane foam 1/2" (a), urethane foam 1", latex sisal pad	Yes	18	4	7.7	1.6	–
Spun rayon cover, urethane foam 1/2" (a), latex foam rubber 2"	Yes	99	64	180	39.9	0.92
Cover, urethane foam 1/4" (a), urethane foam 2"	Yes	32	3	15.1	1.2	–
Cover, urethane foam 2"	Yes	32	6	15.1	2.4	–
Cotton cover FR, urethane foam 3/4", latex sisal pad 1/4"	No	0	0	0	0	–
Cotton cover FR, urethane foam 3/4", latex sisal pad 1/4"	No	4	0	1.6	0	–
Cotton cover FR, urethane foam FR 2"	No	4	0	1.6	0	–
Cotton cover FR, latex foam rubber 2"	No	3	0	1.2	0	–
Cotton cover FR, urethane foam 2"	No	3	0	1.2	0	–
Cotton cover, urethane foam 2"	Yes	35	7	16.8	2.8	1.92
Plastic coated cover, urethane foam 2"	No	3	0	1.2	0	–

(a) Quilted with cover

Table 9. Smoke Produced from Mattress Constructions National Bureau of Standards Test

Material	Exposure		Min. Trans. pct	Specific Optical Density: Maximum, Dm	Specific Optical Density: Final, Df	Time to Ds = 16 min.
Spun rayon cover, urethane foam 3/4" (a), felt batt, latex sisal pad	F		7.62	147.1	5.0	0.46
	NF		0.062	422.1	53.9	0.37
	NF		0.105	392.0	62.0	0.40
Spun rayon cover, felt (a), latex foam rubber 2"	F		0.0145	504.8	43.7	0.82
	NF		0	780	68.8	0.22
Spun rayon cover, urethane foam 1/2" (a), urethane foam 1", latex sisal pad	F		0.147	372.7	8.2	2.75
	NF		0.00906	532.0	60.6	0.49
Spun rayon cover, urethane foam 1/2" (a), latex foam rubber 2"	F		0	780	107.3	0.35
	NF		0	780	131.6	0.26
Cover, urethane foam 1/4" (a), urethane foam 2"	F	(b)	0.920	267.2	12.5	4.50
	F		4.98	171.3	7.3	3.95
	NF	(b)	0	780	84.5	0.30
	NF		0.014	506.8	50.1	0.37
Cover, urethane foam 2"	F		11.8	122.0	6.6	2.87
	NF		0.110	389.0	38.5	0.40
Cotton cover FR, urethane foam 3/4", paper synthetic pad 1/4"	F		6.80	153.6	4.0	0.78
	NF		0.0381	449.9	20.1	0.33
	NF		0.0435	442.3	18.6	0.31
Cotton cover FR, urethane foam 3/4", latex sisal pad 1/4"	F		4.9	172.4	6.0	1.05
	NF		0.0201	486.4	21.4	0.34
Cotton cover FR, urethane foam FR 2"	F		0.0650	419.1	15.7	0.50
	NF		0.0150	502.8	16.8	0.35
Cotton cover FR, latex foam rubber 2"	F	(b)	0	780	74.4	0.20
	F		0	780	61.0	0.13
	NF	(b)	0	780	9.1	0.30
	NF		0	780	20.0	0.37
Cotton cover FR, urethane foam 2"	F		14.2	111.5	4.6	2.65
	NF		0.0150	502.8	20.9	0.36
Cotton cover, urethane foam 2"	F		2.30	215.4	9.8	1.91
	NF		0.0688	415.8	30.8	0.52
Plastic coated cover, urethane foam 2"	F	(b)	0.410	314.0	16.4	0.14
	F		1.71	232.4	17.8	0.15
	NF	(b)	0.460	307.4	27.8	0.50
	NF		0.0010	657.5	103.0	0.33

(a) Quilted with cover
(b) Not wrapped in aluminum foil

The results of the second and third series of tests tended to confirm expectations based on the results of the first series, with the effect of core materials predominating due to their higher weight or volume fraction of the composite. Composites containing latex foam rubber generally exhibited the highest smoke densities.

It is the opinion of the author that the second series of tests, employing methenamine pill ignition in the XP2 smoke chamber, gave the results which most closely approximated the smoke observed in the larger-scale tests. The principal advantage of this group of data is that it tends to incorporate in the results the effect of relative ease of ignition.

The NBS test procedure involves a heat flux level which, by both direct measurement and observation of results, constitutes substantially more severe fire exposure than was developed in the larger-scale tests. For fully developed room fires, the NBS test may be the most appropriate, but data such as time to reach a critical density level become irrelevant, since the occupant of the bed would probably have been killed several times over by lethal levels of temperature and carbon monoxide.

CONCLUSIONS

The use of fire-retardant cotton covering in innerspring and urethane foam core type mattresses has been shown to provide adequate resistance to ignition by small fire sources applied to the mattresses themselves. Use of fire-retardant cotton covering did not prevent the same ignition sources from causing progressive smoldering in latex foam rubber core mattresses.

The presence of sheets, pillows and pillowcases on the mattress, and the involvement of these articles in fire, increases exposure severity, and introduces many interacting variables all affecting the resistance of the mattress to ignition and involvement. Each of the following changes has been shown to improve this resistance:

1. Fire-retardant cotton covering instead of non-fire-retardant cotton or rayon covering in the mattress.
2. Fire-retardant urethane foam instead of non-fire-retardant urethane foam in the mattress.
3. Down-and-feather filling instead of polyester fiberfill in the pillow, and polyester fiberfill instead of latex foam rubber filling in the pillow.
4. Aromatic polyamide sheets instead of non-fire-retardant cotton sheets.
5. Modacrylic or polyamide pillowcases instead of non-fire-retardant cotton pillowcases.

Latex foam rubber cores in mattresses exhibited the characteristics of progressive internal smoldering, vigorous burning when ignited, and production of large quantities of dense smoke.

Currently available materials and technology are capable of giving innerspring and urethane foam core type mattresses adequate resistance to small ignition sources. Resistance to burning sheets and pillows can be achieved in varying degrees by selection of materials.

Fire experience indicates that population habits may be the major factor in bedding fires. The extent to which bedding materials must be improved to attempt to offset human carelessness is an issue which has not been resolved.

It is the opinion of the author that a bedding system cannot be considered resistant to ignition by fire exposure of a given severity unless the necessary level of resistance is possessed by all the components of the system: mattresses, sheets, pillows, pillowcases, blankets, and sleepwear.

REFERENCES

1. Anon., "The Single-Fatality Fire," *Fire Journal*, Vol. 63, No. 1, 34–35 (January 1969).
2. Anon., "Fires and Fire Losses Classified, 1971," *Fire Journal*, Vol. 66, No. 5, 65–69 (September 1972).
3. Anon., "Fires and Fire Losses Classified, 1970," *Fire Journal*, Vol. 65, No. 5, 21–25, 28 (September 1971).
4. Anon., "Fires and Fire Losses Classified, 1969," *Fire Journal*, Vol. 64, No. 5, 65–69 (September 1970).
5. C. A. Hafer, C. H. Yuill, "Characterization of Bedding and Upholstery Fires," Final Report, NBS Contract No. CST-792-5-69, SWRI Project No. 3-2610, Southwest Research Institute, San Antonio, Texas (March 31, 1970).
6. C. H. Yuill, "The Life Hazard of Bedding and Upholstery Fires," *Journal of Fire and Flammability*, Vol. 1, No. 4, 312–323 (October 1970).
7. U.S. Department of Commerce, "Standard for the Flammability of Mattresses," DOC FF 4–72, *Federal Register*, Vol. 37, No. 110, 11362–11367 (June 7, 1972).
8. U.S. Department of Commerce, "Notice of Finding That Flammability Standard for Upholstered Furniture May Be Needed," *Federal Register*, Vol. 37, No. 230, 25239–25240 (November 29, 1972).
9. U.S. Department of Commerce, "Standard for the Surface Flammability of Carpets and Rugs," DOC FF 1–70, *Federal Register*, Vol. 35, No. 74, 6211–6214 (April 16, 1970).
10. U.S. Department of Commerce, "Standard for the Surface Flammability of Small Carpets and Rugs," DOC FF 2–70, *Federal Register*, Vol. 35, No. 251 (December 29, 1970).
11. U.S. Department of Commerce, "Standard for the Flammability of Children's Sleepwear," DOC FF 3–71, *Federal Register*, Vol. 36, No. 141, 14624–14632 (July 21, 1971).
12. W. H. Gloor, "Heat Resistant Textile Fibers," *American Dyestuff Reporter*, (June 17, 1968).
13. N. P. Setchkin, "A Method and Apparatus for Determining the Ignition Characteristics of Plastics," *Journal of Research of the National Bureau of Standards*, Vol. 43, 591–608 (December 1949).
14. D. Gross, J. J. Loftus, A. F. Robertson, "Method for Measuring Smoke from Burning Materials," ASTM Special Technical Publication No. 422, 166–204 (1967).
15. Rohm and Haas Co., NFPA Quarterly, Vol. 57, 276–287 (January 1964).

Carlos J. Hilado

Carlos J. Hilado is a Project Scientist in the Fire Research Laboratory of the Chemicals and Plastics Research and Development Department of Union Carbide Corporation at South Charleston, West Virginia, where he is concerned with evaluating the flammability characteristics and fire safety aspects of materials. He received his B.S. degree in chemical engineering from De La Salle College in Manila, Philippines in 1954, and his M.S. degree in Chemical Engineering Practice from Massachusetts Institute of Technology in 1956. His M.I.T. Practice School experience included the pulp and paper mill of Eastern Corporation at South Brewer, Maine, the chemical plant of Hercules Inc. at Parlin, New Jersey, and the steel mill of Bethlehem Steel Co. at Lackawanna, New York. Since 1956 he has been with Union Carbide Corporation at South Charleston, West Virginia, where he has worked in the fields of vinyl and urethane foams, vinyl polymerization, testing and evaluation, industrial insulation, and fire research.

A registered Professional Engineer in the state of West Virginia, he is a member of the American Chemical Society, American Institute of Chemical Engineers, American Society for Testing and Materials, Society of the Plastics Industry, National Fire Protection Association, and Combustion Institute. He represents Union Carbide Corp. on several ASTM and SPI committees, and is a member of the Ad Hoc Committee on Fire Safety Aspects of Polymeric Materials of the National Materials Advisory Board of the National Research Council. In 1970 he received the Distinguished Service Award of the Cellular Plastics Division of the SPI, and in 1973 he received the Educational Service Award of the Plastics Institute of America.

He is the author or co-author of 42 technical papers, author of the "Flammability Handbook for Plastics" and "Handbook of Environmental Management," editor-in-chief of the Journal of Fire and Flammability, and editor of the Fire and Flammability book series. He has been an invited lecturer at the University of Utah, University of Detroit, Wayne State University, Stevens Institute of Technology, and Brooklyn Polytechnic Institute.

Description of a Fireproof Torso-Form and Ignition System Used as a Testing Technique for Studying Garment Flammability

E. L. Finley and W. H. Carter

Agricultural Experiment Station
Louisiana State University

(Received January 9, 1970)

ABSTRACT

A fireproof torso-form was used to develop an original testing technique for burning garments to simulate a "life situation" in the study of fabric flammability relative to time of burning, temperatures registered at sensor points on the torso-form, and effect of garment form on burning patterns. An original procedure for igniting garments by an open flame was developed which was controlled electrically for instant on and off flame application to fabric.

INTRODUCTION

Present methods for testing fabric flammability neither simulate an "actual life" situation when clothing is ignited, nor do they evaluate flammability characteristics in relation to potential body injury.

According to statistics made available in a Senate report, the Public Health Service has estimated that one million people are burned in the home each year and 150,000 persons suffer injuries as a result of the ignition of clothing. The Health Service further estimates that 2,000 to 3,000 persons die each year from fire and that deaths from burning (all fire) rank third—behind accidents and falls—as the chief causes of accidental injuries and deaths in this country [1, 2]. Textile fibers commonly used for apparel purposes are inherently susceptible to decomposition by fire [3, 4].

In 1967, 79 percent of all blended fabrics were made of cotton and polyester fibers [5]. These fabrics are highly combustible.

Consumers of textile products, the medical profession, textile producers and finishers, and research groups would benefit from an investigation of clothing flammability to determine the interrelationships between fiber, fabric finish, area density, and shape-form of the fabric when it comes in contact with a flame.

A study was undertaken to develop experimental equipment, consisting of a fireproof torso-form and an automatically controlled ignition device to ignite

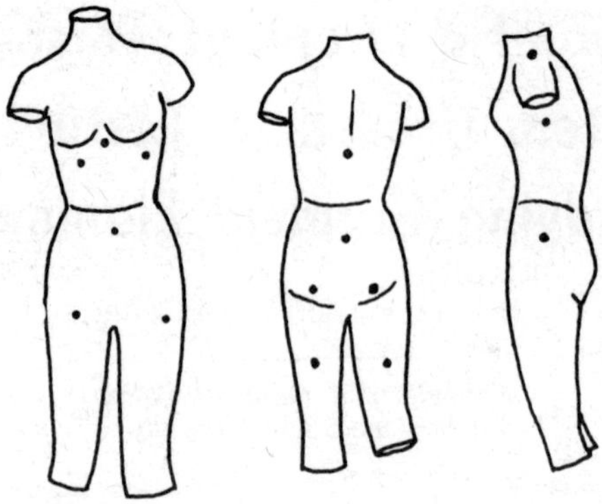

Figure 1. *Location of thermocouples on the torso-form.*

a garment used to simulate conditions of a garment afire on a human body.* A simple A-line dress was the type of garment selected for burning on the torso-form to measure flame temperature at designated locations on the torso, surface temperature on the torso in relation to possible degree of burn injury, effect of convection of gases on ignition of fabric, rapidity of fabric ignition and burn, and the comparative effect of flame retardant finishes on the fabrics.

The simple A-line dress was selected because this represents a current style dress and night dress. The dresses were made of all-cotton, and 50/50 and 70/30 cotton-polyester fabrics, plain weave, and light weight, 3 to 4 oz/yd^2. The all-cotton and cotton-polyester fabrics were treated with each of the three fire retardant finishes of a phosphorus base type. A designated amount of each fabric was retained in the untreated state. The fabrics were manufactured into dresses by a commercial firm.

The purpose of the present paper is to describe the new technique applicable to the study of clothing flammability which may contribute useful information to the development of protective standards for clothing.

MATERIALS AND METHODS

The fireproof torso-form was made by modifying a commercial dress mannequin. It was modified by removing the extremities including head, arms 3 inches below the shoulder point, and legs 3 inches below the knee. The exterior surface was coated to a depth of ¾ inches with a wet mixture of ground

* Work in progress at Louisiana State University and Contributing to Regional Textile Project, SM-38, work undertaken by Alabama, Louisiana, Missouri, North Carolina, Oklahoma, and Tennessee.

Figure 2. *Arrangement of base and burner platforms with copper conduits and thermocouples into torso-form.*

asbestos and calcium sulfate. The position of eighteen individual thermocouple sites for taking temperature measurements were located on the torso as indicated in Figure 1A front, 1B back, and 1C side.

Nine copper tubes, ⅜ inch diameter, were drawn into the hollow interior of the torso through each leg and positioned slightly below the asbestos surface. The copper tubes served as conduits through which two sets of chromel-alumel thermocouple wire, insulated with stranded glass, were drawn for positioning onto the surface of the torso at designated sites, as shown in Figure 2.

One set of eighteen thermocouples on the torso surface was used to record the flame temperature as the garment burned. The component ends of each wire were spot welded together and extended ¼ inch above the torso surface. To measure total heating at each location, each of the other set of eighteen thermocouple wires was welded to an individual disc made of cold rolled copper, 1 inch in diameter, 1/32 inch thick and treated with a temperature stable black paint. The discs were mounted by embedding them flush onto the torso surface to record surface temperature in relation to potential burn injury. The thermocouple points and discs were located in adjacent positions on the torso-form.

The torso-form was mounted in a rigid upright position by metal support braces attached at the knee and bolted to a 24 × 24 inch steel plate supported on casters.

The ends of the chromel-alumel components of each set of thermocouple units extended out of the copper tubes. The chromel-alumel components of each set were separately spot welded to a No. 18, 2 lead copper Bell wire. The chromel component (magnetic) was welded to a red lead and the alumel component (negative) was welded to a white lead of the Bell wire. The thermocouple cables at the welded junctions were immersed in a mixture of ice and water to obtain a zero reference point during burning experiments. The range of temperature actuation of the thermocouple wire was 0-1000°F. An individual recording instrument was connected to a Bell wire-thermocouple lead from the torso. A record of the temperature for an experiment at each set of the eighteen thermocouple sites was inscribed by an individual recorder on a Type A chart at a speed of 8 inches per minute and at a rate of 4 strikes per second with a gear train speed of 18 rpm. The temperature range was 0-1000°F. ±2% for the recorders.

A steel plate platform, 19 × 24 inches, supported 6 bunsen type burners positioned around the torso to ignite the garment. A vertical side adjustment on the burner platform was provided to raise or lower it according to the length of the garment. A 7 × 12 inch rectangle was cut out in the center of the burner platform plate to fit it around the torso base. The gas supply tubes were threaded up through this opening to the burners. Heavy asbestos paper was used to cover the burner platform to protect the exposed tygon tubes from burning debris.

A diagramatic sketch of the ignition device is shown in Figure 3. The key names the parts identified by a capital letter on the diagram. The pilot lead to ignite each burner was made from a 5 inch length of ⅛ inch diameter steel tubing. These leads were clamped to the stem of each burner. The burners and pilots were supplied with chemically pure butane from separate cylinders, 1 and 2. Two independent fuel sources permitted adjustment of the small pilot flames that burned continually as the fuel from cylinder 1 to the burners was turned on and off for garment ignition.

The gas under high pressure from cylinder 1, to the main burners, was reduced to a pressure of 5 inches of water column by the Fisher valve. The pressure gauge controlled the gas pressure when the apparatus was used. Tygon tubing conducted the low pressure gas to the magnetic valve which controlled the uniform flow of gas. The magnetic valve, normally closed, was energized with 120 volt AC current by a toggle switch which instantly opened the valve and permitted gas to flow to the burners. A six-outlet manifold located on the torso base received gas from the magnetic control valve and divided it equally to each burner through 3-foot long, 5/16 inch diameter tygon tubes.

Gas for the pilots was supplied from cylinder 2. Gas pressure was reduced by the pressure regulator, M, and the needle valve, N, provided an adjustment to accurately control the flow of gas to stabilize a very small pilot flame. Gas flowed through a 5/16 inch tygon tube to a small 6-outlet manifold placed on the torso base beside the burner manifold. Three-foot lengths of ⅛ inch rubber tubing conducted the gas from the pilot manifold to each burner pilot.

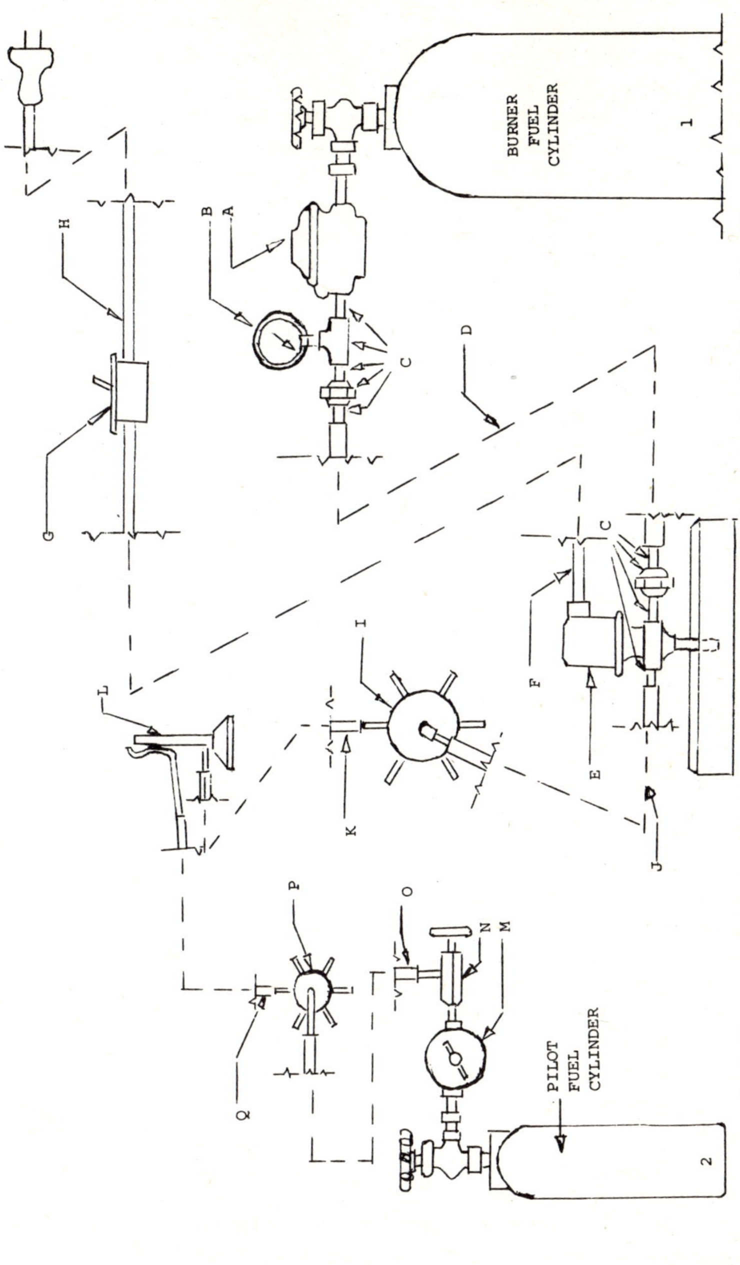

A. Fisher valve No. 912, B. Pressure gauge, C. Pipe fittings, D. Gas lead to solenoid, E. Solenoid valve, F. Electric cable from toggle switch to magnetic valve, G. Gas control toggle switch, H. Electric cable from outlet to toggle switch, I. Burner manifold, J. Gas lead to manifold, K. Gas lead from manifold to burner, L. Burner, M. Pressure reducing valve, N. Needle valve, O. Gas lead from needle valve to pilot manifold, P. Pilot manifold, Q. Gas lead from pilot manifold to pilot.

Figure 3. *Diagramatic sketch of the ignition device.*

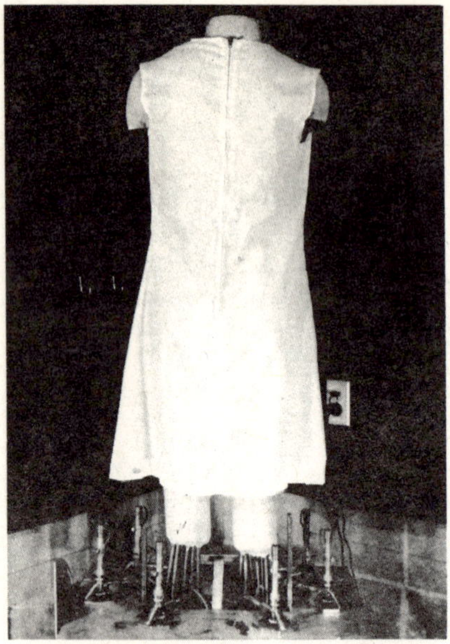

Figure 4. *Garment on the torso-form.*

EQUIPMENT APPLICATION FOR TESTS

Operation of the equipment as illustrated in Figure 4 shows a garment positioned on the torso-form with the six burners positioned below the hem—back, side, and front.

Figure 5 shows the six burners lighted as they function to ignite a garment. The flame simulated an open flame situation where clothing is accidently ignited. The garment was ignited at the hem because clothing is usually ignited where there is freedom to swing such as the lower part of a skirt, long full sleeves, or loose ties. Igniting the garment at the hem imposed the most severe test of the fabric flammability which would indicate its fire resistance when a flame is applied in a vertical direction.

Figure 6 illustrates ignition of a garment 2-3 seconds after flame was impinged at hem edge of the garment from the six burners.

Mean surface temperature from individual thermocouple points on the torso form is shown by location for leg (below crotch) and body position in Table I. The garments used were all-cotton and 70/30 cotton-polyester, without any finish and each having the same flame retardant finish. The untreated garments were consumed; whereas, the treated garments were charred at points of flame impingement. It was concluded that the temperatures observed for the treated fabrics was due to heat rise from the burners.

Further work is in progress and future papers are planned to report in detail on the testing of various types of untreated and treated fabrics in garment form using the equipment and technique described in this paper.

Figure 5. *Lighted burners positioned around torso-form.*

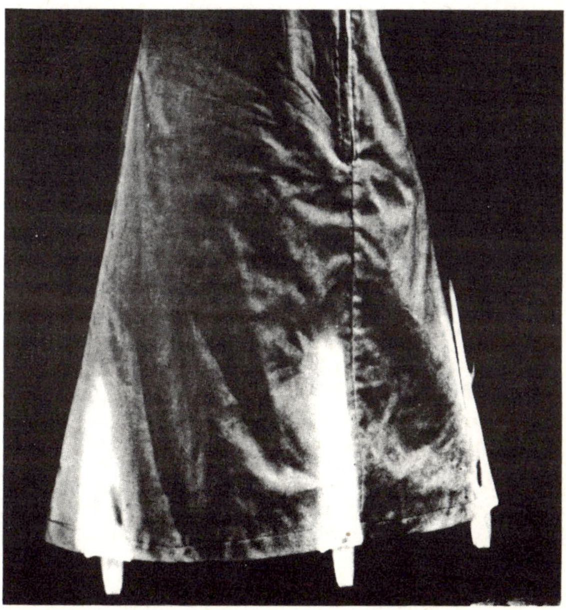

Figure 6. *Burning garment 2-3 seconds after flame applied at hem.*

Table 1. Summary of Temperatures on Torso Surface by Position, Fabric and Flame Retardant Finish.

Garment Identification	Position			
	Legs		Body	
	Temperature			
Fabric/Finish	F°	C°	F°	C°
All-Cotton—0	231.0	120.0	189.0	87.2
All-Cotton—1	149.0	65.0	135.0	57.2
70/30 Blend—0	219.0	113.5	190.0	87.8
70/30 Blend—1	150.0	65.6	135.0	57.2

Code:
70/30 Blend: Cotton-Polyester
Finish: 0 no finish; 1, Same Flame Retardant Finish

CONCLUSIONS

Uniform ignition of a garment was obtained by impinging flame on the hem edge from the six burners. Instant on and off electrical control of burner flame was manipulated by a toggle switch connected to the magnetic control valve. Flame height was regulated by the gas pressure (Fisher) valve. The micro-pilot flames did not constitute any hazard in the dressing of the torso-form in changing garments.

The fireproof torso-form was serviceable and evidenced no chipping or cracking after ignition of many garments.

The three component arrangement of equipment, torso-form-ignition device-thermocouple system, provided a satisfactory and unique technique to study flammability characteristics of garments.

ACKNOWLEDGEMENTS

Grateful acknowledgement is made to Personnel of the Southern Utilization Research and Development Division, USDA, for modification of commercial type mannequins to meet the equipment specifications.

Sincere appreciation is expressed to Mr. Albert S. Cooper, Jr. and to Mr. George L. Drake, Jr., Southern Utilization Research and Development Division, USDA, for their helpful consultation on the project.

REFERENCES

1. Senate Report No. 407, 90th Congress, 1st Session, Flammable Fabrics Act Amendments of 1967.

2. Papers of "Burn Conference," New York, N. Y., 1966.
3. Evelyn E. Stout, *Introduction to Textiles.* Second Edition. New York: John Wiley & Sons, Inc. 1965, pg. 318.
4. Cameron A. Baker, "The Pyrotechnics of Textile Materials" *Modern Textiles,* Vol. XXXXIX, No. 11, (November 1968) pp. 57-62.
5. *Textile Organon,* Vol. XXXVIII, No. 1, (August 1967) pg. 148.

E. L. Finley

Etta Lucille Finley is an Associate Professor at Louisiana State University and Mechanical College in Baton Rouge, Louisiana. She received her B.S. and M.A. degrees from Texas Woman's University in 1931 and 1943, respectively, and did her sabbatical study at the University of Tennessee in 1966. Before joining the faculty at Louisiana State University in 1955, she was a teacher, did cotton fiber research for the Cotton Research Committee of Texas, and worked for Anderson, Claton, and Co. She has experience in textiles and physical properties research.

W. H. Carter

William Hobart Carter is a Professor of Agricultural Engineering at Louisiana State University and Mechanical College in Baton Rouge, Louisiana. He received his B.S. and M.S. degrees in Agricultural Engineering from Iowa State University in 1929 and 1934, respectively. Before joining the faculty of Louisiana State University in 1930, he was employed by the U.S.D.A. Division of Agricultural Engineering. He has experience in farm power and machinery, farm and home utilities, and household equipment.

Temperature and Heat-Flux Measurements on Life-Size Garments Ignited by Flame Contact

E. L. Finley

School of Home Economics

and

W. H. Carter

Department of Agricultural Engineering
Louisiana State University
and
Agricultural Mechanical College

(Received June 3, 1971)

ABSTRACT

A life-size, fireproof mannequin or torso-form has been used to test a garment of a single thickness of fabric made of selected cotton, and cotton-polyester blend fabrics, with no finish, and treated with APO–THPC, THPOH–MM–Urea, and THPOH–NH3 finishes. Flame from Bunsen type burners was impinged on the garment at the hem edge. Surface heat and flame temperature of fabrics which ignited was measured at thermocouple sites located on the torso-form.

INTRODUCTION

The annual economic loss in terms of mortality and morbidity caused from burn injuries cannot be accurately assessed, but this has been estimated at 349.1 to 660.0 million dollars in the United States [6]. The highest incident of burn injuries has been cited as occurring among two groups, for persons under 15 years of age and for persons between the ages of 25-44 years [10].

According to information released on a series of case studies, made by the Consumer Protection and Environmental Health Service, the home was identified as the place where burns from clothing fires occur most frequently [6]. The National Burn Information Exchange cites flame ignition as the cause of two out of three flame burns [6]. The use of open type heaters in the Southwest is the frequent cause of ignition to clothing by flame.

Ignition of clothing occurs most frequently in the outer garment as revealed in a study made on individual textile fires. The types of outer garments most frequently mentioned were shirts, blouses, night dress, and trousers. Positive identification of fiber content of the garments was known for only 38 per cent of the cases, and of these 75 per cent were reported as made of cotton with 8 per cent identified as nylon and 8 per cent identified as other man-made fiber [6]. There is further evidence of the lack of such information in the literature concerning specific fibers used in fabrics in garment form when ignited by flame contact.

The potential flame hazard of a fabric evaluated under laboratory conditions may not exhibit the same flame hazard as it would under an actual life accident situation. There is a lack of data relative to generic type of fiber, fabrics composed of fiber blends, style of the garment and its construction in relation to ignition, flame spread, and heat generated. The wide consumer use of blend fabrics poses unresearched problems in the flame behavior of these fabrics and their finishing for effective flame retardant performance. Approximately 80 per cent of all blended fabrics used for apparel wear are made of a blend of cotton and polyester fiber [3]. Between 1960 and 1970, the production of cotton-polyester blends increased from 200 to 2,250 million yards [13].

The behavior of garments in a situation of accidental ignition is a needed area for exploration. The present work was undertaken to evaluate the performance of garments made of selected cotton and cotton-polyester fabric before and after treatment with flame retardant finishes, and to measure thermocouple point temperature and thermoplate surface heat located at certain sites on a fire-proof torso-form clothed in the single garment.

TORSO–FORM AND IGNITION SYSTEM

Equipment for the experimental work was developed to simulate a human figure clothed in a garment consisting of a single layer of fabric and ignited at the hem edge by controlled application of flame. A fire-proof mannequin, designed as a torso-form, was made by modifying a size 12 to 14 commercial dress mannequin. These modifications were made to facilitate putting on and removing the test garments and included removing the head, the arms 4 inches below the shoulders, and the legs 4 inches below the knees. Nine copper tubes were inserted into the mannequin through each leg to the designated surface locations for temperature sensors. Separate wires were used for thermocouples and plates and were drawn into each copper tube to the designated sites on the surface of the form as shown in Figures 1 and 2, and each thermocouple component was connected to an automatic thermal recorder as shown in Figure 3.

An arrangement of six Bunsen-type burners were modified after an original design by adding a shop-built pilot light tube. The burners were used to ignite the garment on the torso-form by spacing them uniformly around and under the hem edge of the garment. A magnetic valve, operated by a toggle switch, controlled the flow of C.P. butane to the burners. Butane gas was used because it produced a visible flame which was easily adjusted.

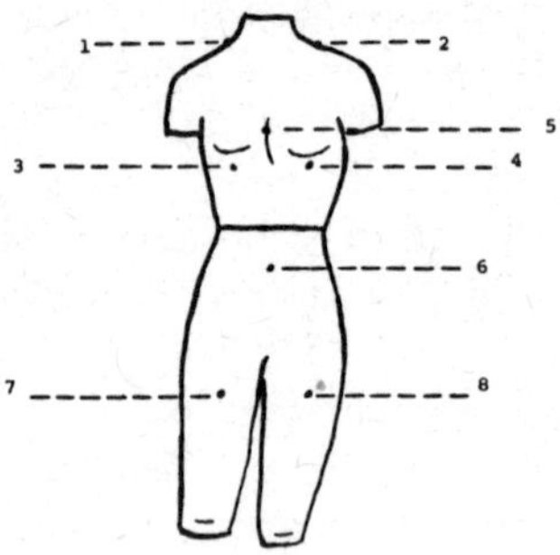

Figure 1. *Location of temperature sensors on front torso-form.*

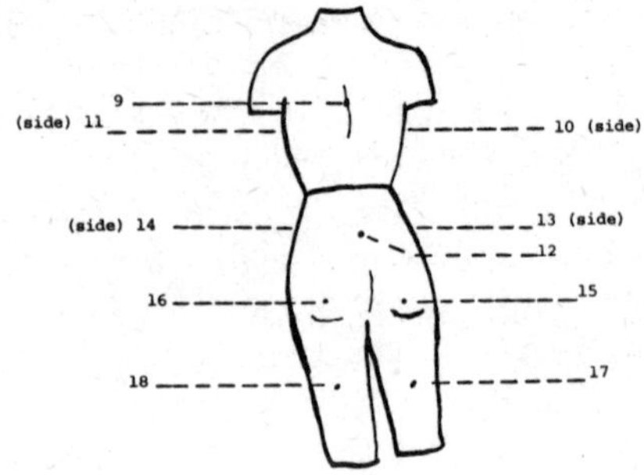

Figure 2. *Location of temperature sensors on back torso-form.*

Figure 3. *Arrangement of individual recorder instruments.*

Fuel was supplied from standard butane cylinders, and the flow was regulated by appropriate pressure reducing valves and gauges. The six pilots and six main burners received fuel from separate shop-built manifolds through equal lengths of tygon tubing. A complete description with specifications for the torso-form and total system equipment was presented in an earlier paper [3].

TEMPERATURE RECORDING INSTRUMENTATION

The experimental time-temperature data from burning experiments were obtained by using individual temperature recording instruments within a temperature range of 0-1000 F, operating at a chart speed of 8 inches per minute and gear train powered from a primary source of 105-125 volts, alternating current. Appropriate type chart paper was used as determined by specifications for the recorders which timed the temperature data in increments of 1.875 seconds.

The thermocouple leads used on the torso-form were made from type K, 18 gauge chromel-alumel, non-oxidizing components, with woven fiberglass insulation. The tips of the two components were spot welded at the surface of the torso-form. Range of activation for thermocouples was 0-1000 F (± 2% accuracy). Surface heat was measured in degrees F by using thermoplates made of copper discs, 1 inch in diameter, 0.025 inches thick, with a weight of 10 grams, and finished with Sicon (R) Flat Black and S744 reducer. Twenty-eight gauge chromel-alumel thermocouple wire with fiberglass insulation was silver soldered to the center rear side of each disc. The discs were mounted flat at the designated locations on the torso-form. The discs were calibrated as follows:

> A radiant panel was used as a heat source for energy response measurements of copper disk/asbestos board radiometer submitted for calibration by 411.06 and reported to be used for clothing/mannequin studies. Measurements were obtained by a substitution method, *i.e.,* by substituting the radiometer for a blackened copper disc calorimeter (421.02) of known Mass, Specific Heat and Area. The energy response for the 421.02 slug-type calorimeter was determined by the energy relationship.

$$E = \frac{Mc}{\alpha A}\frac{dT}{dt} \times 4.186 = 1.00\,\frac{dT}{dt}$$

When E = Incident heat flux $\frac{\text{Watt}}{\text{Cm}^2}$

M = Mass of Copper Slug = 27.51 gm

c = Specific Heat = 0.0918 Cal/gm°C

α = Absorptivity = 0.93

A = Area of Copper Slug = 11.36 Cm^2

$\frac{dT}{dt}$ = Rate of Temperature Rise = °C/sec

1 cal/sec = 4.186 Watt

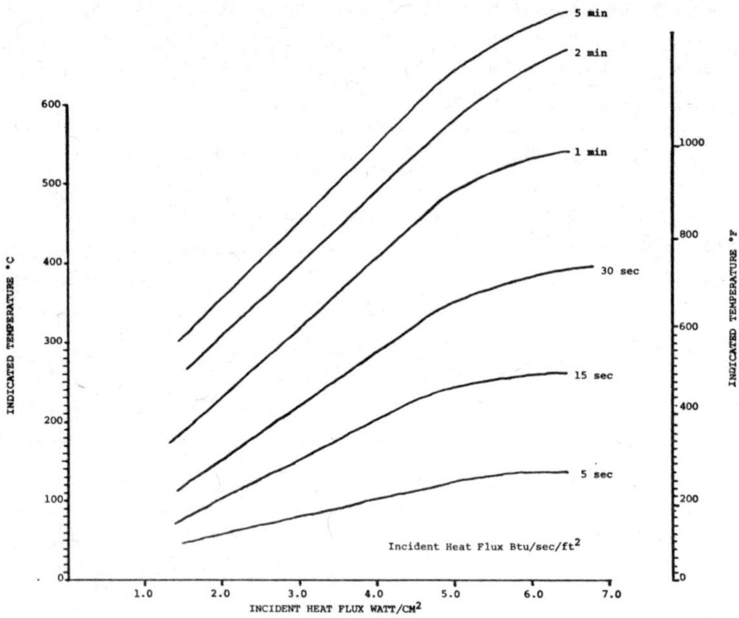

Figure 5. *Copper disk/asbestos board radiometer.*

Figure 5 shows a series of curves for the 411.06 radiometer representing (Watt/CM^2) versus indicated temperature °C, the values used to obtain the curves were obtained. . .for time intervals from 5 seconds to 5 minutes. Also shown as an abscissa value are Btu/ft^2/sec values, of which (1) unit is equivalent to 1.14 Watt/Cm^2 [11].

The temperature recording instruments were checked for calibration prior to experimental testing of garments. Thermocouple wire of alumel-chromel was spot-welded to No. 18 Bell wire in lengths equivalent to the components used to wire the torso-form as prepared for testing. An ice and water reference junction was used. The spot-welded tips of the thermocouple wires were sealed in a micro-Tek, GC-1607, oven and the Bell wire unit was attached to a recorder. The oven temperature was checked by an ASTM calibrated thermometer, 20-760 F, with temperature increased by 50-degree increments. A calibration curve was then prepared for each recorder to correct minor differences in operational function between instruments.

TEST SPACE DESCRIPTION

The experimental work was carried out in a room having an overall area of 64 square feet, without a window opening and having a concrete floor and concrete block walls. A 4-foot square metal hood was installed at the ceiling. A duct, 8 inches in diameter, opened into the hood and was extended through the attic to roof vents. Two 3/4 h.p. motors each operated a fan which pulled a strong suction to exhaust smoke and fumes from the room after each burning experiment was completed. During a burning experiment still air conditions were maintained in the room.

FABRIC DESCRIPTION

The all-cotton carded print cloth had 77 x 58 yarns per inch and a bleached weight of 3.54 oz/yd^2. The 50/50 per cent cotton-polyester broadcloth was made of combed cotton and Kodel fiber with 136 x 58 yarns per inch and a bleached weight of 3.04 oz/yd^2. The 70/30 per cent cotton-polyester broadcloth was made of 50/50 per cent combed cotton and polyester warp and 100 per cent combed cotton filling with 136 x 60 yarns per inch and a bleached weight of 3.40 oz/yd^2. The fabrics were processed at the mill according to their routine bleaching and light mercerizing procedures. A light P.V.A. sizing treatment was applied to the cotton-polyester blend fabrics.

CHEMICAL FINISHES

The fabrics were shipped by the mill to the Laboratory where they were treated with certain flame retardant finishes under controlled conditions (1A). A designated amount of each fabric was treated with APO–THPC [12], THPC–MM–Urea [12], and THPOH–NH_3 [12] in sufficient yardage for the required number of garments. Some of the fabric was retained without any finish. The flame retardant finishes were applied to the fabrics by the following procedure:

APO–THPC was applied by padding for 2 dips and 2 nips through a 30 per cent solids solution of APO–THPC under a tight squeeze roll pressure, with wet pick-up between 70-80 per cent, dried on tenter frame for 3 minutes at 85 C, and then cured on frame for 4 minutes at 140 C. The THPC–MM–Urea finish was applied to the fabrics by padding through a 40 per cent solids solution of THPC–MM–Urea, for 2 dips and 2 nips under a tight squeeze roll pressure with wet pick-up of 70-80 per cent, dried at 85 C, for 3 minutes on a tenter frame, and cured at 150 C for 3 minutes on frame. The THPOH–NH_3 finish was applied to the fabrics by padding through a 40 per cent solids solution of THPOH for 2 dips and 2 nips under a tight squeeze roll pressure, with wet pick-up of 80-90 per cent, dried on a tenter frame at 85 C to a moisture content of 20 per cent. The fabrics were then exposed to ammonia gas in an enclosed jig for 10 ends. All of the fabrics were after-washed on jig and dried on frame [12].

GARMENT DESIGN

The type of garment burned on the torso-form was a simple A-line dress. The dresses were made of three fabrics typical of those used in volume yardage by consumers for light weight apparel. This type was also similar in style to a free hanging night dress. Each fabric was procured directly from the same mill in order to authenticate fiber content and finish and to have fabric with other desired specifications. A designated amount of each fabric was made into dresses by a commercial manufacturer to obtain uniformity in the garment design and construction. The garment was sleeveless with normal type armhole, normal round neckline without a collar, and closed with a zipper at the center back which could be opened to dress the torso-form. A metal zipper opened down 21 inches from the neckline. The zipper tape was cotton. The plain seams were stitched with cotton thread, and a 3/4 to 1 inch single edge was turned up for a hem. The fabric edge around the neck and armhole was turned once and stitched. The dress front was made in three sections, and the dress back was made in two sections with a center back seam below the zipper. The dresses were made of one all-cotton fabric and two cotton-polyester blend fabrics.

The experimental tests were made on garments with and without flame retardant finishes as received. Data were obtained on maximum temperature, thermocouples, and total heat absorbed, thermoplates. To determine the effect of washing on flame retardant durability, tests were also made on garments made of the same fabrics and finishes after 20 laundering cycles.

ORDER OF TEST PREPARATION FOR BURNING

The procedure followed for each burning experiment required an orderly series of preparations. The safety precautions, including a fire hose attached to a water line and a fire extinguisher, were available in case of uncontrolled flame. The temperature and relative humidity in the room was recorded before and after each experiment. The interior temperature of the torso-form was also recorded for each experiment by inserting a thermometer into a small opening bored into the top of the torso-form at the neck. The chart on each recorder was coded in order to identify the particular experiment when the roll of chart was removed from the instrument. Each recorder had a glass visor flap which could be raised in the front window for coding charts. Each recorder was plugged into the power receptacles and checked for "running" operation. The security of Bell wire leads was checked at the recorder poles. The cables of thermocouple – Bell wire at the junctions were submerged in a bucket of ice and water for reference temperature control. The gas supplied through the pilot manifold to pilot burners was turned on, lighted with a safety match, and gas pressure was adjusted for proper height of flame at the pressure reducing valve and the needle valve. The valve at the main tank which supplied gas to the master burners was opened and the gas pressure adjusted to 5 inches water column by a Fisher valve. The toggle switch was closed and each master burner was then ignited automatically from its pilot. The toggle switch was activated several times to control the "on" and "off" supply of gas through the magnetic control valve to exhaust air out of the line until the burners sustained a blue flame of uniform height which would just tip the hem edge of a garment to produce a vertical burn. The master burners were then turned off. A garment was placed over the torso-form, and adjusted for drape and position of hem edge over the burners, as shown in an earlier paper [3]. The control switches for the master burners and the recorders were then activated simultaneously. As flame contacted and ignited the garment hem, the burner switch was immediately turned off. The recorders were operated for three minutes to obtain data via the thermoplate or the thermocouples on the torso-form. After an experiment was completed any charred remnants of fabric and/or resinous material, from the blends, was removed from the surface of the torso-form. Smoke and fumes were exhausted by operating the suction fan. The entrance into the testing area was fitted with a plexiglass door through which the experiment could be observed by the operators and provide them with some protection from smoke and fumes.

WASHING PROCEDURE

The garments were washed in an agitator-type washer and tumble dried in an electric dryer, both of which were Kenmore (R) Model 600. A 4-5 pound load of garments with the same flame retardant finish was washed together using 20 grams of regular Tide (R) for a total cycle of 34 minutes, at high water level and temperature of 140 F, with rinse temperature at 105 F. The garments were tumble dried on normal cycle.

STATISTICAL PLAN

The experimental temperature-time data were statistically treated by least squares analysis of variance, totally uncorrected, using an unbiased estimate of means and weighted means for missing data. Experiments were replicated five times for each garment type.

When the garment was positioned on the torso-form it lay flat against the torso-form across the upper back, over the shoulder, above breast, and at breast prominence, hips, side hip position, and abdomen. The garment was not in flat contact with the torso-form at approximately sternum position, upper chest, at waist, and legs below crotch position.

TEMPERATURE RECORDS

A "temperature curve" was registered by a series of fine points imprinted on the chart by the needle as the flame temperature of the garment was transmitted from the thermocouple points or thermoplates on the torso-surface to the recorder. The "temperature plateau" represented the highest stable temperature at the thermocouple points generated by heat rising from the burners for treated garments that did not burn. Figure 4 is a schematic illustration of a "temperature curve" and "temperature plateau" to illustrate the type of data that were obtained for an individual test. The "temperature curve" was a record from one 50/50, cotton-polyester garment with no finish. The temperature plateau was record from a fabric with a flame retardant finish.

Maximum temperature in degrees F, as registered on the torso-form, was calculated from the chart by plotting the increase in temperature increments from the

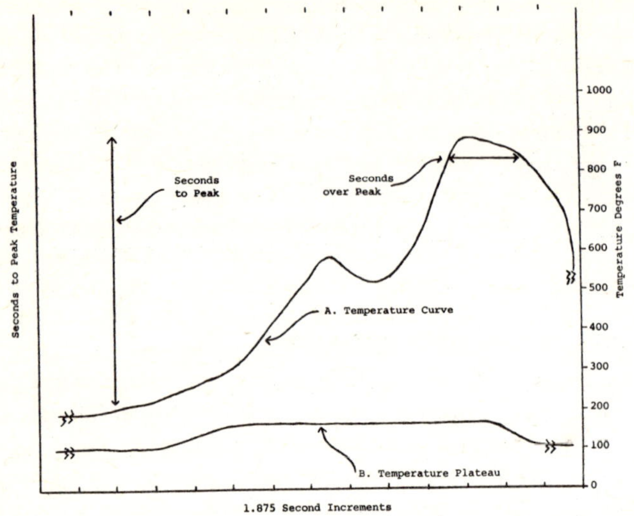

Figure 4. *Temperature diagrams.*

base of the curve to the peak point of the curve. The time required for the temperature to rise from the base point to the peak point was determined for each experimental burn by determining the increments from base point to peak point. The total time in seconds for peak temperature sustained at an individual thermocouple point was determined by locating an arbitrary position 2.5 increments below the maximum strike point for temperature. A base line was then drawn through the located position to intersect the two sides. Each time increment on the chart was equivalent to 1.875 seconds. Each temperature increment was equivalent to twenty degrees. Recorders were operated for 3 minutes for each experiment to obtain complete data on a burn at each sensor location on the torso-form.

THERMOPLATE – Cal/CM^2/15 Sec

The thermoplate or surface temperature was measured in °F on the torso-form by thermocouples welded to the thermoplates mounted on the surface of the form. These data were then converted into equivalent Cal/CM^2/15 Sec according to the calibration formula, Figure 5. In Table 1 a summary mean in Cal/CM^2/15 Sec for the three fabrics with no flame retardant finish shows that the fabrics were different due to fiber composition. The differences were significant according to the calculated statistical value as shown in Table 2. It would appear that the fabrics without a fire retardant finish produced sufficient heat to cause equivalent third degree burn injury.

The specially treated copper discs had an absorption coefficient similar to human skin. The United States Naval Air Development Center conducted studies to determine the tolerance of human tissue to incident heat for second degree burn injury, or the point of threshold pain and blister using human subjects. Their data indicated that in $Cal/Cm^2/Sec$, as time (sec) ratio, threshold pain occurred at 0.125 $Cal/Cm^2/10.1$ Sec; threshold blister occurred at 0.275 $Cal/Cm^2/9.5$ Sec; and full blister at $0.205/Cm^2/15.1$ Sec [2].

The mean data in Tables 2 and 3 would seem to indicate that the fabric having a higher content of polyester fiber produced a lower incident of heat. However, there was less difference (0.159 $Cal/Cm^2/15$ Sec) between the all-cotton fabric vs. the 50/50 cotton-polyester fabric than the difference (0.694 $Cal/Cm^2/15$ Sec) between all-cotton vs. the 70/30 cotton-polyester fabric. It was concluded from visual observations of the experiments that the difference in heat absorbed was due to the fiber composition of the garments under conditions of full flame. Although heat radiation from the burning garments was not measured, it was subjectively apparent that the radiant heat was more intense from the blend fabrics. Upon ignition, the flame quickly flashed over the surface of the all-cotton fabric leaving a surface char after

Table 1. Summary of Mean $Cal/Cm^2/15$ Sec, For Three Unwashed Fabrics With No Flame Retardant Finish.

Fabric Number	Fabric Fiber Content	$Cal/Cm^2/15$ Sec
1	All-Cotton	3.986
5	50/50, Cotton-Polyester	3.827
9	70/30, Cotton-Polyester	4.521

Table 2. Calculated F Values by $Cal/Cm^2/15$ Sec, For Three Unwashed Fabrics With No Flame Retardant Finish.

Source	d.f.	F
Reps	4	3.437**
Fabric	2	9.108**
Thermoplate	13	7.332**
Fabric x Thermoplate	26	1.555
Remainder (Error)	64	–

**Significant at 1% level.

which afterglow or flameless combustion continued slowly and the progressive edge flame consumed the char. As the initial flame flashed over the all-cotton garment it shrank against the form, Figure 6B. The fabric was entirely consumed except for fragments across the upper back shoulder where the fabric was flat against the slightly convex surface, Figure 6C. Smaller fragments occasionally remained on the flat surface of the front torso-form just below the shoulder and above the breast prominence. In the room environment very little smoke was produced from the decomposition of cotton fabric into carbon dioxide and water. The relative humidity rose 1 to 2 per cent after an experiment. The testing room was not conditioned for controlled temperature and relative humidity. The residue from the burned cotton fabric was white, and extremely fine and none remained on the torso-form.

The characteristic burning performance of the cotton-polyester blend fabrics was quite different from the all-cotton especially for the 50/50 cotton-polyester blend fabric. These differences are shown in Figure 6D and 6E. Figure 6D shows rapid flame advance and emission of fumes and flame between the garment and torso-form has ignited an area (dark) slightly above waist position. The fabric was ignited as easily as the all-cotton fabric, and burned with a vigorous, rapid flame advance which generated an up-draft current of air next to the torso-form. The convection current caused the garment to "billow" away from the torso-form. As this phenomenon occurred, sections of the flaming fabric fell away to the burner platform below and continued to burn until consumed. In Figure 6E a section of flaming fabric was held

Table 3. Mean Cal/Cm^2/15 Sec, For Three Unwashed Fabrics With No Flame Retardant Finish.

		Fabric x Thermoplate		
Thermoplate			Fabric	
Position	Temperature	1	5	9
3	4.02	4.28	3.51	4.27
4	4.13	3.79	4.01	4.59
5	3.48	3.15	3.40	3.91
6	3.64	3.52	3.44	3.97
8	6.09	4.29	6.45	7.54
10	3.82	3.72	3.37	4.37
11	3.83	3.48	3.40	4.62
12	4.00	3.81	3.52	4.69
13	3.75	3.72	3.58	3.94
14	3.37	3.62	3.51	2.97
15	4.03	4.23	3.47	4.41
16	4.97	5.80	4.33	4.76
17	4.59	4.95	4.01	4.81
18	3.82	3.44	3.56	4.44
	Means of Means	3.99	3.83	4.52

Fabric:
1. Cotton, 100%
5. Cotton-polyester, 50/50%
9. Cotton-polyester, 70/30%

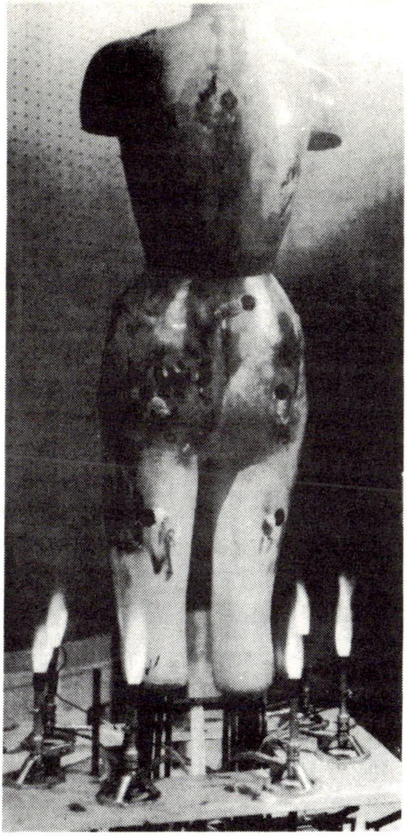

Figure 6A. *Torso-form and lighted burners.*

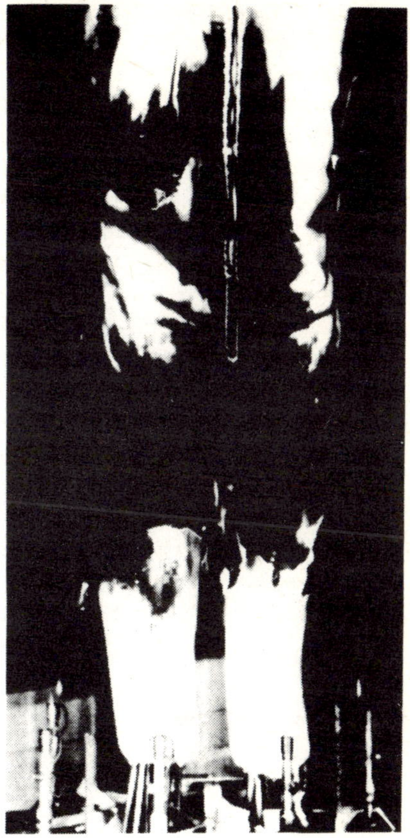

Figure 6B. *Afterglow or flameless combustion of all-cotton fabric.*

by the seam momentarily before falling in flames to the platform below. Some black residue formed which hardened as it cooled. Some of the residue also adhered to the torso surface. Therefore, between each experiment it was necessary to carefully clean the surface of the torso-form to remove the gummy deposits. Considerable smoke and fumes were produced.

The 70/30 cotton-polyester fabric burned similarly to the other blend and quite differently from the all-cotton. However, less of the gummy residue formed on the torso-form and burner-platform, although the burning fabric was prone to flake away and drop in full flame. The bright yellow-orange flame appeared equally as vigorous and produced the same type of strong convection current as the 50/50 blended fabric.

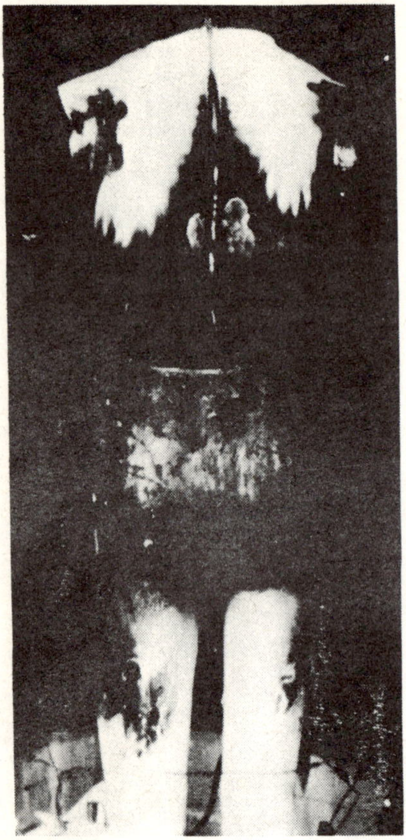

Figure 6C. *Remnants of all-cotton fabric across shoulder area with afterglow.*

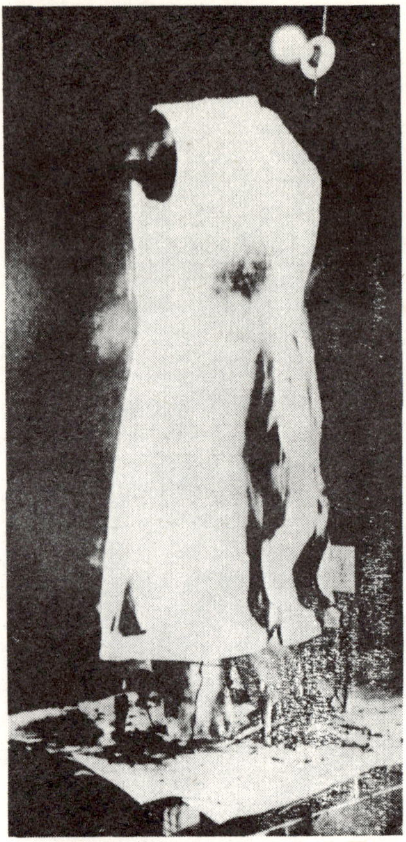

Figure 6D. *Ignition of cotton-polyester fabric showing propagation of flame and liberation of fumes (burners off).*

The thermoplate data in Cal/Cm^2/15 Sec by area of the torso-form are given in Tables 4 and 5. Mean Cal/Cm^2/15 Sec data were higher for the all-cotton and 70/30 cotton-polyester blend fabric than for the 50/50 cotton-polyester blend except for the thermoplate at position 8 on the lower torso-form. No explanation other than characteristic burning pattern can be given for this exception. The reading for underarm positions and sternum (Nos. 3, 4, 5) were similar. These positions represented trapped heat situations due to the short ledge-like overhang of the arm, and the concave area at the sternum and under the breast. The upper torso-form positions at Nos. 10, 11, 5, 4, 3 were similar to but slightly higher than No. 6, 13, 12, 14. Here the fabric lay rather flat against the surface of the torso-form. The side hip and mid-back positions below the waist registered lower readings than buttock and legs below the buttock positions. The sensor positions on the lower leg were farther away from the flaming fabric than the buttock positions and the flame advance passed very quickly over these sensor positions on the leg.

THERMOCOUPLE POINT TEMPERATURE

Temperature of the flaming fabric at thermocouple points differed among the three unwashed fabrics with no flame retardant finish. The temperature-time relationship of the data is shown in Table 6. A higher fabric temperature showed a longer interval for the peak to occur and a longer time sustained over the peak. The all-cotton fabric ranked first, or higher in temperature-time, followed in order by 70/30 cotton-polyester and 50/50 cotton-polyester, respectively. Mean data are shown in Table 8 for the unwashed fabrics without any flame retardant finish. These data were analyzed statistically, as shown in Table 7, and differences in mean temperature by fabric and thermocouple were highly significant. The difference in time to peak was highly significant for fabric and thermocouple. The difference in time over peak was significant by thermocouple.

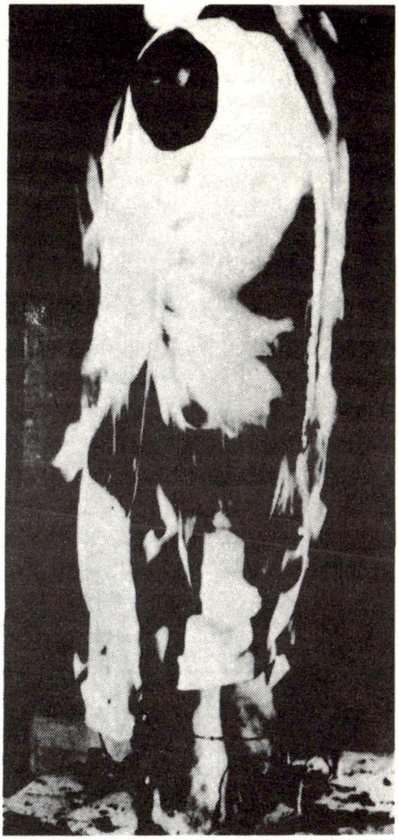

Figure 6E. *Rapid flame spread over cotton-polyester fabric. Section of fabric separation on lower left. (burners off).*

FLAME RETARDANT FABRICS

The finishes were effective in making the three fabrics flame retardant prior to washing. When the flame was applied to the garment hem edge for three minutes the treated fabric charred within ten seconds at the point of direct flame contact for a few inches above the hem edge. Regardless of the duration of flame contact with fabric there was no further change caused by the flame. The garment edge glowed while flame was in contact with the fabric, but the glow extinguished when the flame was turned off. Excessive fumes were liberated by the contact of flame with the chemically treated fabrics because THPC—MM—Urea, APO—THPC, and THPOH—NH_3 were volatilized during burning. A burst of flame sometimes appeared between the fabric and the torso-form and well above flame tip of burner. This was apparently caused by the ignition of trapped flammable gases which would burn for a few seconds and then extinguish leaving a charred and stained area. The ignition of gases often occurred near the waist area. The condition of char, scorch, stain differed by fiber content and finish. Although charred areas at the six flame contact points were

Table 4. Mean Cal/Cm² /15 Sec, For Front And Back Of Torso-Form.

Thermoplate Position	Mean	Cal/Cm² /15 Sec Fabrics 1	5	9
FRONT				
5		3.150	3.400	3.906
3		4.284	3.512	4.270
4		3.794	4.014	4.586
11		3.474	3.404	4.616
6		3.516	3.440	3.974
13		3.722	3.584	3.942
8		4.286	6.452	7.538
	Area Mean of Means	3.747	3.972	4.690
BACK				
10		3.724	3.368	4.368
12		3.808	3.516	4.692
14		3.624	3.512	2.972
15		4.226	3.470	4.406
16		5.802	4.334	4.762
17		4.946	4.014	4.814
18		3.444	3.556	4.444
	Area Mean of Means	4.225	3.824	4.351
	Overall Mean	3.986	3.898	4.251

Table 5. Mean Cal/Cm² /15 Sec, By Section And Thermoplate Position On Torso-Form.

Thermoplate Position	Cal/Cm² 15 Sec Fabrics 1	5	9	Mean
		Torso Above Waist		
Underarm Nos. 10, 11	3.601	3.386	4.492	3.829
Front Nos. 5, 4, 3	3.743	3.642	4.281	3.899
		Torso Below Waist		
Front Nos. 6, 13	3.619	3.512	3.958	3.696
No. 8	4.286	6.452	7,538	6.092
Back Nos. 12, 14	3.716	3.514	3.832	3.684
Nos. 16, 15	5.064	3.902	4.584	4.516
Nos. 18, 19	4.195	3.785	4.629	4.203

extremely fragile, a representative number of garments were cut apart at the center back and carefully removed for measuring length of char, scorch, stain and visual assessment.

OBSERVATIONS BY GARMENT

Garment 2 – All-cotton garments with APO–THPC finish did not ignite. The total char length occurred within 10 seconds with some scorch above the char but no appreciable discoloration of the fabric. Edge glow was observed with "eating away" at flame contact points. At the moment of contact of flame with the fabric a "flame-up" occurred on the garment and this was followed by liberation of fumes and formation of char and stain above the char.

Garment 3 – All-cotton garments with THPC–MM–Urea finish were charred above the six burners within 10 seconds. The fabric scorched in advance of the flame both on the top and underside of the garment. "Eating in" at flame contact points with edge glow occurred. There was no discoloration above the char and the fabric was soft around the char area. Fumes were liberated.

Table 6. Mean Data For Temperature-Time As Measured At Thermocouple Points For Unwashed Fabrics With No Flame Retardant Finish.

Fabrics		Temperature	Seconds	
Code	Fiber	Peak °F	To-Peak	Over-Peak
1	All-Cotton	399.59	37.02	14.07
5	50/50, C-P	360.59	25.28	7.74
9	70/30, C-P	386.96	30.22	9.01

C-P, Cotton-Polyester

Table 7. Calculated F Values For Three Unwashed Fabrics With No Flame Retardant Finish.

		F Value		
		Temperature	Seconds	
Source	d.f.	Peak	To-Peak	Over-Peak
Reps	4	1.561	1.149	1.337
Fabric	2	6.703**	7.364**	2.689
Thermocouple	14	22.840**	7.025**	4.321**
Fabric x Thermocouple	28	2.254	1.260	1.035
Remainder (Error)	176	–	–	–

**Significant at 1% level.

Garment 4 – All-cotton garments with $THPOH-NH_3$ finish had a broader area of char at the base, more scorch above the char areas, and heavy yellow discoloration indicating an inadequate treatment. Heavy fumes were liberated and the fumes ignited and produced isolated spots of char and scorch on the underside of the fabric above the flame contact. The fabric was soft around the charred areas.

Garment 6 – The 50/50 cotton-polyester garments with APO–THPC finish were charred further up the garment than the all-cotton garments with the same finish. There was no further charring from the burner flame after 10 seconds. Visible fumes were liberated and frequently ignited spontaneously under the garment. The discoloration was a translucent grey rather than yellow.

Garment 7 – The 50/50 cotton-polyester garments with THPC–MM–Urea finish charred within 5 to 10 seconds. Secondary flame-up occurred from liberation of fumes. After-glow and shrinking in at the hem edge also occurred. Fumes were heavier than for the same finish on the all-cotton fabric. The discoloration was a translucent smoky grey. The length of char averaged 21 inches from the hem. Isolated char and scorch spots on the fabric were caused from ignition of liberated fumes.

Garment 8 – The 50/50 cotton-polyester garments with $THPOH-NH_3$ finish showed visible flame-up at higher positions above the burner points than the other

Table 8. Maximum Thermocouple Point Temperature, Degrees F, For Three Unwashed Fabrics With No Flame Retardant Finish.

Thermocouple		Fabric x Thermocouple		
		Fabric		
Position	Temperature	1	5	9
1	269.07	214.40	288.20	304.60
3	398.33	456.20	305.80	433.60
4	454.80	460.60	464.20	439.60
5	362.27	390.20	336.40	360.20
6	451.80	509.40	408.40	437.60
7	258.27	263.60	240.20	271.00
9	287.93	226.60	318.40	318.80
10	392.33	382.00	384.40	410.60
11	308.20	311.40	300.20	313.00
12	343.20	387.00	311.60	331.00
13	422.53	417.00	363.40	487.20
14	481.67	452.20	469.80	523.00
15	475.87	506.80	429.80	491.00
16	543.07	615.00	518.20	496.00
17	300.53	326.20	319.40	256.00
18	368.20	474.80	311.00	318.80
	Mean of Means	399.59	360.59	386.96

Fabric: 1. Cotton, 100%
5. Cotton-Polyester, 50/50%
9. Cotton-Polyester, 70/30%

finishes on this fabric. Very heavy fumes were liberated which contributed to further momentary fabric ignition above the area in contact with the flame. Fumes were observed to vent at the open armhole of garments. The fabric shrank at the edges of char points and the fabric glowed at flame points as long as the flame remained in contact. Minute "shiny" particles were observed in the char. Appreciable areas of discoloration were deposited around the charred areas and these varied from translucent grey to deep yellow. In addition, there was a heavy deposit of black, smoky material on the under side of the fabric. There was a pronounced rippling or puckering of the fabric around the charred areas. The charred area was bordered by a translucent band, then by a yellow discoloration. Isolated spots of char appeared on the body of the garment. These were caused by ignition of liberated fumes. Frequently the charred areas arched across the garment from one area to connect with an adjacent area.

Garment 10 – The 70/30 cotton-polyester garments with APO–THPC finish, liberated heavy fumes and the char formed above the six burner points within 10 seconds. There was no yellow discoloration on the fabric but the charred areas were bordered by a deep orange ring and then a translucent band which was stiff and slightly puckered. Re-ignition of the fabric at the side seams on the underside of the garment occurred along side folds of the garment. This accounted for the small isolated spots of independent char.

Garment 11 – The 70/30 cotton-polyester garments with THPC–MM–Urea finish produced heavy fumes with re-ignition on the underside of the garment. Fumes exited at the open armhole. The char length was formed within 5 to 10 seconds. Edge glow appeared while the flame was in contact with the fabric. The charred fabric was more crumbly and easily cracked. The discoloration was translucent grey with pronounced dark orange to brown scorch around the char.

Garment 12 – The 70/30 cotton-polyester garments with THPOH–NH_3 finish produced heavy fumes which re-ignited on the underside of the garment. Isolated spots of char and scorch appeared on other areas of the garment. The charred areas spread into adjoining char areas. The charred areas were stiff and caused puckering of the adjacent fabric. The charred areas were banded by a narrow yellow-orange scorch and then a band of yellow discoloration.

The inches of char, scorch, and stain per garment represent a mean measurement per garment for the six burner points as shown in Table 9.

PLATEAU TEMPERATURE

Heat rising inside the garment from the three-minute duration of flame was sufficient to be monitored by the thermo sensors and register as a temperature increase on the chart. After the initial temperature elevation occurred a plateau or straight line record was maintained, as shown in Figure 4. There was no temperature curve recorded. Tables 10 and 11 present mean temperature data by fabric, finish and by individual thermocouple points.

INTERIOR TORSO-FORM TEMPERATURE

Interior temperature of the torso-form was measured by boring a hole through the solid surface at top of the neck and inserting a thermometer. All-cotton garments with no flame retardant finish produced a mean interior heat rise of 24 F as measured just before and just after a burn. The two blends, 50/50 and 70/30, cotton-polyester with no flame retardant finish, produced a mean interior heat rise of 8 F and 11 F, respectively. The temperature remained relatively stable with 2-3 degrees change as measured just before and just after the completion of a burn for the garments with flame retardant finishes where flame contact was made with the garment edge for 3 minutes.

Table 9. Mean Inches Of Char, Scorch, And Stain By Fabric And Finish Per Garment For Six Burner Points.

	Total Inches Per Garment		
Identification	Char	Scorch	Stain
Fabric			
All-Cotton	35.64	26.66	25.67
50/50, P-C	35.87	24.85	32.87
70/30, P-C	35.96	28.51	31.50
Finish			
APO–THPC	35.84	27.06	31.31
THPC–MM–Urea	35.68	27.62	27.87
$THPOH-NH_3$	35.95	25.34	30.86

Table 10. Mean Temperature At Plateau Level By Thermocouple For Unwashed Flame Retardant Fabrics.

Thermocouple Position	Temperature Degrees F
1	111.6
3	158.1
4	187.2
5	188.4
6	207.2
8	173.7
9	113.5
10	184.9
11	182.4
12	173.4
13	143.3
14	249.4
15	208.8
16	311.1
17	209.8
18	169.1

FLAME RETARDANT FABRICS AFTER WASHING

The garments treated with flame retardant finishes were washed for twenty cycles in the laboratory using a typical home process. Fabrics with the APO–THPC and THPC–MM–Urea finishes showed no appreciable difference in the type or amount of charring formed on the fabrics before and after washing when a forced three-minute flame was applied at the hem edge of the garment. After twenty washings the three fabrics with the THPOH–NH_3 were no longer flame retardant. The flame was self-propagating from the six contact points and spread over the entire garment, confirming the inadequate treatment as applied.

The flame was then self-extinguished and the charred garment shape remained intact until touched whereupon the material crumbled into ash and fine particles. The application of THPOH–NH_3 used in this particular experiment was neither an optimum nor a representative finish of this type. The results are reported however, to illustrate the dangers of inadequate flame resistant finishes. Mean temperature data for fabrics with THPOH–NH_3 after washing are shown in Table 12. A comparison of

Table 11. Mean Plateau Temperature For Unwashed Flame Retardant Fabrics.

Fabrics/Finish	Temperature Degrees F
2 Cotton – APO–THPC	168.9
3 Cotton – THPC–MM–Urea	167.1
4 Cotton – THPOH–NH_3	178.4
Cotton – Polyester	
6 50/50 – APO–THPC	179.4
7 50/50 – THPC–MM–Urea	178.0
8 50/50 – THPOH–NH_3	197.0
10 70/30 – APO–THPC	194.4
11 70/30 – THPC–MM–Urea	189.7
12 70/30 – THPOH–NH_3	225.2

Untreated Fabrics Excluded.

Table 12. Means for Temperature-Time For Fabrics With THPOH–NH_3 Retardant Finish After 20 Washes.

Fabric		Temperature	Time Seconds	
No.	Fiber	Degrees F	To-Peak	Over-Peak
4	All-cotton	350.18	18.36	8.14
8	50/50, C-P	355.14	25.31	8.57
12	70/30, C-P	328.28	22.49	13.05

C-P, Cotton-Polyester.

the data for the three fabrics, unwashed, with the three fabrics treated with $THPOH-NH_3$ after 20 washes, Tables 6 and 12, respectively, show that the treated, washed fabrics exhibited a lowered peak temperature. The change in time to peak and over peak were very slight for the 50/50 blend treated with $THPOH-NH_3$ after 20 washes. The 70/30 blend treated with $THPOH-NH_3$ reached the peak temperature more quickly and sustained the peak temperature for slightly longer after 20 washes.

SUMMARY

Data obtained from the burning experiments were measured by inscribing a temperature-time ratio curve on temperature chart records.

Burning temperature and heat-flux, $Cal/Cm^2/15$ Sec, was higher for fabrics with higher cotton fiber content. However, during burning the fabrics with a blend of polyester fiber burned more vigorously which caused the garment to "billow-away" from the torso-form. This appeared to dissipate heat away from the torso-form and thermo-sensors. Heat-flux measured in $Cal/Cm^2/15$ Sec exceeded 3.0 cal. for each of the three fabrics. Heat-flux measured by fabric x thermocoplate ranged between 3.0 – 5.0 + cal. except at one thermoplate position for the 70/30 cotton-polyester blend. This was a side hip position. This level of heat generated by the burning fabrics measurements exceeded the level which would produce third degree burns. Severe body injury or third degree burn damage occurs in 30 sec. at 1.5 $Cal/Cm^2/Sec$ or 2 – 3 sec at 2.0 $Cal/Cm^2/Sec$. Ignition was instantaneous and flame advance was very rapid for the untreated cotton-polyester fabrics. From observations of the experiments it would appear extremely unlikely that a person could succeed in removing such a flaming garment. Although, the all-cotton fabric flashed, the flash heat would not cause a serious burn. The flameless combustion which followed was relatively slow in comparison. The melt product from the burning blend fabric would contribute additional hazard in that it adhered to the torso-form or dripped. Thermoplate heat was not significantly different among sensor positions on upper, lower, front, or back areas of the torso-form.

Flame temperature registered at thermocouple points, by fabrics, ranged between 360-400 F, being higher for the fabrics with higher cotton content. Likewise, the time required to reach the peak temperature and the time of temperature was greater for the fabrics with higher cotton fiber content.

The fabrics with flame retardant finishes did not ignite and support combustion with 3-minute flame contact. A brief flame-over occurred at the burner points due to volitization of the finishes with char formation. The chemical finishes were volatilized and collected under the garment causing momentary spontaneous combustion and residual formation of char and scorch on the fabric well above the point of flame contact. After twenty washes the fabrics finished with APO–THPC and THPC–MM–Urea were not ignited by 3-minute forced flame contact. The fabrics finished with $THPOH-NH_3$ were not flame retardant. The poor performance of this finish could only be assessed for this particular experiment.

ACKNOWLEDGMENTS

1A. Grateful acknowledgment is made to Personnel of the Southern Utilization Research and Development Division, USDA, for applying the chemical finishes.

Individual appreciation is expressed to Mr. Albert S. Cooper, Jr. and to Mr. George L. Drake, Jr., Southern Utilization Research and Development Division, USDA, New Orleans, Louisiana, for their invaluable guidance and assistance with the work.

2A. Sincere appreciation is expressed to Mr. James V. Ryan, National Bureau of Standards, Washington, D. C., for his helpful consultation on the work and for supplying the calibrated copper disc sensors.

3A. Other phases of the regional research project under SM-38 were conducted at Auburn University, University of Missouri, University of North Carolina (Greensboro), Oklahoma State University and University of Tennessee, respectively.

REFERENCES

1. John V. Beninate, Eileen K. Boylston, George L. Drake, Jr.; and Wilson A. Reeves, Better Flame Resistant Finish for Cottons, *Textile Industries,* Vol. 131, No. 11, November 1967, pp. 110-118.
2. W. P. Behnke and R. E. Seaman, Development of Clothing for Protection From Thermal Hazards. Paper Delivered to the ASTM Committee, D-13, Belmont Plaza Hotel, New York, March 1969.
3. E. L. Finley and W. H. Carter, Description of a Fireproof Torso-Form and Ignition System Used as a Testing Technique for Studying Garment Flammability. *The Journal of Fire & Flammability,* Vol. 1, April 1970, pp. 166-174.
4. R. Bruce LeBlanc, The Present Status of Fire Retardance, *Textile Chemist and Colorist,* Vol. 2, No. 7, April 8, 1970, pp. 31-33.
5. A. R. Markezich, L. N. Bosworth, C. H. Haydel, and G. L. Drake, Jr., The Effect of Certain Variables in the Vertical Flame Test. *Textile Chemist and Colorist,* Vol. 2, No. 18, September 9, 1970, pp. 13-15.
6. K. McDonald, R. Dardis, and B. F. Smith, Accidental Burn Injuries: A. Review. *Textile Chemist and Colorist,* Vol. 3, No. 4, April 1971, pp. 33-37.
7. Bernard Miller and Charles H. Meiser, Jr., Steady State Burning of Textiles in Flowing O_2/N_2 Mixtures, *Textile Chemist and Colorist,* Vol. 2, No. 12, June 12, 1970, pp. 35-38.
8. E. B. Neilson and H. R. Richards, Measuring the Rate-of-Burn of Thermoplastic Fabrics, *Textile Chemist and Colorist,* Vol. 2, No. 4, February 25, 1970, pp. 35-40.
9. J. M. Sather and H. R. Richards, Efficiency of Phosphorous Containing Compounds for Inhibiting Afterglow in Cellulose Fabrics. *American Dyestuff Reporter,* Vol. 59, No. 3, March 1970, pp. 21-26.
10. Stanley M. Suchecki, Flammability of Textiles: Progress is Accelerating, *Textile Industries,* Vol. 133, No. 2, February 1969, pp. 69-72, 97.
11. Personal Communication. Report of Text of Calibration of Copper Disk/Asbestos Board Radiometer Work Order Number 6057, 411.06, March 11, 1969, U. S. Department of Commerce, National Bureau of Standards, Washington, D. C., 20234.
12. Personal Communication. Albert S. Cooper, Jr. and George L. Drake, Jr. Southern Utilization Research and Development Division, USDA, New Orleans, Louisiana.
13. Single Dye System for Polyester-Cotton Blends. *American Dyestuff Reporter,* Vol. 60, No. 3, March 1971, p. 25.

E. L. Finley

Etta Lucille Finley is an Associate Professor in the School of Home Economics at Louisiana State University and Agricultural and Mechanical College in Baton Rouge, Louisiana. She received the B.S. and M.A. degrees from Texas Woman's University in 1931 and 1943, respectively. She has done sabbatical study at the University of Tennessee. Before joining the faculty at Louisiana State University, her experience was in teaching followed by cotton fiber research for the Cotton Research Committee of Texas, and for Anderson, Clayton & Co. of Houston, Texas. She has extensive experience in physical properties research in textiles.

W. H. Carter

William Hobart Carter is a Professor of Agricultural Engineering at Louisiana State University and Agricultural and Mechanical College in Baton Rouge, Louisiana. He received his B.S. and M.S. degrees in Agricultural Engineering from Iowa State University in 1929 and 1934, respectively. Before joining the faculty of Louisiana State University in 1930, he was employed by the U.S.D.A. Division of Agricultural Engineering. He has experience in farm power and machinery, farm and home utilities, and household equipment.

W. H. Carter

Department of Agricultural Engineering

E. L. Finley

School of Home Economics

B. R. Farthing

Department of Experimental Statistics
Louisiana State University and
Agricultural and Mechanical College
Baton Rouge, Louisiana 70803

FLAME TEMPERATURE, HEAT RADIATION AND HEAT CONTENT MEASURED FOR ALL-COTTON AND COTTON-POLYESTER FABRICS*

(Received November 17, 1972)

ABSTRACT: This research measured the heat radiation rate when full size A-line dresses made of all-cotton, 70/30 cotton-polyester and 50/50 cotton-polyester were burned on a life-size mannequin form. Radiation rates were evaluated in relation to surface temperatures measured on the mannequin and to the total heat content in cal/gram for the three fabrics. An Eppley pyrheliometer placed at waist height and 18 inches out from the mannequin surface was used to measure heat radiation rate. The mannequin surface temperatures decreased as the polyester fiber content increased; whereas, the total heat contents and radiation rates increased with increase in polyester fiber.

INTRODUCTION

Significant differences were found between flame temperature data for fabrics composed of all-cotton and cotton-polyester fiber and made into A-line dresses which were burned on a fire-resistant mannequin. Heat radiation was not measured in this study [1]. Flame temperature was highest for all-cotton fabric, lower for 70/30 cotton-polyester, and lowest for the 50/50 blend. On the basis of flame temperature data obtained, the fabric with a higher polyester content appeared to produce a lower incidence of heat. However, it was observed that combustion of the blend fabrics produced very intense heat which seemed greater than an all-cotton fabric.

*Results of the Regional USDA-CSRS Project No. S-86 entitled, "Performance of Selected Fabrics Treated with Flame Retardant Finishes."

The burning pattern for the all-cotton fabric was described as an instant flash-over of the fabric surface which produced surface char followed by relatively slow flameless combustion. At this stage the fabric shrank to the mannequin. This was substantiated by the longer time required to reach the maximum flame temperature (Table 1).

Table 1. Mean Values of Flame Temperature, Radiation and Heat Content as Measured for Three Fabrics

	Mannequin			Pyrheliometer			Parr Calorimeter
Fabrics	Flame (°F)	To Peak (Sec)	Over Peak (Sec)	Time During Radiation (Sec)	Recorder Gain (M/v)	Calculated Radiation Rate (Cal/cm² /min)	Heat Content (Cal/Gm)
1 All-Cotton	522.8	39.0	16.6	106.0	1.6	0.7	3952.4
5 50/50, C–P	380.5	31.5	16.5	74.6	4.2	1.8	4716.6
9 70/30, C–P	426.1	31.5	12.8	74.4	3.9	1.7	4398.3

Note: C–P Percent Cotton/Polyester Fiber Content

In contrast to the behavior of the all-cotton fabric, the cotton-polyester blend burned with a very vigorous flame, rapidly advancing over the dress and between the dress and the mannequin causing the dress to billow away from the mannequin. Less time was required to produce the maximum flame temperature which was sustained for a shorter period of time as indicated in Table 1. The burning blend fabrics produced noxious fumes.

It was hypothesized by the investigators that the total heat content generated by the dresses made of three fabrics and their different burning characteristics influenced the flame temperatures as measured on the mannequin. This factor accounted for the apparent differences in heat as combustion occurred. Hence the work reported here was undertaken.

MATERIALS AND METHODS

The same type of A-line dresses made of all-cotton (print cloth type) and two blends, 70/30 and 50/50 cotton-polyester broadcloth fabric, were used. Five replications of each type of fabric were tested in these experiments using the same fire-resistant mannequin, ignition system, data recording instruments and testing procedures described in an earlier paper [2]. Flame temperature and time data were measured at each of the 18 thermocouple points on the fire-resistant mannequin. The Parr calorimeter and Eppley pyrheliometer were used for these experiments.

In these experiments determination was made of the total heat content for each fabric, flame temperature on the mannequin surface, time to reach peak surface temperatures, heat radiation rate under normal atmospheric conditions,[1] and the relationship among these factors. Dresses were burned in the same manner as reported in an earlier paper [1]. As each garment burned, surface flame temperatures and radiation rate measurements were taken simultaneously. One radiation rate measurement was made for each replication.

RADIATION RATE

A single Eppley pyrheliometer was used to measure radiation rate. The constant for this instrument was 2.31 millivolts per cal/cm^2/min. Its response was recorded on a Minneapolis-Honeywell millivolt recorder, calibrated to a range of 0–5 millivolts. The radiation in cal/cm^2/min was calculated by dividing the chart reading in millivolts by the constant factor of 2.31 millivolts per cal/cm^2/min.

The pyrheliometer was mounted on a vertical board so that the surface plane of the sensor was parallel with a vertical center line position through the mannequin. The board was attached to a stand and then positioned at waist height on the garment, approximately 18 inches horizontally out from the waist position of the mannequin and directed so that a line perpendicular to the sensor face would strike the mannequin at about a 45° angle, with line through the larger dimension of the waist and intersecting the vertical center line of the mannequin. This position was an arbitrary selection since there were many positions at which the pyrheliometer could have been located and this was approximately the position an investigator stood when differences in radiation among fabrics were observed. While radiation rate data were collected at only one position, it was assumed that the same trend would be obtained wherever the pyrheliometer might be placed around the mannequin. No attempt was made to account for the total heat liberated by a burning garment.

Before making radiation measurements, the recorder was turned on about 20 seconds preceding dress ignition and allowed to run for about the 20 seconds after burning was completed. This established a base line from which the deflection of the recorder needle was measured. In this paper, the deflection of the recorder needle from this base line, read in millivolts, is termed "gain." The time during burning was scaled from the recorder chart paper and was the time when deflection from the base began, through peak rate to where it returned again to a horizontal base line. The radiation rate used in this paper was maximum radiation rate. It was calculated by dividing maximum deflection in millivolts by 2.31, a constant for the pyrheliometer.

---

[1] Kwan-Nan Yeh and Barker [3] have reported on work which measured the total heat generated by burning fabrics under normal atmospheric conditions.

TOTAL HEAT CONTENT

A Parr oxygen bomb calorimeter was used to determine total heat content for each of the three fabrics. Each sample of fabric weighed approximately 1 gram, including the thin strip used to tie the fabric after rolling it into a tight, compact cylinder. The cylinder of fabric was placed in the calorimeter crucible for the determination. Standard procedures for operating the calorimeter were followed. Five replications of each fabric were burned and a mean value for replications was used in this paper.

DISCUSSION OF RESULT

The data presented in Table 1 substantiate the hypotheses made as a basis for this research. The trent in mannequin flame temperature among fabrics as reported previously [1] concurred with results of this investigation. Mean flame temperature for all-cotton fabric was 522.8° F, the mean for 70/30 polyester blend fabric was 426.1° F, and the mean for 50/50 cotton-polyester blend was 380.5° F. The time required for the flame temperature to peak decreased as the polyester fiber content increased. These differences in lower peak flame time and flame temperature resulted when the cotton-polyester blend fabrics billowed away from the mannequin during burning. The vigorous burning of the blend fabrics created strong convection currents between the dress and the mannequin. The all-cotton fabric dresses burned more slowly and did not create strong convection currents.

The time to reach peak flame temperature decreased as the percent of polyester fiber content increased. The concurrent radiation readings showed that maximum radiation rate, increased as the polyester fiber content increased. The radiation time decreased as percent of polyester fiber content increased. Recorder gain was higher for the two-blend fabrics with the 50 percent polyester fiber (1.8 cal/cm^2/min) higher than the 30 percent polyester fiber (1.7 cal/cm^2/min). The all-cotton fabric registered a relatively low radiation rate (0.7 cal/cm^2/min) compared to the blends with polyester fiber.

These differences among fabrics for peak time, radiation time, radiation rate, and flame temperature show that heat was dissipated away from the mannequin, which accounted for an apparent lower temperature measurement for the blend at the surface thermocouple points.

The Parr calorimeter heat content determination was a measure of the fiber potential for heat liberation during combustion. The total heat content was found to be higher when polyester fiber was present in the fabric and to increase in direct ratio to the polyester fiber content. It is interesting to note that as total heat content increased, the radiation rate increased, likewise. The nature and rate of burning of the blend fabrics observed in an earlier research [1] coincide with the results of this experiment; and this explains the trends which are characteristic for all-cotton and cotton-polyester fabrics.

Significant differences were found among fabrics and thermocouple readings for flame temperature, to peak time, and over peak time (Table 2). Fabrics were inherently different. Differences among thermocouple readings occurred since a garment positioned on the mannequin could actually be a greater distance from some thermocouple points than other points.

Statistically significant differences were found among fabrics for radiation rate and total heat content (Table 3).

Table 2. Calculated F Values for Flame Temperature as Measured on Mannequin Form

Source	d.f.	F VALUE		
		Flame	To Peak	Over Peak
Fabric	2	21.8**	4.5**	5.4*
Thermo	17	15.6**	18.4**	11.0**
Fabric x Thermo	33	2.4	1.1	2.4

*Significant at 5% level
**Significant at 1% level

F test is the ratio of two mean squares used for testing the null hypothesis of no treatment differences.

Table 3. Calculated F Values for Radiation Rate Measured by Pyrheliometer and Parr Calorimeter

		Parr Cal.	Pyrheliometer		
Source	d.f.	Heat Content (Cal/Gm)	Time During Radiation (Sec)	Recorder Gain (M/v)	Radiation Rate ($Cal/Cm^2/Min$)
Fabric	2	131.7**	9.2*	110.8**	90.7**
Residual	12	–	–	–	–

*Significant at 5% level
**Significant at 1% level

CONCLUSIONS

The presence of polyester fiber in a fabric increases the potential heat content and radiation rate if the fabric is ignited. Fabrics containing blends of polyester and cotton fiber are highly flammable and burn with a startingly vigorous flame advance. Garments made of cotton-polyester fabrics burn more rapidly than all-cotton as shown by radiation time and time to peak flame temperature. Blend fabrics leave a gumming residue on the mannequin surface. Radiation data and total heat content were lower for all-cotton fabric, which accounted for the higher flame temperature. The all-cotton higher flame temperature was due to limited heat radiation, whereas the lower flame temperature generated by the blend fabric was due to heat radiation away from the mannequin surface. Although the all-cotton fabric had a higher flame temperature, the phenomena of rapid burning and gumming residue of the polyester-cotton blends seem to balance the potential burn hazard between the two kinds of fabrics. In fact, the slower flameless combustion second stage burning of the all-cotton would offer a brief time period in which to remove the charred garment, whereas the rapid burning and melting behavior of the cotton-polyester blend fabrics offer almost no time margin for escape from the burning garment.

REFERENCES

1. E. L. Finley and W. H. Carter, Temperature and Heat-Flux Measurements on Life-Size Garments Ignited by Flame Contact, *J. Fire & Flammability*, Vol. 2, October 1971, pp. 298–319.
2. E. L. Finley and W. H. Carter, Description of a Fireproof Torso-Form and Ignition System Used as a Testing Technique for Studying Garment Flammability, *J. Fire & Flammability*, Vol. 1, April 1970, pp. 166–174.
3. Kwan-Nan Yeh and Robert H. Barker, Pyrolysis and Combustion of Cellulose Part IV: Thermochemistry of Cotton Cellulose Treated with Selected Phosphorus-Containing Flame Retardants, *Textile Research Journal*, Vol. 41, No. 11, November 1971, pp. 932–938.

W. H. Carter

William Hobart Carter is a Professor of Agricultural Engineering at Louisiana State University and Agricultural and Mechanical College in Baton Rouge, Louisiana. He received his B.S. and M.S. degrees in Agricultural Engineering from Iowa State University in 1929 and 1934, respectively. Before joining the faculty of Louisiana State University in 1930, he was employed by the U.S.D.A. Division of Agricultural Engineering. He has experience in farm power and machinery, farm and home utilities, and household equipment.

E. L. Finley

Etta Lucille Finley is an Associate Professor in the School of Home Economics at Louisiana State University and Agricultural and Mechanical College in Baton

Rouge, Louisiana. She received her B.S. and M.A. degrees from Texas Woman's University in 1931 and 1943, respectively. She has done sabbatical study at the University of Tennessee. Before joining the faculty at Louisiana State University, her experience was in teaching followed by cotton fiber research for the Cotton Research Committee of Texas, and for Anderson, Clayton & Company of Houston, Texas. She has extensive experience in physical properties research in textiles.

B. R. Farthing

Barton R. Farthing is Professor and Head of the Department of Experimental Statistics at Louisiana State University and Agricultural and Mechanical College in Baton Rouge, Louisiana. He received his B.S. degree from Wake Forest College, and M.S. and Ph.D. degrees from North Carolina State University at Raleigh, North Carolina. Before joining the faculty at Louisiana State University in 1959, he was a staff member of the Animal Science Department at North Carolina State University. His specialty is application of statistics to Agricultural Research.

E. L. Finley and C. T. Butts

School of Home Economics
Louisiana State University and
Agricultural and Mechanical College

GARMENT CONFORMATION ON A MANNEQUIN CHANGES FLAMMABLE CHARACTERISTICS

(Received January 26, 1973)

ABSTRACT: The conformation of the A-line dress on a fire-resistant mannequin was changed by tieing a self-fabric belt at the waist position. The dresses were made of light weight all-cotton fabric and cotton-polyester fabric in two blend levels. These experiments were conducted to obtain information on visual burning characteristics, maximum flame temperature, and flame temperature-time relationship. The fabrics with and without flame-retardant finishes were tested on the mannequin before laundering. Those with flame-retardant finishes were also tested after laundering. The selected flame-retardant finishes included APO-THPC, THPC-MM-Urea, and THPOH-NH_3. The dresses were tested and flame temperature-time data were obtained, using the methods described in previous issues of This Journal.

Conformation of the dresses on the mannequin affected burning characteristics and flame temperature. Folds and gathers were formed by the belt at the waist position and mean flame temperature was higher for the belted-in dress than the A-line dress with no flame-retardant finish before laundering.

Conformation of the A-line dress produced higher flame temperature for the laundered A-line dress finished with THPOH-NH_3. The belted-in dress was held closer to the mannequin. After the burning fabric was extinguished, the form of the dress remained on the mannequin as a layer of char which appeared to shield thermocouple points from flame.

Among fabrics, finished with THPOH-NH_3 after laundering, flame temperature was higher for fabric with the higher polyester fiber content (50/50) than for the other two fabrics for both types of dress conformation.

The burning time was faster for the two fabrics containing polyester fiber than all-cotton fiber and between the two blends burning rate increased as the content of polyester fiber increased.

INTRODUCTION

The conformation or drape of a garment on a mannequin (body form) affects its flammable characteristics [7]. These flammable characteristics were cited in recent literature as including propagation of combustion, flame intensity, surface heat, radiation temperature, and transfer of heat to the surface of the skin [10, 13]. A recent study reported that temperature and heat-flux measurements were different for simple A-line dresses made of fabrics differing in fiber content when these were tested on a form designed to simulate an accidental clothing fire situation [7]. It was, therefore, hypothesized that the flammable characteristics of an A-line garment could be altered by changing its drape or shape in relation to the way it fit against the contours of the mannequin form. Therefore, the A-line shape of the simple dress, as a single layer of fabric over the mannequin surface, was changed by adding a narrow self-fabric tie belt at the waist position of the fire resistant mannequin. The A-line garment shape was selected, originally, because this shape corresponded to a silhouette common to both day and night type dress. The waist is recognized as a natural position on the human figure for the logical division of its proportion and a focal point in designing and fitting clothing. Many basic types of garments, as well as popular fashions in clothing, include a belt for either structural or decorative purposes.

McDonald, et al. [10] reviewed case histories of individual textile fires which cited the type of garments most frequently ignited, the source of garment ignition, and the estimated annual monetary losses due to clothing fires. Outer garments were most frequently ignited and the types of outer garments associated with clothing fires included shirts, trousers, blouses, and night dress. In the home, two out of three clothing fires occurred from igniting by actual contact of fabric with an open flame. The use of open type heaters in the Southwest was the principal cause of flame ignition of clothing. The annual coverage for economic, mortality and morbidity losses in the United States was estimated to be from 349.1 to 660.0 million dollars.

It is generally recognized that many of the generic textile fibers appropriate for clothing fabrics are susceptible to fire. The flammability of a fabric includes the ease of ignition and the rate of burning. Conversely, fire resistance of a fabric refers to the inherent qualities of the fabric, or to the presence of a chemical treatment [9]. The current use of fabrics made of a blend of different fibers has posed unresearched problems with respect to their flame behavior.

Light weight cotton-polyester blend fabric is currently produced and used in great volume for clothing and other purposes. In 1971, a total of 2777.8 million linear yards of blend fabrics were used. Blends made of polyester-cotton accounted for 2002.5 million linear yards of the total. In percentage of the total consumption this amounted to 72.02 percent for polyester-cotton, and 27.91 percent for other blends, or 258.29 percent as much for polyester-cotton [15]. Reeves, et al. [12] found that it was more difficult to flameproof polyester fiber than cotton fiber in a cotton-polyester blend fabric. The amount of chemical flame retardant needed to make a cotton-polyester blend flame-safe increased in direct proportion to the percent of polyester fiber.

STATEMENT OF PROBLEM

This research problem was undertaken to study the flammable characteristics of the A-line dress after it was belted-in at the waist position on the mannequin. This study was a logical sequence to the work on Temperature and Heat-Flux Measurements on Life-Size Garments Ignited by Flame Contact [7, 8], [B].

The general objective of this research was to measure flame temperature and flame time for selected fabrics, used in a single-layer, "belted-in" type of dress, at 18 thermocouples located on a fire-resistant mannequin. The position of each of the 18 thermocouples on the mannequin was described by Finley and Carter [8]. The specific objectives were: 1), to observe the burning pattern of an A-line type dress "belted-in" at the waist position by a narrow belt made of self-fabric; and 2), to measure the maximum flame temperature, and the time that a maximum flame temperature was sustained at the designated thermocouple sites for a) laundered and unlaundered fabrics with no finish, and b) laundered and unlaundered fabrics with three flame-retardant finishes.

MATERIALS AND METHODS

The modification of shape of the simple dress from an A-line shape to a belted shape is shown in Figure 1 and Figure 2. A narrow belt of self-fabric was placed around the mannequin at the waist position and tied in a square knot at the back of the dress. Above and below the waist the fabric formed gathers due to the belt at the waist. The garment lay flat or next to the mannequin above the waist on the shoulder, upper breast, upper center back, and on the side hips and over the abdomen below the waist.

Figure 1. *Front and back view of an A-line dress without belt.*

The A-line, belted and unbelted, dresses were made by the same commercial manufacturer to obtain uniformity in their design and construction. The dress design was sleeveless with a normal shape armcye, normal round shape neckline without a collar, and finished with a zipper at the center back. The metal zipper

Figure 2. *Front and back view of an A-line belted dress.*

opened the dress 21 inches from the neckline for dressing the mannequin. The zipper tape was cotton. The plain seams were stitched with untreated cotton thread. A three-fourths-to one inch-wide single edge was turned once and stitched. The dress front was made in three sections, and the back was made in two sections with a center seam below the zipper.

The selected fabrics were typical of those used in volume yardage for many types of light weight apparel. One fabric was an all-cotton print cloth and two fabrics were cotton-polyester broadcloth in 70/30 and 50/50 blends. The fabrics were purchased from a mill to authenticate the fiber content and other specifications. Yarns per inch and fabric weight were matched as closely as possible in a commercial fabric. No finishes were applied to the fabrics at the mill other than the routine processing of bleaching and light mercerizing. The fabrics were shipped directly from the mill to a laboratory where selected flame-retardant finishes were applied under controlled pilot plant conditions [A].

A sufficient amount of each fabric for the dresses was treated with APO-THPC, – THPC, tetrakis hydroxymethyl phosphonium chloride reacted with APO, tris (l-aziridinyl) phosphine oxide; THPC-urea MM tetrakis hydroxymethyl phosphonium chloride reacted with urea-methylomelamine; THPOH-NH_3, tetrakis hydroxymethyl phosphonium chloride reacted with sodium hydroxide. A sufficient amount of each type of untreated fabric was retained for use as controls.

The reaction of APO with THPC produced a flame-retardant finish which made light weight all-cotton fabrics glow resistant, rot resistant, and flame resistant without appreciable change in hand, texture and strength [5]. The reaction of aqueous THPC and aqueous NaOH, in the presence of ammonia vapor during processing formed a flame-resistant polymer suitable for use on cotton fabrics between 2.8 and 8.5 oz/sq. yd. and having only slight change in tensile strength and hand, satisfactory wrinkle recovery, and durable through 40–50 washes [3, 4, 11]. Moisture pick-up is a desirable property for a flame-treated fabric when the fabric is used as apparel. Drake, et al. [6] found that fabric treated with THPOH-NH_3 picked up more moisture than fabric treated with THPOH-amide.

Equipment and procedures for the experimental burning tests were developed to simulate a figure clothed in a single-layer garment which might be ignited at the hem edge by flame [8]. Aenishanslin [2] has stated that procedures are needed for testing clothing flammability which have a close relationship to the way clothing is actually worn. Fabric in the long dimension of a garment is easily ignited at a lower edge and flame spread is generally rapid in the upward direction [7]. The fire-resistant mannequin and burner system for applying flame to the hem edge of the dresses were the same as that used by Finley and Carter [8].

Flame was applied to the fabric of the dress for 5 seconds. The fabric ignited within the 5-second interval and flame at the burners was then turned off by an automatic switch control. Once the fabric was ignited it was allowed to burn until the flame and afterglow combustion ceased. The temperature-time data were inscribed by a recording device connected to the end of each thermocouple cable which extended several feet from the mannequin. Flame temperature was measured in degrees Fahrenheit and duration of flame time was measured in increments of 1.875 seconds [8].

The dresses with and without flame-retardant finishes were laundered in an automatic washer by a procedure considered equivalent to general home laundering practices. Since home laundering is important to the consumer in the care of much of today's wearing apparel, a flame-retardant finish should retain its flame-retardant properties for a reasonable service life of the apparel. An agitator type washer and an electric dryer, according to AATCC Test Method, 124-1967, were used for laboratory laundering [14]. A four- to five-pound load of dresses with the same flame-retardant finish was washed together using 20 grams of regular Tide® for a total cycle of 34 minutes, at high water level, and at a temperature of 140°F, with rinse temperature at 105°F. The dresses were tumble dried in a normal temperature-time cycle.

Thermal data were analyzed using a split plot arrangement in a completely randomized design. The laundered and unlaundered fabrics with no flame-retardant finishes were analyzed under objective 2a where the main plot was a 3×2 factorial (3 fabrics, 2 treatments – laundering) with 3 observations per treatment combination. The sub-plot factor was 18 thermocouples located at selected sites on the mannequin. Significant differences among fabrics, treatments, and thermocouple positions were determined by applying the F test to the experimental data (Table 1). The F test is the ratio of two mean squares used for testing the null hypothesis of no treatment differences.

The experimental flame temperature-time data, reported earlier by Finley and Carter [7] were examined by a different statistical arrangement. These data for the A-line shape dresses were treated statistically by least squares analysis of variance, totally uncorrected, using an unbiased estimate of means and weighted means for missing data. Experiments were replicated five times for each garment by type of fabric and finish.

Table 1. Analysis of Variance for Peak Flame Temperature of Fabric with No Finish Before and After Laundering Treatment

Source of Variation	d.f.	Mean Squares	"F"
Fabric	2	158,236.34	5.2650*
Treatment	1	24,981.56	< 1
Fabric X Treatment	2	58,018.60	1.9304
Error A	12	30,054.48	
Thermocouple	17	762,692.52	35.4309**
Fabric X Thermocouple	34	94,476.70	4.3889**
Treatment X Thermocouple	17	33,349.21	1.5492
Fabric X Treatment X Thermocouple	34	22,732.81	1.0560
Error B	204	21,526.21	
Total	323		

* Significant at .05 level
** Significant at .01 level

RESULTS AND DISCUSSIONS

Description of the burning phenomena observed for the belted dresses without flame-retardant finishes is pertinent to an interpretation of the statistical data. Flame was impinged onto the fabric for 5 seconds. Each of the three fabrics was ignited immediately when flame from the six burners contacted the fabric. The self-supported flame spread rapidly upward. The self-fabric belt confined the garment to the mannequin immediately below the waist. This limited the rapid flame spread over the abdomen and hip area. The flame spread was halted at the waist position until it burned through the belt. Release of the garment at the waist position generally occurred in about three minutes and then vigorous flame-up occurred on the fabric when the constricted fabric was released.

Combustion of the all-cotton fabric with no flame-retardant finish occurred in two stages. The first stage burning was a flame-over of the total lower portion of the dress. This produced surface char on the fabric and the fabric shrank toward the mannequin. In the second stage the surface flame was extinguished and a slow flameless combustion occurred at the edges of the charred fabric until it was finally consumed except for typical unburned "lobe" shaped fragments over the back of each shoulder and slightly over the front shoulder and above the breast area. The single layer of fabric lay flat against the mannequin over these areas.

The visual burning pattern of the all-cotton fabric used in this study was consistent with the findings on the combustion of cellulose fabrics as reported by Sather and Richards [13]. They found that the immediate flame-over caused formation of char over the fabric, and the char was then consumed by secondary or afterglow burning.

The all-cotton zipper tape, at the center back, flamed slowly on all the dresses made of fabric without a flame-retardant finish. The flame advance on the zipper tape was retarded because the zipper tape was made of a narrow strip of closely woven fabric.

The 50/50 and 70/30 percent cotton-polyester blend fabrics burned differently from the all-cotton. These blended fabrics were ignited instantly. Flame advance was rapid in an upward direction and complete combustion of the fabric occurred as flame advanced. Areas of the mannequin became visible as the flame spread upward. Flaming sections of the blend fabrics fell away from the dress and continued to burn on the platform beneath the mannequin. The 50/50 cotton-polyester fabric appeared to burn more rapidly than the 70/30 cotton-polyester blend. The belt was generally burned through in one minute for the 50/50 blend and within two minutes for the 70/30 blend. A greater deposit of gummy residue formed on the mannequin from the flaming 50/50 blend fabric than from the 70/30 blend fabric. Considerable black smoke and pungent fumes were liberated for both blended fabrics. Little smoke was liberated from the flaming all-cotton fabric and only traces of grey ash formed on the surface of the mannequin.

Flame was applied for one minute to the hem edge of unlaundered dresses treated with flame-retardant finishes, APO-THPC, THPC-MM-Urea, and THPOH-NH_3. Scorching and charring of the fabric occurred directly above the point of flame contact. The char occurred within 10 to 17 seconds. The char did not burn. Pungent fumes were liberated prior to the formation of char on the fabric. In some instances spontaneous flame appeared on both the top and under surfaces of the fabric due to the ignition of liberated gases well above the burner flame. No temperature data were recorded for the flame-retardant fabrics since the area of flame in contact with the fabric was well below any of the thermocouple points, and the fabric did not ignite to sustain combustion. Dresses made of the three fabrics treated with APO-THPC, and THPC-MM-Urea flame-retardant finishes were not ignited after 20 launderings. The laundered fabrics treated with APO-THPC and THPC-MM-Urea produced results very similar to the unlaundered fabrics with respect to char formation and the liberation of gases.

A symmetrical arrangement of thermocouple sites on the right and left sides, and the front and back of the mannequin, allowed comparison of the thermal data recorded of the garments at identical locations on the mannequin. Mean peak flame temperatures recorded on the right and left position of the front and back side of the mannequin produced closely similar burning patterns and data.

There were no observed differences in the burning patterns for the dresses laundered twenty times and unlaundered dresses made of the three fabrics without flame-retardant finishes. Peak flame temperature was not affected by laundering of dresses without the flame-retardant finishes.

The mean peak flame temperature as shown in Table 2 was higher for the unlaundered belted dresses than for the A-line dresses. For each type of dress, the higher mean peak flame temperature was, 1) all-cotton, 2) 70/30 cotton-polyester, and 3) 50/50 cotton-polyester. The higher flame temperature and the longer time required to reach peak flame temperature for the all-cotton fabric used in this study supported findings reported by Sather and Richards [13] on multistage process in the combustion of cellulose fibers. The mean peak flame temperature was higher at thermocouple sites where the fabric in an unlaundered belted dress formed folds or gathers, than it was where the fabric was in close surface contact with the mannequin at side, hip, center bodice back, and shoulder areas. Mean peak flame temperature was approximately 50°F lower at thermocouple positions on the upper torso above the "belted-in" waist area than at thermocouple positions on the lower torso below the belt.

Table 2. Mean Peak Flame Temperature for Three Fabrics in Degrees Fahrenheit

Treatment Dress Shape	All-Cotton	70/30, C-P	50/50, C-P
Unlaundered			
No Finish			
Belted	608.05	531.11	487.31
*A-line	399.59	386.96	360.59
Laundered			
THPOH-NH_3			
Belted	162.96	157.96	278.33
*A-line	350.18	328.28	355.14

*Data from Finley-Carter study
C-P, Cotton-Polyester Blend

In Table 3, mean time-temperature data are shown for seconds to peak and seconds over peak for the belted and A-line dresses made of three fabrics. The time to peak for unlaundered dresses was longer for the A-line shape than the belted shape. The mean time to-peak flame temperature was longer for the unlaundered all-cotton fabric than for the blended fabrics in the belted and A-line dresses. The time to-peak flame temperature for the 70/30 cotton-polyester fabric was longer than for 50/50 cotton-polyester fabric in both belted an A-line dresses.

The change in dress shape on the mannequin increased the time for over peak flame temperature for the belted dresses made of 70/30 and 50/50 cotton-polyester fabrics. The time in seconds for over-peak flame temperature was longer for the A-line dress made of all-cotton fabric than for the other two fabrics. The time in seconds for over-peak flame temperature was longer for the belted dress made of 50/50 cotton-polyester than for the other two fabrics. The over-peak time was longer for belted dresses made of the blend fabrics in the order of 50/50 and 70/30 cotton-polyester, respectively.

Table 3. Mean Seconds To-Peak and Over-Peak Temperature for Three Fabrics

Treatment Dress Shape	All-Cotton	70/30, C-P	50/50, C-P
Unlaundered			
No Finish			
To-Peak			
Belted	33.02	28.16	24.43
*A-line	37.02	30.22	25.28
Over-Peak			
Belted	11.96	13.43	16.34
*A-line	14.07	9.01	7.74
Laundered			
THPOH-NH_3			
To-Peak			
Belted	8.85	12.86	18.30
*A-line	18.36	22.49	26.31
Over-Peak			
Belted	12.29	15.78	13.55
*A-line	8.14	13.05	8.57

*Data from Finley-Carter study
C-P, Cotton-Polyester Blend

The time over peak for flame temperature was longer for the laundered belted dress than for the A-line dress. The all-cotton fabric for both types of dress showed a shorter over-peak time than either of the cotton-polyester blends. The over-peak time was longer for the 70/30 cotton-polyester blend fabric than for the 50/50 cotton-polyester blend.

After 20 launderings the belted and A-line dresses made of all fabrics finished with THPOH-NH_3 ignited and burned on the mannequin. The fabrics were not flame retardant. Why this reaction occurred for the THPOH-NH_3 finish in this experiment was not known.

A difference in mean peak flame temperature was found for all fabrics and both types of dresses finished with THPOH-NH_3 after laundering. The mean peak flame temperature was higher for the A-line dress than for the belted dress. It was also higher for the 50/50 cotton-polyester fabric than for either of two other fabrics. The all-cotton fabric produced a higher flame temperature than the 70/30 cotton-polyester.

The time in seconds to peak flame temperature for both dress shapes was less for all-cotton than for either of the two other fabrics. The 50/50 cotton-polyester fabric required the longer time to peak temperature than the other two fabrics.

CONCLUSIONS

Conformation of the dress shape on the mannequin affected its flammability. Belted-in dresses produced higher mean flame temperature than the A-line dresses due to conformation of the dress into folds, and gathers in certain areas.

The percent of polyester fiber in the two blend fabrics affected flame temperature. These findings agree with work reported by Reeves et al [12]. A difference in flame temperature occurred between A-line and belted-in dresses. Polyester fiber caused a char residue to be deposited on the mannequin surface as the blend fabrics burned. The blend with higher percentage of polyester fiber produced more char deposit. As the cotton-polyester blend fabrics burned, molten residue and flaming fragments of fabric fell away to the platform below.

Mean peak flame temperature occurred in less time and was sustained longer for the belted-in dresses than for the A-line dresses. Flame advance above the waist was retarded by the belt at the waist. Higher flame temperature in the area of folds and gathers showed that burning was intensified by the extra fabric and entrapped air. The belt (or a band) would function as a temporary barrier to flame advance.

The flame-retardant properties of APO-THPC and THPC-MM-Urea were not affected after 20 launderings. The flame-retardant performance of these finishes were consistent with findings by Aenishanslin [1] and Reeves, et al [12].

ACKNOWLEDGMENTS

A. Appreciation is expressed to the Southern Regional Research Laboratory, Agricultural Research Service, United States Department of Agriculture, for applying the experimental flame-retardant finishes.

B. Research under regional research project SM-38 was supported by the Louisiana Agricultural Experiment Station, Louisiana State University, Baton Rouge, Louisiana. Other phases of the regional project were developed at Auburn University, University of Missouri, University of North Carolina (Greensboro, N. C.), Oklahoma State University, and University of Tennessee, respectively.

REFERENCES

1. R. Aenishanslin, Flame-Retardant Finishes for Cellulosic Fibers. *Ciba Review*, 1969, 4, 35–43.
2. R. Aenishanslin, Flame-Retardant Finishes, *Textile Industries*, Vol. 133, No. 11, (November, 1969) pp. 99–104.
3. John V. Beninate, Eileen K. Boylston, George L. Drake, Jr. and Wilson A. Reeves, Application of a New Phosphonium Flame Retardant, *American Dyestuff Reporter*, Vol. 57, No. 25 (December 2, 1968), pp. 981–985.
4. John V. Beninate, Eileen K. Boylston, George L. Drake, Jr., and Wilson A. Reeves, Conventional Wrinkle Resistance with Tetrakis (Hydroxymethyl) Phosphonium Hydroxide (THPOH), *Textile Research Journal*, Vol. 38, No. 3 (March, 1968), pp. 267–272.
5. K. M. Decossas, B. H. Wojcik, A. deB. Kleppinger, W. A. Reeves, and H. L. E. Vix,

Flame-Resistant Cottons, *Textile Industries*, Vol. 130, No. 7 (July, 1966), pp. 128–136, 141–143.

6. George L. Drake, Jr., et al. Members of the Gulf Coast Section's Research Committee, How Moisture Affects Flame Retardancy, *Textile Chemist and Colorist*, Vol. 4, No. 12 (December, 1972), pp. 287/43–50/294.
7. E. L. Finley and W. H. Carter, Temperature and Heat-Flux Measurements on Life-Size Garments Ignited by Flame Contact, *The Journal of Fire & Flammability*, Vol. 2 (October, 1971), pp. 298–320.
8. E. L. Finley and W. H. Carter, Description of a Fireproof Torso-Form and Ignition System Used as a Testing Technique for Studying Garment Flammability, *The Journal of Fire & Flammability*, Vol. 1 (April, 1970), pp. 166–174.
9. R. B. LeBlanc, Flammability and Fire Resistance of Textiles, *American Dyestuff Reporter*, Vol. 57, No. 27, (December 30, 1968), pp. 1093 1096.
10. K. McDonald, R. Dardis, and B. F. Smith, Accidental Burn Injuries: A Review, *Textile Chemist and Colorist*, Vol. 3, No. 4 (April, 1971), pp. 33–37.
11. S. James O'Brien, New Trends in Fire Retardants, *Textile Chemist and Colorist*, Vol. 2, No. 12 (June 17, 1970), pp. 201/31–34/204.
12. W. A. Reeves, G. L. Drake, Jr., J. V. Beninate, and R. M. Perkins, A Comparison of Four Flame Retardants, *Textile Chemist and Colorist*, Vol. 1, No. 17 (August 13, 1969), pp. 24–28.
13. J. M. Sather and H. R. Richards, Efficiency of Phosphorus Containing Compounds for Inhibiting Afterglow in Cellulose Fabrics, *American Dyestuff Reporter*, Vol. 59, No. 3 (March, 1970), pp. 21–26.
14. *Technical Manual* of the Association of Textile Chemists and Colorists, Voi. 45, 1969, pp. 181–182.
15. *Textile Organon*, Vol. XLIII, No. 5, (May, 1972), p. 60.

E. L. Finley

Etta Lucille Finley is an Associate Professor in the School of Home Economics at Louisiana State University and Agricultural and Mechanical College in Baton Rouge, Louisiana. She received her B.S. and M.A. degrees from Texas Woman's University in 1931 and 1943, respectively, and did her sabbatical study at the University of Tennessee in 1966. Before joining the faculty at Louisiana State University in 1955, she was a teacher, did cotton fiber research for the Cotton Research Committee of Texas, and Anderson, Clayton & Co. She has experience in textiles and physical properties research.

C. T. Butts

Christine Thomas Butts was a graduate student in the School of Home Economics, Louisiana State University when this study was completed. She received the Bachelor of Science and Master of Science degrees from the School of Home Economics, Louisiana State University in 1966 and 1972, respectively. Her professional experience has been as a Home Service Advisor with Mississippi Power Company, Laurel, Mississippi and Gulf States Utilities Company, Baton Rouge, Louisiana.

Flammability and Heat Transfer Characteristics of Cotton, Nomex and PBI Fabric*

W. D. Freeston, Jr.

Fabric Research Laboratories, Inc.
Dedham, Mass.

(Received August 10, 1970)

ABSTRACT

The ignition temperatures and burning rates in air of similar PBI, Nomex and flame-retardant-treated cotton fabrics are given and the thermal shrinkage of PBI and Nomex fabric are compared. The thermal conductivity of a series of fabrics at ambient and elevated temperatures is given. The temperature rise in a fabric-covered skin simulant under flame impingement is investigated for a range of fabric constructions. The flame-impingement heat-transfer experimental data are compared with theoretically predicted results. A parametric study of the material properties which govern heat transfer through a fabric under flame impingement is presented. The results indicate that one type of fibrous structure which will provide improved thermal protection is moderate weight, tightly woven PBI fabric over a PBI tricot mesh.

INTRODUCTION

Apparel for use in applications where under emergency conditions a hazardous thermal environment might exist must satisfy the following four requirements: the clothing system must be an effective thermal barrier and prevent severe skin burns; the garment itself must not contribute to the wearer's injury by burning, melting or disintegrating in the presence of heat or open flames; it must exhibit good dimensional stability, since large thermal shrinkages might restrict limb mobility; it must also be durable and sufficiently comfortable to encourage its use.

Flame-retardant-treated cotton is used in applications such as aviators' flying suits which must provide thermal protection in emergency situations. Recently, Nomex**, a synthetic fiber exhibiting improved heat resistance has replaced cotton in many special applications. However, a need still exists for a fiber with thermal characteristics superior to those of Nomex. Test data

* Part of several research programs sponsored by the Air Force Materials Laboratory under U. S. Government Contracts F33615-67-C-1731, F33657-69-C-0256, F33657-70-C-0152 and F33615-69-C-1167.

** E. I. du Pont de Nemours & Co., Inc.

indicate that polybenzimidazole (PBI), a new fiber developed by the Air Force Materials Laboratories and Celanese Research Corp., fulfills this need. It has good apparel textile characteristics and superior thermal properties.

The response of fibrous materials to heat proceeds in stages:

1. Temperature rise, heat transfer through the material.
2. Decomposition, change in physical form.
3. Ignition.
4. Combustion.

The rate of heat transfer through a material depends on the heat source temperature and heat flux, and the characteristics of the material, e.g., specific heat, thermal conductivity, heat of chemical reaction, heat of fusion, heat of vaporization, structural geometry, etc.

Decomposition can involve the formation of various types of products: combustible or noncombustible gases, liquids, smoke (entrained solid particles), carbonaceous char residue. Changes in the physical form of the structure are also important for many applications. In apparel applications, melting and shrinkage can result in serious burns.

Ignition is a function of the temperature at which the gases evolved from the material can be caused to flame by a spark or flame or the temperature at which the reactions within the material result in spontaneous flaming.

The most important characteristic of the combustion of a material is whether or not it is self-supporting. Some materials will smolder without a visible flame; other materials will burn with a visible flame. The heat of combustion, flame temperature and rate of flame propagation are also important material characteristics. The rate of flame propagation will depend upon the material, environment, geometry of the fibrous structure, etc.

The construction of the fabrics whose response to heat is investigated herein are outlined in Table 1; the three fabrics are similar. The flame-retardant-treated cotton and spun Nomex fabrics are currently used in aviator flight suits.

Flammability

The ignition temperatures and burning rates in ambient air (14.7 psia, 70°F, 65% RH) of treated cotton, spun Nomex and spun PBI fabric are given in Tables 2 and 3 respectively.

A preliminary determination of ignition temperature is made by holding a vertically oriented fabric test specimen against a Calrod-type heater to which a thermocouple has been bonded and increasing the electrical power until fabric flaming is readily visible. The temperature of the surface of the heater at the instant before initiation of burning is the approximate fabric fire point. Additional tests are then made by preheating the Calrod to temperatures above and below this level by means of a proportioning controller, moving the fabric test specimen against the heater and noting whether or not ignition occurs. The time from contact to initiation of burning is also noted. The ignition temperature is thus determined approximately within 50°F.

Table 1. Fabric Construction.

Fabric Identification	Weight (oz/sq yd)	Yarn Size[2]/Ply Warp	Yarn Size[2]/Ply Fill	T.M.[3]	Fiber Denier	Fabric[5] Thickness (inch)	Weave Type	Construction Ends/Inch	Construction Picks/Inch
Cotton, Flame-Retardant-Treated[1] (grey)	4.5	68/2	68/2	—[4]	~1	0.010	3 × 3 twill	108	97
Spun Nomex (sage green)	5.0	35/1	35/1	5.1	2	0.014	2 × 2 herring-bone twill	92	70
Spun PBI	5.4	21/1	21/1	4.6	1½	0.017	1 × 2 twill	70	74

[1] Hooker Chemical Corp. THPC treatment.
[2] Cotton count.
[3] Twist Multiplier (singles)—turns per inch twist/$\sqrt{\text{cotton count}}$.
[4] T.M. of warp yarn (singles) = 4.3, T.M. of filling yarn (singles) = 3.2.
[5] Under 0.9 psi transverse pressure.

This test procedure was developed originally for the evaluation of fabric flammability characteristics because it simulates roughly the type of environment in which apparel must occasionally perform. In the laboratory test there is a variable temperature heat source (Calrod heater) on one side of the fabric and a heat sink (room-temperature air) on the other. In an emer-

Table 2. Fabric Ignition Temperature.

Material	Calrod Temperature (°F) 1400	1450	1500	1550	1600	1650	1700	Ignition Temp (°F)
Cotton, flame-retardant-treated	DNI 2 min†	I 5 sec	I 5 sec					1450
Spun Nomex		DNI 2 min		DNI 2 min	I 5 sec	I 5 sec		1600
Spun PBI				DNI 5 min	DNI 5 min	DNI 5 min	I 2 min	1700

DNI = did not ignite.
I = ignited.
† = approximate contact time of specimen with heat source.

Table 3. Fabric Burning Rate.

Material	Ignition Method	End Ignited	Burning Rate (inch/sec)	Remarks
Cotton, flame-retardant-treated	burning tissue	top	—	did not ignite
	burning tissue	bottom	—	charred surface but no apparent ignition; ignition may be masked by burning tissue flame.
	Calrod	bottom	—	ignites and immediately self extinguishes
Spun Nomex	burning tissue	top	—	did not ignite
	burning tissue	bottom	0.61	self extinguishes after 20 sec; 3-4 inches of fabric burned
Spun PBI	either	either		does not burn

gency situation, clothing would have, in fact, flame or hot air on one side and either air or underclothing as well as human tissue on the other.

The burning rates of vertically oriented specimens, in inches of fabric consumed per second, were measured using two ignition techniques: 1) the NASA burning-tissue ignition method and, 2) that used to determine ignition temperatures, i.e., the Calrod-type heater. In the first method a 1-½ inch wide by 3-inch long piece of paper facial tissue is wound around a straight length of Nichrome wire located ⅛ inch from one end of the 2-inch wide specimen. The paper is ignited by electrically heating the wire. The temperature at the hottest point in the flame, approximately one inch above the Nichrome wire, is 1440°F. The flame temperature is considerably less at other locations: approximately 1000°F ⅛ inch above the wire, which is the location of the edge of the test specimen during ignition from the bottom. Specimens were ignited from both the top and bottom.

In the second method, a Calrod-type heater whose temperature can be held at any desired level up to approximately 2000°F is used to ignite the fabric specimen.

All Nomex and PBI fabric test specimens were scoured in Triton X-100 and hot-water, rinsed and air-dried prior to testing.

Investigation showed that the same fabric ignition temperature and burning rate are obtained in both flowing and quiescent atmospheres demonstrating that the chamber used is sufficiently large that constant atmosphere replenishment is unnecessary. Consequently all tests were performed in a static air environment.

It should be noted that at the ignition temperature reported, PBI requires minutes of contact with the heat source to produce ignition, and the flame is very small and of short duration (the sample is blackened.). This contact time significantly exceeds the 5 seconds required for ignition of the Nomex and treated cotton fabrics. The ranking in order of increasing ignition temperature is treated cotton, Nomex, PBI.

As the data in Table 3 show, both flame retardant-treated-cotton and Nomex fabric can be ignited and do burn in air. However they are self extinguishing. The PBI fabric does not burn in air.

The treatment of PBI fabric with various commercial finishes to prevent generation of static electricity and to provide water and oil repellency and crease resistance was found not to adversely affect the fabric flammability in air. Treatments were evaluated alone and in combination.

Shrinkage

As noted in the Introduction, the shrinkage of clothing at elevated temperatures is an important consideration in determining the degree of protection it affords the wearer. High thermal shrinkage can result in intimate contact of the clothing with the skin thereby increasing heat transfer to the skin. This could then cause a severe burn. Excessive shrinkage might also restrict limb mobility or result in fabric tearing with subsequent exposure of the skin to the flame or heat source.

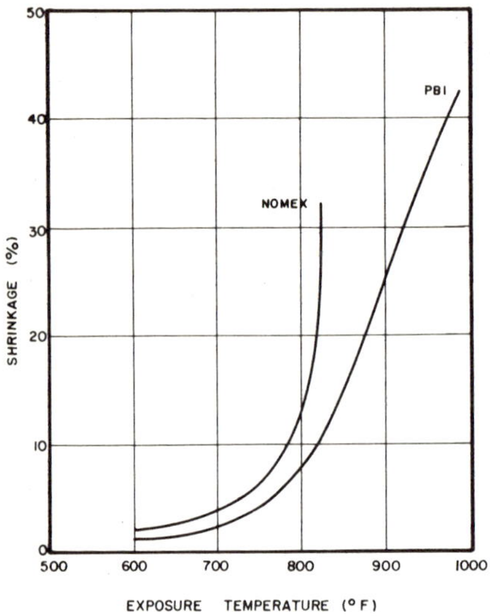

Figure 1. *Fabric thermal shrinkage.*

The lineal shrinkages of Nomex and PBI fabric are plotted in Figure 1 as a function of exposure temperature. These data are for ten-minute exposures of 6-inch by 6-inch specimens suspended freely in a circulating hot-air oven. The measurements were made five minutes after the specimens were removed from the oven. At temperatures above 800°F the Nomex fabric was badly charred and curled, making an accurate measurement of shrinkage very difficult. At 900°F most of the PBI samples exhibited curling and embrittlement.

As the data show, PBI fabric exhibits considerably less shrinkage upon exposure to hot air than does Nomex. However, the thermal shrinkages of PBI may be too great for many potential applications.

Because of the foregoing shrinkage test results, a thermal stabilization process was developed for PBI fabric [1]. Laboratory tests indicate that it reduces the fabric shrinkage from approximately 25% at 900°F to approximately 4-½% and from approximately 45% at 1000°F to 23-½% (ten-minute exposure) without degrading the mechanical and textile characteristics of the fabric at ambient conditions.

Thermal Conductivity

The thermal conductivity and specific heat capacity of all polymeric fibers are of the same order of magnitude [2]. Consequently, fabric thermal conductivity or insulation is largely independent of the fiber type. It has been well established that the heat insulation property of a fabric is a function of its thickness and ability to maintain that thickness; the insulation performance increases linearly with thickness [2]. Because all fibers have considerably higher thermal conductivities than air, the thermal insulation of a fabric is largely dependent upon the entrapment of still air within the fabric structure.

The thermal conductivities of the cotton, Nomex and PBI fabrics described in Table 1 were measured so that the theoretical temperature rise through them under flame impingement could be calculated. However, prior to determining the thermal conductivity the fabric thicknesses were measured as a function of transverse pressure to determine the most appropriate pressure at which to measure thermal conductivity. It was found that the thicknesses of the fabrics are very sensitive to small changes in pressure at pressures below 1 psi but at higher pressures are less sensitive. A pressure of 1 psi therefore appeared to be the lowest pressure at which the thermal conduc-

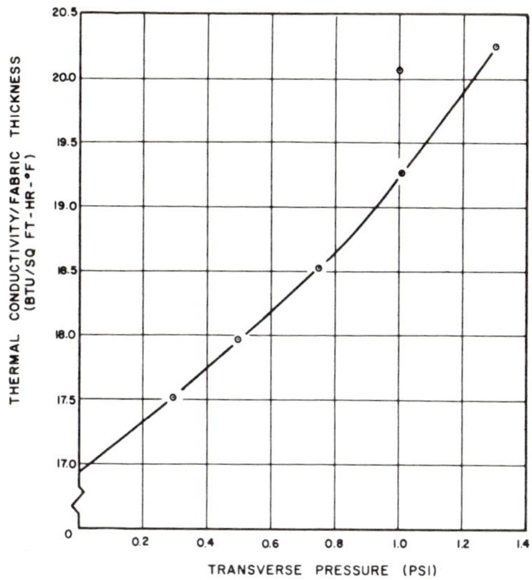

Figure 2. ***Heat flow through spun PBI fabric.***

tivity of the fabrics could be reproducibly measured and was the pressure at which measurements were made.

The thermal conductivities of the fabrics were determined by the guarded hot-plate method [3]. The test conditions and results are given in Table 4. Also given is the ratio of fabric thermal conductivity to thickness; i.e., the quantity of heat passing through the fabric per unit of fabric surface area per unit time per degree difference in temperature between the two fabric faces. As the results show, fabric thermal conductivity increases with increasing temperature.

The thermal conductivity of the PBI fabric was also measured at 70°F and transverse pressures ranging from 0.3 to 1.3 psi. The resulting ratio of fabric thermal conductivity to thickness is plotted in Figure 2 as a function

Table 4. Fabric Thermal Conductivity.

	Thermal Conductivity (Btu-inch/sq ft-hr-°F)		Thermal Conductivity/Thickness (Btu/sq ft-hr-°F)	
Material	at 70°F	at 500°F	at 70°F	at 500°F
Cotton, flame-retardant-treated, 4.5 oz/sq yd	0.39	—	40.2	—
Spun Nomex 5.0 oz/sq yd	0.29	0.45	22.3	34.6
Spun PBI 5.4 oz/sq yd	0.35	0.51	20.1	29.3

of the transverse pressure. As expected, the diagram shows that the heat flow through the fabric increases with increasing pressure; that is, the insulation protection provided by the fabric decreases with increasing transverse pressure (decreasing thickness).

Flame-Impingement Heat Transfer

The heat transfer through fabrics under flame-impingement was measured by determining the temperature rise through fabric specimens exposed to a controlled open flame [4]. The apparatus used is shown schematically in Figures 3a and 3b and is described below. Briefly, the fabric test specimen is clamped over a simulated skin device (see Figure 3b) which is attached to a movable carriage. The carriage is connected to a solenoid controlled by a timing device, thereby allowing the specimen to be moved into and out of the flame in a precisely timed manner. A relatively large diameter flame of constant temperature is achieved with a Meker burner connected to a propane gas source through a gas flow meter. The latter permits setting of the gas flow rate at a specific level.

The simulated skin device was developed by the Naval Material Laboratory and is composed of a resinous compound with thermal and optical properties similar to human skin [5]. The temperature within the compound is measured with a cooper-constantan ribbon thermocouple with an effective thickness of 0.01 mm located 0.49 mm below the surface of the skin simulant. The distance between the top of the Meker burner and the skin simulant is 1-⅞ inches.

The flame temperature was determined using thermocouples of successively finer diameters; the temperature was found to be approximately 1180°C (2156°F).

The severity of burns depends on the integral of the time-temperature profile [6, 7, 8, 9]. Therefore, in order to determine a precise correlation between burn severity and temperature history, the temperature history within the skin simulant must be compared to that obtained for situations for which the burn severity is known. However, such an elaborate procedure is not necessary for screening materials. The maximum temperature reached in the skin simulant under specified flame-impingement conditions is sufficient.

The heat flux caused by the flame impinging on the simulant was determined by exposing the simulant to a 3-second flame exposure. An average temperature rise of 61.4°C (142.5°F) in the simulant, equivalent to a heat flux of 1.69 calories/sq cm-sec (2240 Btu/sq ft-hr) was obtained.

Most of the testing was carried out using a square-wave flame exposure of 3-seconds duration. This length of time was chosen since it is the exposure duration used in mannequin fire pit testing. (This test is discussed in a later section.)

A typical skin-simulant temperature-rise vs time plot is shown in Figure 4. The sequence is initiated at time zero; the shutter opens to begin the flame exposure at approximately one-half second and closes at approximately 3-½ seconds. These time intervals are established by pre-set timers. The peak temperature is reached at approximately 5 seconds, after which the

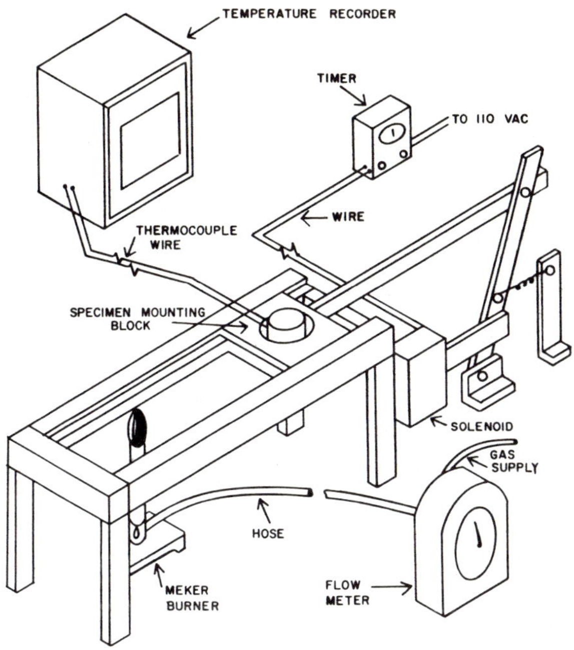

Figure 3a. *Flame-impingement heat-transfer tester.*

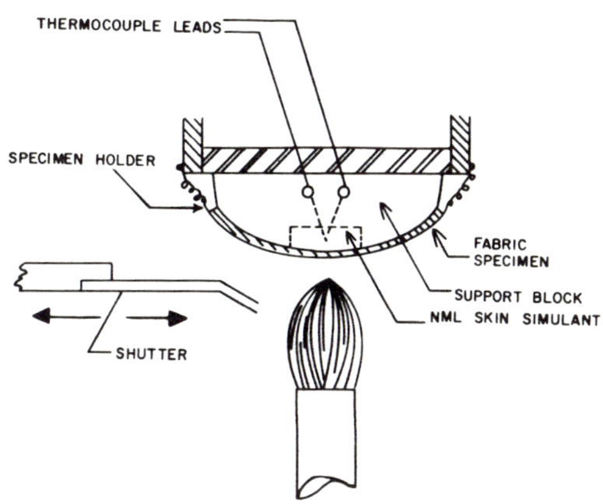

Figure 3b. *Specimen mounting fixture.*

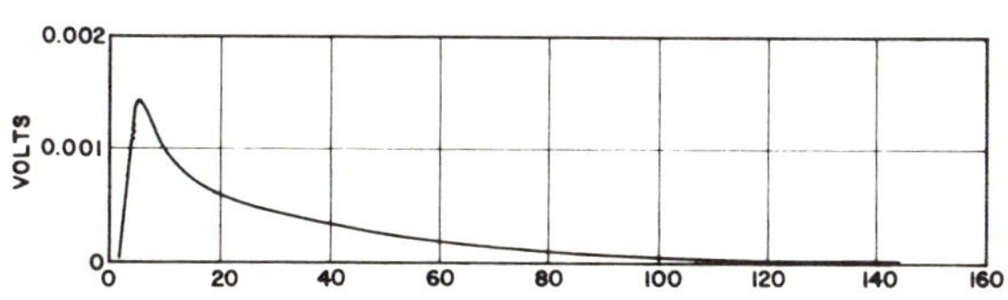

Figure 4. *Temperature rise vs time for 3-second flame exposure of spun PBI fabric.*

temperature drops as the system loses heat to the surroundings. Generally similar curves are obtained for all fabrics, differing only in the slope and peak amplitude.

The average maximum temperature rises within the NML skin simulant covered with single and multiple layers of various fabrics for 3-second flame exposures are tabulated in Table 5. All of the test specimens were stored in a conditioned room at standard temperature (70°F) and relative humidity (65%) before testing. Since the PBI fabrics were virtually undamaged by 3-second flame exposures, all samples were deliberately exposed repetitively. Higher temperatures in the skin simulant were recorded during the second and third exposure, with a general leveling-off of peak temperature after three, four or more exposures. Presumably this is due to the presence of moisture, or other volatile material, in the specimen which is evaporated during the initial flame exposures [6]. (The moisture regain of PBI fabric is 11-12% at 70°F, 65% RH, approximately twice that of Nomex or treated cotton fabrics.) This demonstrates that the thermal protection potential of a fabric depends significantly on its initial moisture content.

The results presented in Table 5 are for heat transfer through fabric specimens under essentially zero transverse pressure. The effects of pressure on flame impingement heat transfer were evaluated by making tests with the fabric test specimens under a tensile load. The results show that the temperature rise through the fabric increases moderately with increasing tensile load.

Table 5. Temperature Rise in NML Skin Simulant for 3-Second Flame Exposure.

		Maximum Temperature Rise (°C)	
		First Exposure (3-sec)	After Multiple 3-Sec Exposures
Cotton, flame-retardant-treated,			
single layer		29	destroyed
double layer	(zero spacing)	20	30
	(1-mm spacing)	20	29
	(2-mm spacing)	15	25
Nomex,			
single layer		35	39
double layer	(zero spacing)	23	28
	(1-mm spacing)	22	25
	(2-mm spacing)	18	19
PBI,			
single layer		24	34
double layer	(zero spacing)	13	23
	(1-mm spacing)	11	25
	(2-mm spacing)	8	20

Mathematical Analysis of Heat Flow

In an effort to gain a better understanding of the heat transfer through fabric under flame impingement, the measured temperature rise in the skin simulant was compared to that predicted using the Griffith & Horton analysis [4, 10, 11] for uniaxial transient heat flow through a two-layer wall. This analysis assumes a wall comprised of a uniform layer of infinite extent but finite thickness in intimate contact with an infinitely thick base. It also assumes that the surface of the wall absorbs a constant, uniform heat flux. If the heat input and the thermal properties of the two layers are known, the temperature rise at any location in either layer can be calculated as a function of time.

It must be emphasized, however, that this mathematical analysis is only applicable while the flame is in contact with the fabric. As previously noted, the temperature remains at a high level long after the flame is removed. Since

Table 6. Fabric and Skin Simulant Properties.

Property	Spun Nomex	Spun PBI
H — heat flux perpendicular to surface (cal/sq cm-sec)	1.7	1.7
α — fabric thickness at 1 psi (cm)	0.0330	0.0417
k_1 — fabric thermal conductivity at 1 psi (cal/cm-°C-sec) at approximately		
22°C	9.6×10^{-5}[1]	11.7×10^{-5}[1]
260°C	15.7×10^{-5}[1]	17.7×10^{-5}[1]
538°C	21.1×10^{-5}[2]	23.6×10^{-5}[2]
k_2 — skin simulant thermal conductivity (cal/cm-°C-sec)	13.1×10^{-4}[3]	13.1×10^{-4}[3]
ρ_1 — fabric density (gm/cu cm)	0.51[4]	0.44[4]
ρ_2 — skin simulant density (gm/cu cm)	1.8[3]	1.8[3]
C_{p1} — fabric specific heat (cal/gm-°C)	0.29[6]	0.30[5]
C_{p2} — skin simulant specific heat (cal/gm-°C)	0.36[3]	0.36[3]

[1] See Table 4.
[2] Obtained by a linear extrapolation of the values at 22° and 260°C.
[3] Reference 5.
[4] Calculated from weight and thickness.
[5] Assumed.
[6] E. I. du Pont de Nemours & Co., Inc.

the severity of a burn is a function of the total cumulative heat flow into the skin, in order to obtain an exact estimate of the amount of protection provided by a specific fabric under flame impingement, the mathematical model must also take into account the heat flow after the flame is removed. Such a model has yet to be formulated.

As noted previously, the heat flux imposed on the skin simulant by the flame in the flame-impingement heat transfer apparatus is approximately 1.7 cal/sq cm-sec. The properties, and their symbolic designation, of the spun PBI fabric, spun Nomex fabric, and the NML skin simulant required for the mathematical analysis are given in Table 6. Using this data, the temperature rise at the surface of the fabric nearest the flame U_1 ($x = 0$), at the fabric-skin simulant interface U_1 ($x = \alpha$), and at the depth below the skin simulant surface at which the thermocouple is located U_2 ($x = 0.049$ cm $+ \alpha$) were calculated for times t from 1 to 10 seconds using the following expressions:

$$U_1 = \frac{H}{k_1}\left\{ \left[2\sqrt{\frac{D_1 t}{\pi}}\, e^{x^2/4D_1t} - x\left(1 - \operatorname{erf}\frac{x}{2\sqrt{D_1 t}}\right)\right]\right.$$

$$- \frac{1}{\gamma}\sum_{n=0}^{n=\infty}\left(-\frac{1}{\gamma}\right)^n \left[2\sqrt{\frac{D_1 t}{\pi}}\left(e^{-\{x+2\alpha(n+1)\}^2/4D_1t} + e^{-\{x-2\alpha(n+1)\}^2/4D_1t}\right)\right.$$

$$- \{x + 2\alpha(n+1)\}\left\{1 - \operatorname{erf}\left(\frac{x + 2\alpha(n+1)}{2\sqrt{D_1 t}}\right)\right\}$$

$$\left.\left. + \{x - 2\alpha(n+1)\}\left\{1 - \operatorname{erf}\left(-\frac{x - 2\alpha(n+1)}{2\sqrt{D_1 t}}\right)\right\}\right]\right\}$$

$$U_2 = \frac{2H\lambda\sqrt{D_1}}{\gamma}\sum_{n=0}^{n=\infty}\left(-\frac{1}{\gamma}\right)^n\left\{2\sqrt{\frac{D_2 t}{\pi}}\, e^{-\{x-\alpha[1-\sqrt{D_2/D_1}(2n+1)]\}^2/4D_1t}\right.$$

$$- \{x - \alpha[1 - \sqrt{D_2/D_1}(2n+1)]\}$$

$$\left.\left(1 - \operatorname{erf}\frac{x - \alpha\{1 - \sqrt{D_2/D_1}(2n+1)\}}{2\sqrt{D_2 t}}\right)\right\}$$

where

$$D_1 = k_1/S_1,\quad D_2 = k_2/S_2$$
$$S_1 = \rho_1 C_{p1};\quad S_2 = \rho_2 C_{p2}$$
$$\lambda = k_2\sqrt{D_1} - k_1\sqrt{D_2}$$
$$\gamma = \frac{k_2S_2 + \sqrt{k_1S_1k_2S_2}}{k_2S_2 - \sqrt{k_1S_1k_2S_2}}$$

Because of the complexity of these expressions and the large number of repetitive calculations to be made, they were programmed for a digital computer. The temperature rises computed for the 538°C thermal conductivities,

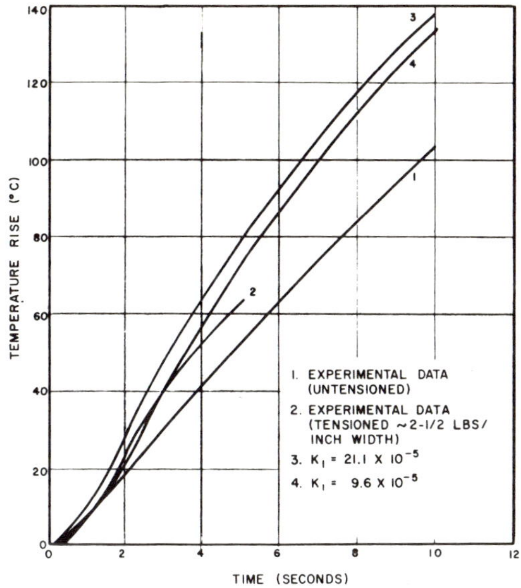

Figure 5. *Temperature rise in skin simulant covered with spun Nomex fabric.*

k_1 can only be considered as rough approximations, since, as noted in Table 6, these values were obtained by linear extrapolation of the data at 22° and 260°C. However, calculations were made using fabric thermal conductivity values for elevated temperatures in an effort to use material properties appropriate at the actual temperature of the fabric under flame impingement.

As expected, the results of the computations show that the temperature rise increases with increasing time of exposure, and decreases with increasing distance through the assembly, i.e., increasing values of x. The temperature rise at the fabric-skin simulant interface and the thermocouple location behind the surface of the skin simulant also increase as expected with increasing fabric thermal conductivity. However, the temperature rise at the fabric surface upon which the flame impinges decreases with increasing fabric thermal conductivity. Since the impinging heat flux is constant, the greater the fabric thermal conductivity the faster the rate of heat conduction away from the fabric surface nearest the heat source and thus the lower the fabric surface temperature.

The temperature rises in the skin simulant determined experimentally using the flame-impingement heat transfer apparatus and those predicted from the above analytical expressions are compared in Figures 5 and 6. As shown, the predicted temperature rise is considerably greater than that determined experimentally with the specimen under no tension. However, as also shown in the figures, when the predicted results are compared to experimental data for specimens under a tension of 2-½ lbs/inch width the agreement is considerably better. Thus, it appears that the Griffith & Horton analysis for uniaxial, transient heat flow through a two-layer wall predicts

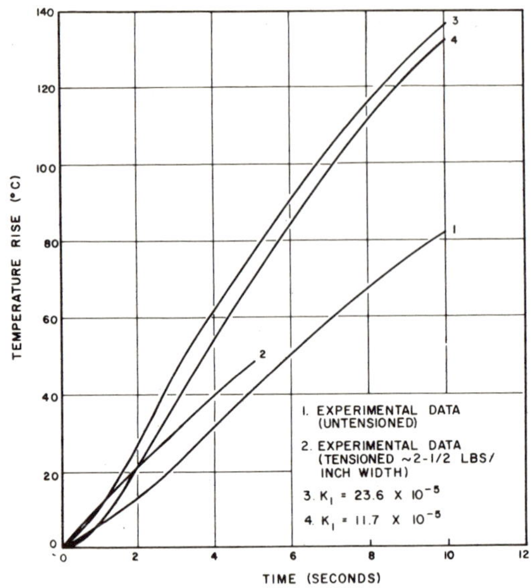

Figure 6. *Temperature rise in skin simulant covered with spun PBI fabric.*

reasonably well the heat transfer through fabric in intimate contact with a skin simulant under flame impingement. The more intimate the contact, the closer the agreement between the measured and predicted temperature rises in the simulant.

The effects of moderate changes in the values of input heat flux and the thermal properties of the fabric and skin simulant on the heat transfer calculations was also investigated. The magnitude of the effect of a systematic variation of each of the parameters was determined by solving the theoretical expressions, again with the assistance of a digital computer, for the following seventeen sets of data:

α = fabric thickness = 0.0417 cm
k_1 = fabric thermal conductivity = 17.7×10^{-5} cal/cm-sec-°C
k_2 = skin simulant thermal conductivity = 13.1×10^{-4} cal/cm-sec-°C
ρ_1 = fabric density = 0.440 gm/cu cm
ρ_2 = skin simulant density = 1.8 gm/cu cm
C_{p1} = fabric specific heat = 0.30 cal/gm-°C
C_{p2} = skin simulant specific heat = 0.36 cal/gm-°C
H = heat flux perpendicular to surface = 1.69 cal/sq cm-sec

followed by varying each of these parameters separately by 0.8 and 1.2 times the values given above. These data are the properties of the spun PBI fabric. The k_1 value is the thermal conductivity of the fabric at 260°C. The theoretical temperatures at the fabric surface, at the fabric-skin simulant interface, and 0.049 cm below the surface of the simulant were calculated for times t from 1 to 10 seconds. The percent changes in temperature rise resulting from the calculations for 3- and 10-seconds are summarized in Tables 7 and 8.

These results show that small deviations from the values used for the thermal properties of the fabric and skin simulant would not have a large effect on the calculated temperature rises. As also shown in the tables, a 20% change in the incident heat flux results in a 20% change in temperature at the three points at which the calculations were made. This is as anticipated since the temperature rise is directly proportional to the incident heat in the analytical expressions used.

The above analysis shows that the only parameters which significantly affect the tempeature rise 0.049 cm below the surface of the skin, and thus the severity of resulting burns a human might suffer, over which an engineer can exercise control are those fabric properties related to fabric construction. It is doubtful that fibers for apparel can be developed with intrinsic thermal properties significantly different from those presently available.

The data in Tables 7 and 8 indicate that a thick fabric, with a low thermal conductivity, high density and high specific heat would provide maximum personnel protection. However, although a material with a higher density and specific heat would reduce the skin temperature during flame impingement, part of the stored heat would be transmitted to the skin after the flame is

Table 7. Flame-Impingement Heat Transfer Through 5.4 oz/sq yd Spun PBI Fabric (t = 3 sec).

								Change in Temperature (%)		
α	k_1	k_2	ρ_1	ρ_2	C_{p1}	C_{p2}	H	Fabric Surface	Fabric-Skin Interface	0.049 cm Within Skin
B	A	A	A	A	A	A	A	−14	+9	+13
C	A	A	A	A	A	A	A	+12	−11	−14
A	B	A	A	A	A	A	A	+19	−5	−7
A	C	A	A	A	A	A	A	−13	+3	+5
A	A	B	A	A	A	A	A	+2	+10	0
A	A	C	A	A	A	A	A	−1	−8	−1
A	A	A	B	A	A	A	A	+2	+6	+8
A	A	A	C	A	A	A	A	−2	−6	−8
A	A	A	A	B	A	A	A	+2	+10	+20
A	A	A	A	C	A	A	A	−1	−8	−14
A	A	A	A	A	B	A	A	+2	+6	+8
A	A	A	A	A	C	A	A	−2	−6	−8
A	A	A	A	A	A	B	A	+2	+10	+20
A	A	A	A	A	A	C	A	−1	−8	−15
A	A	A	A	A	A	A	B	−20	−20	−20
A	A	A	A	A	A	A	C	+20	+20	+20

A = original value.
B = original value × 0.8.
C = original value × 1.2.

extinguished, and thereby possibly cause almost as severe a burn as with a material having a lower density and specific heat.

Consequently, it appears that the most promising approach for decreasing heat transfer through apparel is to use a thick, fibrous structural system with a low thermal conductivity [12]. Since air is the best insulator, having a thermal conductivity of approximately 10.3×10^{-5} cal/cm-sec-°C at 260°C, clothing systems which are thick and contain many small air cells should exhibit superior performance. Additionally, the structure must be made from a nonflammable fiber that is not degraded upon exposure to a flame to the point where its thermal properties are altered detrimentally or the material becomes non-functional, e.g., stiff and brittle. Of course, apparel structures must also meet many other criteria associated with comfort, e.g., a sufficient degree of flexibility and air permeability.

A clothing system that appears to meet all these requirements is a moderate weight, tightly woven PBI fabric over a PBI raschel or tricot mesh. The low porosity outerlayer would minimize convective and radiation heat transfer and, in combination with the mesh, create the desired air cells. Furthermore, this type of assembly provides warmth in cold climates due to its low thermal conductivity and has also been found to be comfortable in the tropics. The mesh underlayer keeps the perspiration or rain-soaked outerlayer off the skin.

***Table 8. Flame-Impingement Heat Transfer Through 5.4 oz/sq yd Spun PBI Fabric* (t = 10 sec).**

								Change in Temperature (%)		
α	k_1	k_2	ρ_1	ρ_2	C_{p1}	C_{p2}	H	Fabric Surface	Fabric-Skin Interface	0.049 cm Within Skin
B	A	A	A	A	A	A	A	−13	+3	+4
C	A	A	A	A	A	A	A	+12	−3	−4
A	B	A	A	A	A	A	A	+16	−1	−1.5
A	C	A	A	A	A	A	A	−12	+½	+1.5
A	A	B	A	A	A	A	A	+3	+11	+7
A	A	C	A	A	A	A	A	−3	−9	−5
A	A	A	B	A	A	A	A	+½	+2	+3
A	A	A	C	A	A	A	A	−1	−2	−2
A	A	A	A	B	A	A	A	+3	+11	+16
A	A	A	A	C	A	A	A	−3	−9	−11
A	A	A	A	A	B	A	A	+½	+2	+3
A	A	A	A	A	C	A	A	−1	−2	−2
A	A	A	A	A	A	B	A	+3	+12	+16
A	A	A	A	A	A	C	A	−3	−9	−11
A	A	A	A	A	A	A	B	−20	−20	−20
A	A	A	A	A	A	A	C	+20	+20	+20

A = original value.
B = original value × 0.8.
C = original value × 1.2.

Figure 7. *Fire-pit test facility.*

The data in Table 9 give an indication of the effectiveness of fabric-mesh-type assemblies. The maximum temperature rise in the NML skin simulant covered with the PBI fabric alone and with a fabric-cotton mesh assembly is given for a 3-second flame exposure. (Cotton mesh was used because no PBI mesh was available.) As shown, the use of the mesh underlayer reduces the temperature rise by 55-65%.

Fire Pit Tests

A series of tests were conducted at the U.S. Army Natick Laboratories Fire Test Facility in Maynard, Mass. (see Figure 7) to compare the performance of PBI coveralls with that of the 100% cotton and 100% Nomex suits of comparable construction. In this test fully clothed mannequins were drawn over a thirty foot long pool of flaming JP-4 aircraft fuel at a rate of 10

Table 9. Temperature Rise Through Fabric-Mesh Assembly.

		Maximum Temperature Rise (°C)	
Outer Layer	Second Layer	Through Outer Fabric Layer	Through Fabric-Mesh Assembly
Spun PBI fabric 2.9 oz/sq yd	cotton tricot mesh 0.04-0.05 inch thick, (0.14-0.16 inch by 0.16-0.20 inch pore size) 8.4 oz/sq yd	43	16
Spun PBI fabric 8.3 oz/sq yd	same as above	22	10

Figure 8. *Fire-pit test results.*

feet per second (total flame exposure time of 3 seconds). Flame temperatures of 1800-2000°F were recorded. Temperature indicating papers were placed at various locations on the mannequins so that the thermal protection effectiveness of the garments could be assessed after the exposure. This test simulates the type hazard aircraft ground and flight personnel occasionally encounter.

From observations made during and after these tests it was apparent that PBI suits were superior to those fabricated from Nomex or cotton. Large portions of the cotton suits and small portions of the Nomex suits were totally destroyed; however, the PBI suit remained completely intact (see Figure 8). Portions of the cotton and the Nomex coveralls were burning as the mannequins came out of the flames. The temperature sensors under the suits indicated that burns covering 50-60% of the body occurred with both cotton (untreated and treated with a flame-proofing agent) and Nomex suits. A maximum of 7% of the body surface was burned under the PBI suits.

The cotton and Nomex suits burn in the JP-4 fuel flame because their ignition temperatures (flame-retardant-treated cotton: 1450°F; Nomex: 1600°F) are lower than the fuel flame temperature (1800-2000°F). The ignition temperature of the PBI fabric (1700°F) is also lower than the fuel flame temperature. However, laboratory tests indicate that PBI must be exposed to high temperatures for considerably longer than 3 seconds before there is any visible sign of ignition. All three materials self-extinguish when removed from a high-temperature environment.

Both the PBI and Nomex suits exhibited considerable thermal shrinkage. However, the shrinkage of the PBI suits was somewhat less than that of the Nomex suits. Shrinkage leads to initimate contact of the fabric with the mannequin and thereby a larger temperature build-up on the mannequin surface. However, a suit fabricated from thermally stabilized PBI fabric exhibited no noticeable dimensional changes and the temperature-indicating devices on the surface of the mannequin were essentially unaffected.

These test results indicate that nonstabilized PBI coveralls provide greater thermal protection than currently used cotton or Nomex coveralls. However, coveralls fabricated from stabilized PBI fabric provide an even greater degree of protection.

ACKNOWLEDGMENT

The author acknowledges the assistance of Mr. J. G. Donovan in obtaining the flammability characteristics of the fabrics, Mr. R. E. Sebring in measuring the flame-impingement heat-transfer characteristics of the fabrics and Mrs. M. M. Schoppee and Dr. M. M. Platt in the preparation of this paper.

References

1. W. D. Freeston, Jr., et al, AFML-TR-70-267 (1970).
2. E. R. Kaswell, "Wellington Sears Handbook of Industrial Textiles," Wellington Sears Co., N.Y.C. (1963).
3. "Physical Methods of Investigating Textiles" Interscience Publishers, Inc., New York (1959).
4. A. M. Stoll, M. A. Chianta, and L. R. Monroe, "Flame-Contact Studies," Transactions of the ASME, Paper No. 63-WA-121 (1963).
5. W. L. Derksen, et al, U.S. Naval Applied Science Laboratory, Final Report SF-001-05-011 (1960).
6. W. L. Derksen, W. J. Bates, and T. I. Monahan, Technical Report AFSWP-834 (1954).
7. F. C. Henriques, Jr., "Studies of Thermal Injury V: The Predictability and Significance of Thermally Induced Rate Processes Leading to Irreversible Epidermal Injury," *Arch Path,* vol 43, p. 489 (1947).
8. A. C. Moritz, and F. C. Henriques, Jr., "Studies of Thermal Injury II: The Relative Importance of Time and Surface Temperature and the Causation of Cutaneous Burns." American Journal of Pathology, vol 23, p. 695 (1947).
9. A. M. Stoll, and M. A. Chianta, NADC-MR-6809 (1968).
10. M. V. Griffith, and G. K. Horton, Proc. of the Phys. Soc., London, England, vol. 58, (1946).
11. A. M. Stoll, and M. A. Chianta, NADC-MR-6216 (1962).
12. R. E. Seaman, Bull. N. Y. Acad. Med. vol. 43, No. 8 (1967).

W. D. Freeston, Jr.

W. Denney Freeston, Jr. is Associate Director of Applied Mechanics at Fabric Research Laboratories, Inc., Dedham, Massachusetts. He received his B. S. degree in Mechanical Engineering from Princeton University in 1957 and his PhD from the same institution in 1961. He has been with FRL® since 1960 specializing in the mechanistic analysis of fibrous structures, evaluation and utilization of high temperature fibers and analysis of the response of fibrous materials at high rates of loading.

START

Flame Retardant Treatments of PBI Fabric

Samuel C. Temin

Fabric Research Laboratories, Inc.
Dedham, Mass.

(Received April 29, 1971)

ABSTRACT

Fabrics knitted or woven from polybenzimidazole (PBI) fibers were treated to reduce flammability in oxygen atmospheres, particularly that of 5 psia oxygen. Bromination to approximately 15% weight gain of such fabrics led to markedly lower burning rates; samples brominated to over 80% weight gain were self-extinguishing in 5 psia oxygen. The loss in tensile strength of fabrics due to bromination was negligible although shrinkage was observed. Free fibers showed negligible losses on bromination. Treatment of PBI fabric with organophosphorus compounds also achieved self-extinguishing character in 5 psia oxygen but the enhanced flameproofing was largely lost on leaching. Reaction with $POC1_3$ in pyridine led to a permanent reduction in flammability.

INTRODUCTION

Fibers and fabrics composed of poly-2,2'-(m-phenylene)-5,5'-bibenzimidazole, or simply polybenzimidazole (PBI) [1], have been observed to possess a high order of flame resistance in air [2] although burning takes place in 100% oxygen atmospheres. The non-flammability of PBI in air, therefore, suggested this material as an attractive starting point to achieve a textile material that would be suitable, from a safety point of view, for use in space missions where atmospheres of 100% oxygen were contemplated.

Chemical modification of PBI appeared to be a promising route to enhanced flame resistance since the structure of the polymer suggested that reaction could be achieved without great difficulty. In particular, bromination [3] and phosphorylation [4] to enhance flame resistance appeared feasible.

EXPERIMENTAL

The PBI fabrics used in this work included (a) a 70 x 70, 2 x 2 twill (200/50

Part of a research program sponsored by NASA Manned Spacecraft Center, Houston, Texas, under Contract NAS 9-8811.

yarns 3Z turns per inch) woven at FRL® and (b) a 5 oz/sq yd fabric (200/50 untwisted yarns) knitted at FRL®.

Bromination

In one series of experiments, the fabric was exposed in a desiccator over a pool of liquid bromine for various times at room temperature. Subsequently, the treated fabric was refluxed in benzene. Other samples, in the form of a roll, were placed inside a vertical, electrically-heated glass tube which was connected to a flask containing bromine. Vapors of the halogen were swept into the tube by bubbling nitrogen through the liquid bromine. Temperature was determined by a thermometer inserted in the fabric roll.

A third bromination procedure consisted of immersing a piece of fabric in a boiling 10% aqueous KBr_3 solution (equal molar concentrations of KBr and Br_2) for 15 minutes.

All brominated fabrics were rinsed in aqueous sodium bisulfite solution, to remove unreacted bromine or any N-Br structures, and then washed in water containing a detergent and plain water before drying.

Analyses (Galbraith Lab) of a few selected samples indicated that elemental analyses and weight gain measurements gave approximately the same values. Consequently, bromine content is generally reported in terms of the weight gain of the fabric in this article.

Phosphorylation

Several chemicals were applied to PBI fabric from suitable solvents. Reactions involving a P-C1 bond were conducted in the presence of an acid acceptor. The most extensively explored reagent, $POC1_3$, was used in pyridine solution. In a typical experiment, a 6-inch long strip of scoured PBI fabric was immersed in 150 ml of a 50% solution of $POC1_3$ in dry pyridine. The solution was heated (under a condenser fitted with a drying tube) at 90-95°C for 20 hours. The fabric was then washed in acetone and dried *in vacuo.* After the weight gain was calculated, the sample was washed in water, redried and the net weight gain noted.

Flammability Testing

The FRL® flammability tester is shown schematically in Figure 1. The apparatus, as currently constituted, is more sophisticated than is apparent from the diagram. However, for the work reported here, ignition comprised wrapping the resistance wire with tissue paper and heating to produce a paper fire. The sample holder held the sample vertically about 1/8 inch above the resistance wire. The holder was adjustable, permitting samples of various widths and/or lengths to be tested. During the course of the work reported here, the width of the sample exposed to the flame was held constant at 1.25 inches.

The sample was clamped in the holder which was then placed in position. After bolting on the transparent door to seal the chamber, the vacuum pump (oil) was

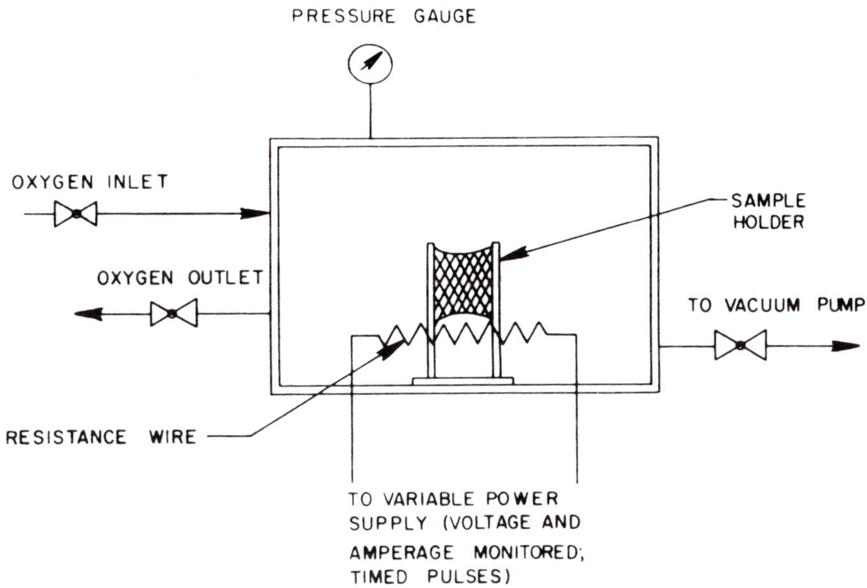

Figure 1. *Schematic of flammability test chamber.*

turned on and the chamber evacuated. Following evacuation, the chamber was repressurized with bone dry oxygen to the desired gauge reading and the tissue paper ignited by heating the resistance wire. The test results were recorded both as time for the flame (or glow) to reach the top of the sample or, if the specimen were shorter than the standard six inches, as rate of burning (inches per second). The 100% oxygen atmospheres used in this work were either 5 psia or 16.5 psia.

RESULTS

Effect of Sample Geometry on Burning Rate

It has been demonstrated that burning times in air increase in a roughly linear fashion with increasing fabric weight for both cellulosic [5] and thermoplastic [6] fabrics. To verify this relationship for combustion in oxygen atmospheres, PBI fabrics of various constructions were burned in 5 psia oxygen. Also burned, in order to compare different sample geometries were PBI films, including one prepared from brominated PBI powder supplied by Whittaker Corporation. Results are given in Table 1.

The data of Table 1 indicate, as expected, that heavier fabrics do take longer to burn but, as evident from a comparison of Samples 1 and 4, a change in the yarn denier is not as important as the relative openness of a fabric. With films, it appears that burning rates are slightly higher than for fabrics of comparable weight, possibly due to better heat transfer in films than in the discontinuous fabric.

Accordingly, strict attention must be paid to the form and dimensions of polymer specimens selected for flammability testing if valid comparisons are to be made. In order to make direct comparisons of two polymers as possible materials for flameproof textiles, it seems expedient that they be compared in the form of similarly constructed fabrics.

Bromination

Bromination of PBI fabrics was straightforward and effective. Elemental bromine substitutes PBI without any added catalyst. Also, simply heating in an aqueous solution of KBr_3 was found quite effective. The mechanism of bromination is not known but probably involves formation of an N-Br intermediate which rearranges to give nuclear bromination.

The fate of the HBr formed in the reaction was not ascertained; presumably salts could be formed with basic nitrogen. To eliminate this possibility, and to remove any N-Br structures or free bromine, the fabrics were rinsed in sodium bisulfite. It was shown that sodium bisulfite (or sodium carbonate) at room temperature does not react with either PBI itself or brominated PBI. In one case, a highly brominated PBI fabric lost 15% weight on treatment with alkaline solutions. This weight loss is roughly equivalent to the loss of a molecule of HBr from a PBI substituted with an average of two bromines per repeat unit.

The rate of bromination was not quantitatively determined. However, the data of Table 2, which shows the net weight gain obtained on exposure to bromine vapors at room temperature, show that the degree of bromine substitution is a function of exposure time. Sample 7 differed from the other of Table 2 in that the fabric was refluxed in benzene following exposure and before bisulfite washing. Since the fabric carried a good deal of bromine into the refluxing liquid, additional bromination could have occurred during this process. Samples 5 and 6 were dried in an 105°C

Table 1. Effect of Sample Weight on PBI Flammability

Sample No.	Type	Fabric Weight (oz/sq yd)	Yarn Denier[a]	Burning Rate In 5 psia Oxygen (in/sec)
1	Woven fabric	4.5	200	0.46
2	Woven fabric	10	200	0.09
3	Woven fabric[b]	14	260	0.04
4	Knit fabric	5[c]	200	0.75
5	Film[d]	7.3	–	0.13
6	Film[d]	11.2	–	0.11
7	54% Br – Film[d,e]	7.2	–	0.05

a. All yarns contained 50 filaments.
b. PBI woven webbing.
c. Not exact; knit fabric is easily stretched so density of material presented to flame may vary according to tension applied by operator in placing of sample in holder.
d. Films were heated to 350 C before testing to ensure removal of solvent (DMF).
e. Brominated PBI supplied by Whittaker Corporation, San Diego, California.

oven just before exposure to the bromine vapors and the higher weight gain obtained with Sample 5 (relative to Samples 3 or 4) may reflect the influence of moisture present in the undried fabrics.

The rate of bromination is markedly accelerated by an increase in temperature. For example, exposing PBI to bromine vapor at 80 C resulted in a 94% weight gain (after bisulfite washing) after only two hours.

Results of burning tests, Table 3, indicated that bromination even at a 14%

Table 2. Room Temperature Vapor Phase Bromination of PBI Fabric[a]

Sample	Bromine Exposure Time (hrs)	Weight Gain %
1	1	1.8
2	1	3.4
3	3	14
4	3	15
5[b]	2	23
6[b]	4.5	34
7[c]	4.5	84
8	21	83

a. Knit fabric, 5 oz/sq yd, prepared at FRL® from 200/50 untexturized, untwisted yarn.
b. Fabrics dried (105°C) before placing in desiccator.
c. Fabric heated for 1 hour in refluxing benzene before washing.

Table 3. Effect of Bromination on Burning Rate of PBI Knit Fabric

Sample No.	Bromine Treatment[a]	Bromine (%)[b]	Burning Rate In Oxygen (in/sec) 5 psia	16 psia
1	V-RT	1.8	0.30	–
2	V-RT	3.4	0.35	–
3	V-RT	14	0.07	–
4	V-RT	23	0.05	–
5	V-RT	34	–	0.18
6	V-RT	83	–	0.04
7	V-Ref.	84	0.02[c]	0.08[c]
8	V-80	94	0.03[d]	0.27
9	None	0	0.75	–

a. V-RT = vapor phase bromination at room temperature; V-Ref = room temperature bromination followed by refluxing in benzene; V-80 = vapor phase bromination at 80°C.
b. Based on weight gain.
c. Self-extinguished in 5 psia oxygen after 150 sec when only half burned. Remaining half then tested at 16 psia oxygen.
d. First sample tested was self-extinguishing.

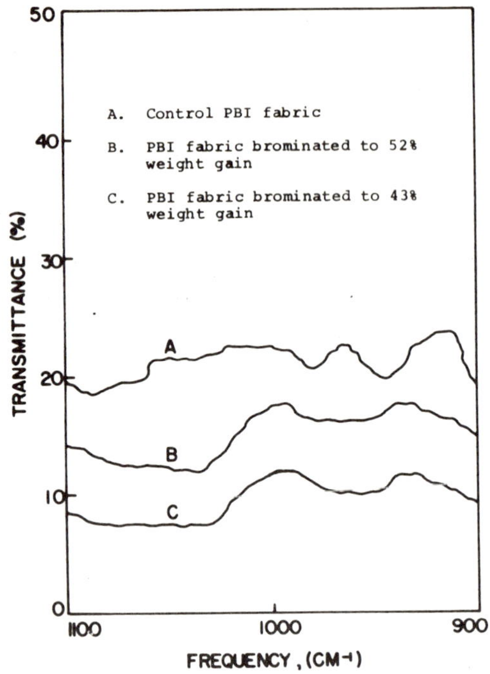

Figure 2. *Infra-red spectra (ATR) of PBI fabrics.*

bromine level effectively reduces the flammability of PBI fabric in oxygen. At higher levels of substitution, the rate of burning becomes extremely slow. At 80-90% bromine content samples are obtained that are self-extinguishing in 5 psia oxygen and with a burning rate in 16.5 psia oxygen reduced to the point where such fabric might provide a wearer sufficient protection to permit escape to a lower oxygen content atmosphere.

It was assumed that nuclear substitution of the arylene units in the polymer had occurred and that stable carbon-bromine bonds were formed. Infra-red analyses, Attenuated Total Reflectance (ATR) using KRS-5 plate, of the brominated samples revealed loss of a doublet at 945, 985 cm^{-1}, which is present in the untreated PBI [7], indicating further nuclear substitution. Other differences in the 900 to 1100 cm^{-1} spectra are shown in Figure 2. Further evidence of the bonding stability of the bromine was afforded by washing studies. Samples of brominated PBI fabric (the treatment of which included the bisulfite wash) were laundered five times in a home washer using a commercial detergent and water at 60°C. There was no weight loss and the burning rate in 5 psig oxygen was unchanged. Fabrics in this study included those brominated by bromine vapors as well as by KBr_3 with bromine contents (based on weight gain) ranging from 40 to 143%.

Brominated fabric was also subjected to a simulated drycleaning with perchloroethylene, using the general procedure of AATCC 85-1968, and again was found to undergo no loss in weight.

Phosphorylation

A number of regents were examined for potential utility in introducing phosphorus into the PBI chain. The most facile route to this goal was believed to be by reaction with the N-H bond in PBI. Accordingly, most of the reagents studied contained a labile halogen bond in a effort to achieve the general reaction shown below:

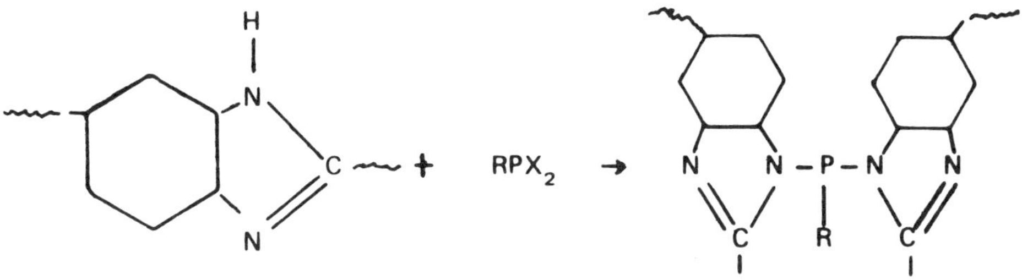

The R in the equation, to satisfy the valence of phosphorus, almost always included oxygen.

Phenyl phosphonic dichloride, C_6H_5P (0) $C1_2$, was reacted with PBI fabric (refluxing isopropyl alcohol with potassium carbonate as acid acceptor) but gave a product with only modestly reduced flammability. A fabric specimen with 15.8% weight again after treatment burned at 0.25 inches per second compared to 0.75 inches per second for unmodified fabric.

Treatment of PBI fabric with either trichloromethyl phosphonic acid, $C1_3C$-P (O) $(OH)_2$, or phosphorofluoridic acid, $FP(O)(OH)_2$, gave initially encouraging results but neither treatment was resistant to water washing. Thus immersing PBI fabric for 24 hours at room temperature in an acetone solution containing phosphorofluoridic acid and pyridine (both present in weights equal to that of the fabric) gave a fabric that was self-extinguishing in 5 psia oxygen. However, the acid was apparently not bound covalently and was lost on water washing. Similar results were obtained with the trichloromethyl phosphonic acid.

A fabric sample reacted with phosphorus pentachloride, $PC1_5$, to 25% weight gain burned at 0.08 inches per second in 5 psia oxygen.

Phosphorus oxychloride treatments (in pyridine as reaction medium and acid acceptor) gave products with weight gains of 30-40% after acetone washing. Such fabrics were self-extinguishing in 5 psia oxygen and one sample self-extinguished in 16 psia oxygen. However, washing with water reduced the add-on to about 15% of (apparently) chemically bound phosphorus-containing entity. One such specimen (15.9% add-on) was self-extinguishing in 5 psia; another piece burned at 0.07 inches per second. In one case, a sample of PBI fabric analyzing 18% bromine was given a subsequent $POC1_3$ treatment (1.9% P) and found to be self-extinguishing in 5 psia oxygen.

The mode of reaction of $POC1_3$ with PBI was not evaluated and thus no explanation can be offered for the observation that the permanent add-on (water resistant) did not exceed 16% even when large excesses of $POC1_3$ were utilized in the reaction.

Effect of Phosphorylation on Rate of Bromination

Attempts to brominate PBI fabric that had been treated with a phosphorylating reagent revealed some inhibition of the bromination reaction. Typical results are given in Table 4 where bromination of a $POC1_3$-treated PBI fabric is compared with a control fabric. It can be seen that the phosphorylation treatment markedly reduces the rate of bromination. This inhibition is not unexpected if the mechanism of bromination involves an initial information of an N-Br bond; a reagent that blocks the reactive nitrogen in PBI will take its toll on the rate of bromination.

Effect of Treatments on Thermal Stability

Since PBI possesses a high degree of thermal stability, it became of interest to determine whether the treatments had affected this property. Assessment was attempted using differential scanning calorimetry (DSC) measurements; samples of fabric were run on the Perkin Elmer DSC-1 at 20°/min under a flow of nitrogen. The curve of the control PBI fabric showed a slight exotherm at between 420 and 440°C but no evidence of decomposition below 500°C. However, the samples after removal from the holder were found to be fused and badly shrunken. In contrast, a PBI fabric brominated to 53% weight gain neither fused nor shrunk on heating to 500°C although it was embrittled. A sample brominated to 143% weight gain also did not shrink or fuse but, further, did not embrittle on heating to 500°C; this sample did exhibit a slight endotherm at 40-460°C and gave off appreciable amounts of yellow-brown material (acetone-soluble) which collected in the sample holder cover. Finally a PBI sample given a $POC1_3$ treatment was found to be remarkably stable, showing no transitions and neither fusion nor embrittlement on heating to 500°C.

Thus it appears that the chemical reactions undergone in the bromination and phosphorylation treatments have resulted in bondings which are both thermally and hydrolytically stable.

Table 4. Bromination of Phosphorylated Fabric

Fabric Sample	$POC1_3$ Weight Gain[a] (%)	Exposure Time[b] (hrs)	Bromination Weight Gain (%)
1A	15.4	2	3.1
1B	15.4	4	8.6
1C	15.4	9	16.2
2A	0	2	8.3
2B	0	4	18.1
2C	0	6	41.9

a. Value refers to weight gain resulting from $POC1_3$-pyridine treatment.

b. To bromine vapors at room temperature.

Effect of Bromination on Fabric Properties

To ascertain whether bromination treatments of PBI caused unacceptable losses in physical properties, measurements were made on yarns and fabrics before and after such treatments.

In one series of experiments PBI yarns (200/50 denier) were subjected to vapor phase bromination, washed with bisulfite, dried and conditioned. The breaking loads and elongations of these specimens were compared, Table 5, with control yarns and specimens given only the bisulfite wash.

Although there was considerable scatter in the Instron data, due primarily to nonuniformity of the yarn, the results indicate no major damage is done to the fiber by 3 hours exposure to bromine at room temperature.

There still remained the possibility that yarns constricted within the fabric might exhibit more noticeable changes. Several brominated fabrics were selected for measurements of shrinkage and strength loss; fabrics given either a room temperature vapor bromination or an aqueous KBr_3 treatments were used. It was found that the KBr_3 technique caused no shrinkage, in contrast to the vapor phase process, and that the strength loss was quite small in both cases. As shown in Table 6, there is no correlation between bromine uptake and strength loss although there appears to be, in the case of vapor treatments, greater shrinkage with extend of reaction. The tensile tests (Table 6) were made on the same number of ends to normalize the results.

Table 5. Effect of Bromination on PBI Yarn Strength

Treatment	Breaking Load (lbs)[a]	Elongation To Break (%)[a]
None	2.1	16.7
Bisulfite wash only	2.3	15.5
Br_2 one hour, bisulfite wash	2.3	19.1
Br_2 three hours, bisulfite wash	2.0	14.0

a. Average of 5 specimens (Instron at 20% extension per minute).

Table 6. Effect of Bromination on Fabric Shrinkage and Tensile Strength

Sample No.	Bromination Technique	Bromination Wt. Gain (%)	Yarn Count Picks	Yarn Count Ends	Tensile Strength Retained (%)[a]
1	0	–	72	74	100
2	vapor	52	80	80	95
3	vapor	143	83	83	100
4	KBr_3	66	72	74	94

a. Based on breaking load of warp strips ravelled to 20 ends.

It appears that even large increases in bromine content can be achieved with little loss in fabric strength. Moreover, even where shrinkage occurs, the handle of the fabric is still acceptable; the KBr_3 brominated samples are essentially unchanged in handle.

CONCLUSIONS

It has been shown (based on burning rates of chemically modified PBI fabric in oxygen atmospheres), that reduced flammability results from incorporating either bromine or phosphorus in the PBI molecule.

Simple, fairly rapid bromination of PBI fabric can be achieved using either bromine vapors or aqueous systems. Self-extinguishing character in 5 psia oxygen can be imparted to a lightweight PBI fabric by brominating to approximately an 80% gain in weight. Bromination of PBI fabric does not seem to cause appreciable loss in fabric properties and the treatment is stable to laundering or drycleaning.

Chemical incorporation of phosphorus-containing substituents or impregnation with organophosphorus compounds is effective in reducing flammability of PBI fabrics.

Brominated and phosphorylated fabrics retain the thermal stability typical of PBI. END

ACKNOWLEDGMENT

The author wishes to acknowledge the assistance of Mr. D. Allen and Mrs. C. H. Park in performing the fabric treatments and of Mr. J. C. Donovan in conducting flammability tests.

REFERENCES

1. R. W., Singleton, Appl. Polymer Sym., *9*, 133 (1969).
2. J. H. Ross, Mod. Text. Mag., *50* (1), 18 (1969).
3. J. K. Jacques, Trans. J. Plastics Inst., Conference Supplement No. 2, "Flame Resistance with Polymers," Jan. 1967, p. 33.
4. P. Volans, Trans. J. Plastics Inst., Conference Supplement No. 2, "Flame Resistance with polymers," Jan. 1967, p. 47.
5. E. B. Nielsen, and H. R. Richards, Textile Chemist and Colorist, *1* (12) 270 (1969).
6. E. B. Nielsen, and H. R. Richards, Textile Chemist and Colorist, *2* (4) 78 (1970).
7. H. Vogel and C. S. Marvel, J. Polymer Sci., Pt. A, *1,* 1531 (1963).

Samuel C. Temin

Samuel C. Temin is Assistant Director, Polymer Chemistry, at Fabric Research Laboratories, Inc., Dedham, Massachusetts. After receiving a B. S. from Wilson Teachers College, Washington, D. C., he received an M. S. (1943) and Ph. D. (1949) from the University of Maryland. Prior to joining FRL®, he was Manager, Exploratory Polymer Chemistry, at Koppers Co., Inc. and Supervisor in the High Polymer Laboratory of Industrial Rayon Corp. He has taught graduate level Polymer Chemistry at both Pennsylvania State (Continuing Education) in Pittsburgh and at Northeastern University (current). Recent research interests center around synthesis of fiber-forming polymers and fiber reactions, including graft copolymerization.

The Status of Research to Develop Flame Retardant Cotton Flote Products for Cushioning Applications

Nestor B. Knoepfler

*Southern Regional Research Laboratory**
New Orleans, Louisiana 70119

(Received October 9, 1970)

ABSTRACT

The Cotton Flote process is reviewed from both mechanical and chemical requirement standpoints. The performance characteristics of Cotton Flote as a cushioning material using a variety of test methods are discussed. The most recent research aimed at developing flame retardant products is described. The addition of such flame retardants as urea phosphate, antimony trioxide, diammonium phosphate and borax, borated amido polyphosphate to Cotton Flote formulations confers acceptable levels of flame retardance with minimum effect upon the other physical characteristics of the products.

The passage of the National Traffic and Motor Vehicle Safety Act, Public Law 89-563, on September 9, 1966, and the amendments to the Flammable Fabrics Act, Public Law 90-189, on December 14, 1967, introduced a new urgency into the development of flame retardant fibrous cushioning products. This urgency has been carried a step further by the announcement of the establishment of Motor Vehicle Safety Standard 302 covering the flammability of automobile interior trim materials by the Department of Transportation on January 8, 1971, and the notice of finding that flammability standards or other regulations may be needed for mattresses, and the institution of proceedings by the Department of Commerce on June 10, 1970, to establish such standards.

It is the purpose of this paper to report the present status of research aimed at the development of flame retardant Cotton Flote products for cushioning applications in furniture, mattresses, and automobile seating.

Product oriented research to develop new and improved cotton batting

* One of the laboratories of the Southern Marketing & Nutrition Research Division, Agricultural Research Service, U. S. Department of Agriculture.

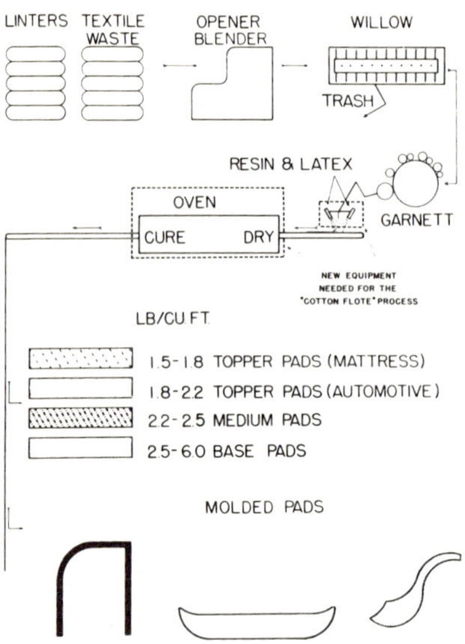

Figure 1. *Flow diagram of the Cotton Flote process.*

products that demonstrate enhanced resilience, dimensional stability, tensile strength, integrity and flame retardance have been researched over the past ten years at the Southern Regional Research Laboratory. This research has been cosponsored by the Textile Fibers and ByProducts Association, The National Cotton Batting Institute, The National Cottonseed Products Association, The Foundation for Cotton Research and Education, Cotton Inc., and the United States Department of Agriculture.

A new high performance cotton nonwoven product for cushioning applications resulted from this research. It has been trade-named Cotton Flote.* The key feature of the process is the simultaneous application of both thermoplastic and thermosetting resins from a single solution/suspension to achieve a concurrent improvement in the inherent resilience of the fibers and to bond the fibers together at points of contact to enhance dimensional stability, tensile strength, and integrity [1, 2, 3, 4, 5, 6, 7].

An engineering evaluation of the processing steps used in the production of nonwoven cotton batting disclosed that the fibers were most openly arrayed, and, consequently, most accessible to a treating formulation when they are in web form as they are discharged from the reciprocating apron of the garnett. Because the properties sought in the finished products seemed to be most readily achieved by resins similar to those used in wash-wear or durable press treatments for cotton goods or in nonwovens, a wet processing system was selected. The garnett webs used in the production of nonwoven cotton batting are extremely fragile and quite difficult to handle in an immersion system. As an alternative, spraying of the chemicals onto the webs was found to be both economical and practical. The introduction of a spray operation into the conventional cotton batting line had little effect on the handling of the material or in the lapping of the web to form a cushion pad.

Inherent in a wet processing system is the requirement for drying. Furthermore, most thermosetting resins must be raised to 300°F and held at that temperature for about 1-½ minutes to chemically react or crosslink them with the cellulose of the cotton. Drying and curing was integrated into the ma-

* The trade name Cotton Flote is registered by the National Cotton Batting Institute and may be used only by manufacturers who adhere to specified standards of performance for products made by use of the process covered by U. S. Patent 3,181,225.

TYPICAL TREATING FORMULATIONS
20% RESIN SOLIDS

METHYLATED METHYLOL MELAMINE	50%	UREA FORMALDEHYDE (MODIFIED)	50%
VINYL ACETATE	50%	VINYL ACRYLATE	50%
BUFFER OR FORMALDEHYDE ACCEPTOR		BUFFER OR FORMALDEHYDE ACCEPTOR	
CATALYST (MAGNESIUM CHLORIDE)		CATALYST (MAGNESIUM CHLORIDE)	
WATER		WATER	

METHYLOL IMIDAZOLIDONE	50%	METHYLATED METHYLOL MELAMINE	50%
VINYL ACETATE-ACRYLATE	50%	VINYL CHLORIDE-ACETATE	35%
CATALYST (ZINC CHLORIDE)		STYRENE BUTADIENE	15%
WATER		BUFFER OR FORMALDEHYDE ACCEPTOR	
		CATALYST (MAGNESIUM CHLORIDE)	
		WATER	

Figure 2. *Typical formulations used to spray the web in the production of Cotton Flote.*

terial flow pattern at the end of the floor apron after the in-process, spray damp product had been laminated or lapped to the desired weight per square foot.

In essence then, Cotton Flote is produced on the same type of equipment that is used for conventional nonwoven cotton batting with the exception that the web is sprayed with thermosetting and thermoplastic resins and dried and cured after it has been laminated to the desired weight or thickness. Figure 1 is a flow diagram of the process.

The process-product patents, U.S. 3,181,225 and 3,350,486 are in the public domain, and available for license on a royalty-free nonexclusive basis within the United States.

During the research, a series of statistically designed experiments were carried out to establish the optimum chemical content of the treating media, the wet add-on needed to achieve product performance characteristics, and the ratio of thermoplastic to thermosetting resin in the formulation. Analysis of the derived data showed that product performance characteristics were best when the concentration of the chemicals in the treating media was 20% by weight and the wet add-on was 100% in terms of the weight of the cotton fibers being processed. The ratio of thermoplastic to thermosetting resin in the formulation varies with the density of the product. For example: 3 parts thermosetting to 2 parts thermoplastic resin is best for products having densities of 1.5 to 2.0 pounds per cubic foot. A ratio of 1 part thermosetting to 1 part thermoplastic resin works best for intermediate density products in the range 2.1 to 3.0 pounds per cubic foot, and a ratio of 3 parts thermoplastic to 2 parts thermosetting resin is optimum for high density products in the range of 3.1 to 6.0 pounds per cubic foot.

Where the recommended wet add-on of 100% of a 20% solids content spray solution is applied, the resulting Cotton Flote products contain about 16.6% resins by weight. Figure 2 shows some typical formulations that have been successfully used in the process.

PERFORMANCE CHARACTERISTICS

The versatility of the Cotton Flote process permits the production of a wide range of products that can be engineered to meet end-use requirements. The density can be controlled over the range of 1.5 to 6.0 pounds per cubic foot by varying (a) the web weight, (b) the speed of the lapping conveyor, (c) the number of webs that are laminated together by adjusting the speed of the floor conveyor, (d) the amount of compression exerted on the products while they are being dried and cured. The density of the product affects its ability to compress under load as well as its initial softness.

The properties of the products such as resilience, tensile strength, dimensional stability and coherence can be readily tailored by the selection of a treating formulation that emphasizes the desired characteristic. Formulations that contain more thermosetting resin than thermoplastic resin give products that exhibit rapid recovery from deformation loading, while formulations that contain more thermoplastic than thermosetting resin give products that have high-tensile strength, dimensional stability, and coherence with somewhat lower resilience.

The hand of the products can be controlled by varying the conditions of drying. Where a harsh skin effect is desired it is achieved by the use of air temperatures in excess of 200°F during the initial phases of drying. Under these conditions the resins tend to migrate to the surface of the pad and are then fixed in that position. Where a soft hand is desired, air temperatures of below 200°F are used. At the lower temperature, practically no resin migration occurs.

One measure of the performance of the products is their ability to recover from repeated compressional loading. The data shown in Table 1 was obtained by subjecting the samples to a load of 1 pound per square inch. The loading and unloading was carried through 50 cycles which has been shown statistically to differentiate sample performance. The ability of the sample to recover is measured by precisely recording its thickness immediately after completion of the 50th cycle and again after 4 minutes. Two sets of conditions were used for the resilience test, the samples were evaluated (1) at 70°F and 65% relative humidity, and (2) at 80°F and 100% relative humidity.

Table 2 shows the tensile strength of typical Cotton Flote products within the density range of 1.5 to 6.0 pounds per cubic foot. These values are much greater than those shown for conventional nonwoven cotton batting.

Cotton Flote products stand up well under use. Samples have been subjected to a cyclic pounding at the rate of 132 strokes per minute at a load of 1.5 pounds per square inch. After 30,000 compression and relaxation cycles, experimental samples in the 2.0 pound per cubic foot density range showed a loss in thickness of less than 10%.

Another measure of cushion comfort for want of a more descriptive terminology, is called the initial softness test. This characteristic is measured, or at least categorized, by the application of sufficient load to deflect the sample 25% of its original thickness. The load is applied to a 50-square-inch pressure foot which is centered on a 144-square-inch sample. Cotton Flote of 2.0

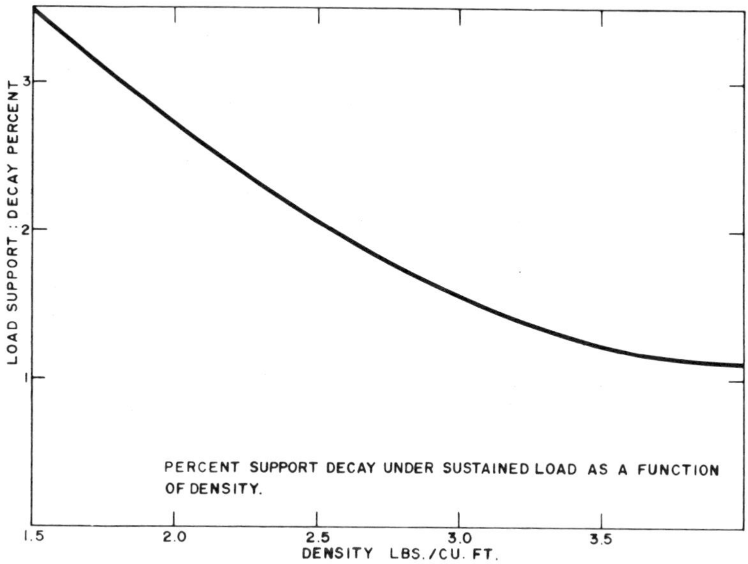

Figure 3. *Percent support decay under sustained load as a function of density.*

Table 1. Typical Performance Data on Cotton Flote Products.

SAMPLE DENSITY, LBS./CU.FT.		1.5	1.8	2.0	2.2	2.5	3.0	3.8	4.2	6.7
PERFORMANCE CHARACTERISTICS										
CYCLIC LOADING RECOVERY, 65% R.H.										
	AFTER 0"	79.1	86.5	90.3	92.0	92.3	95.0	97.2	96.6	96.0
	4"	81.2	89.1	93.2	96.1	96.3	97.0	98.0	100.0	99.5
100% R.H										
	AFTER 0"	53.7	70.2	75.0	77.0	79.2	79.5	79.4	79.3	87.5
	4"	60.0	75.6	79.2	81.2	84.0	85.0	85.0	86.2	91.7
22 HOUR HEAT SET UNDER LOAD AT 158°F										
	AFTER 30"	66.6	60.0	57.9	51.7	48.3	42.2	39.0	38.3	35.5
	120"	59.2	55.5	52.0	42.7	39.0	34.6	28.8	30.5	27.2
COMPRESSION % (1.0 LB./SQ.IN. LOAD)		70.6	66.7	64.8	59.2	52.9	46.7	43.8	40.0	32.0

Table 2. Tensile Strength of 1" Thick Cotton Flote.

DENSITY LBS./CU.FT.	MACHINE DIRECTION LBS. TO BREAK	TRANSVERSE DIRECTION LBS. TO BREAK
1.0 – 1.5	7.5	4.5
1.5 – 1.8	10.0	6.0
1.8 – 2.2	15.0	10.0
2.2 – 3.0	25.0	15.0
3.0 – 4.5	40.0	25.0
4.5 – 6.0	65.0	35.0

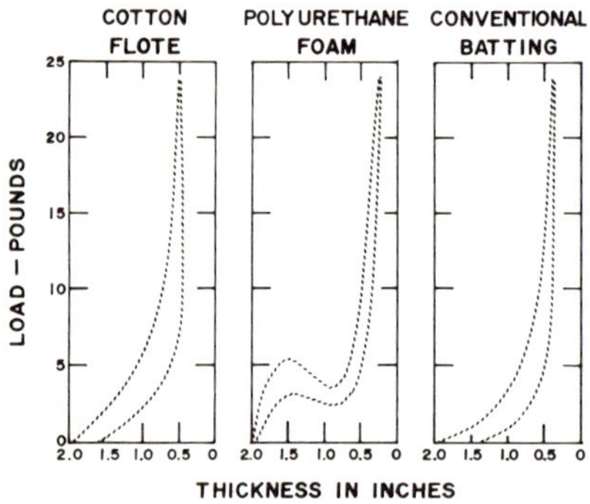

Figure 4. *Comparison of typical energy absorption and recovery characteristics of Cotton Flote, polyurethane foam, and conventional cotton batting.*

pounds per cubic foot density requires a load of 6 pounds to deflect 25%. Comparable density conventional nonwoven cotton batting will deflect 25% under a load of only 3.5 pounds.

The relaxation of cushioning materials under sustained load is also a measure of cushion comfort. To index this characteristic a 16-square-inch sample is subjected to a load of 75 pounds for 5 minutes. While the sample is loaded, the decay in the load over the time interval is measured and recorded. Typical performance data for Cotton Flote products are given in Figure 3.

Another yardstick that describes the performance of a cushioning material is the energy absorption test. It measures the amount of energy that is expended in compressively deforming a sample of cushioning material, and the energy released by the material in its attempts to resume its original configuration.

Figure 4 shows typical energy absorption curves for conventional nonwoven cotton batting, polyurethane foam, and Cotton Flote. The Cotton Flote and the conventional nonwoven cotton batting have smooth curves, whereas the polyurethane foam curve shows a deflection which is characteristic of materials that bottom out under load.

Figure 5. *Die cut Cotton Flote samples.*

SPECIAL PRODUCT CHARACTERISTICS

A number of product characteristics are achieved in addition to resilience and tensile strength. Among these special properties are dimensional stability, integrity, moldability, and flame retardance.

Figures 5 and 6 show samples of Cotton Flote that have been die-cut to selected shapes. Note from the edge view that the samples retain their preselected thickness, and from the front view that the samples retain their shape even when suspended by a corner. It is also possible to precut and shape slab stock Cotton Flote so that conforming cushions can be produced.

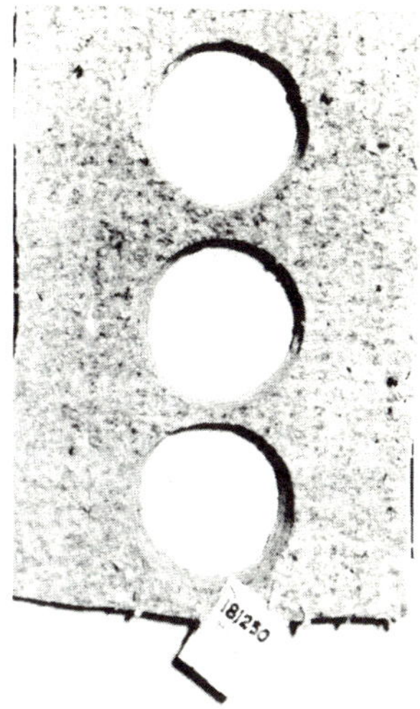

Figure 6. *Die cut Cotton Flote sample.*

The integrity and cohesiveness of the Cotton Flote products is evidenced by the fact that in terms of weight, 1/3 less Cotton Flote is needed to produce a better cushion than conventional cotton batting. Cotton Flote products do not compact in use due to fiber reorientation within the batting structure, nor do fibers or chemicals dust out.

Figure 7 shows molded Cotton Flote products. These are made by placing the spray-damp batting into the molds shown in the exploded view, then drying and curing. The products retain their resilience and conform to the shape imparted to them during manufacture. The molded products demonstrate all of the resilience and tensile strength characteristics that are found in the slab type Cotton Flote products, and in addition the molded products can have densities that vary from point to point to fit the end-use requirement.

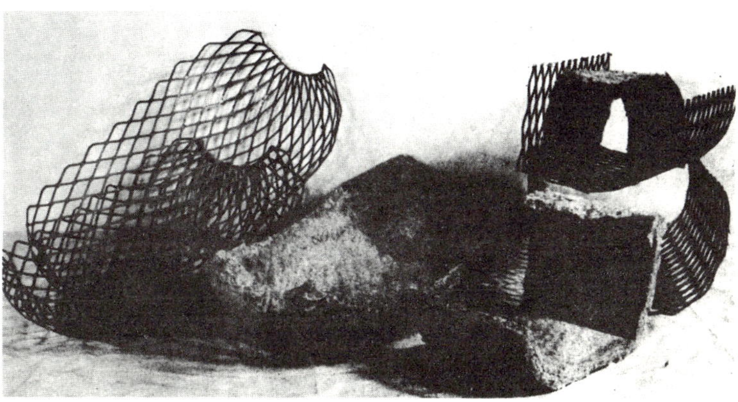

Figure 7. *Experimental samples of molded Cotton Flote products.*

Table 3. Chemical Composition.

SAMPLE	A	B	C	D	CONTROL 1	CONTROL 2
CHEMICAL COMPOSITION OF SPRAY [1]						
THERMOSETTING RESIN	METHYLATED METHYLOL MELAMINE	METHYLATED METHYLOL MELAMINE	UREA FORMALDEHYDE	METHYLATED METHYLOL MELAMINE	METHYLATED METHYLOL MELAMINE	–
% OF SOLIDS	45.2	39.7	50.0	40.0	50.0	–
THERMOPLASTIC RESIN	VINYL CHLORIDE	VINYLIDENE CHLORIDE	VINYL CHLORIDE	ACRYLIC	ACRYLATE	–
% OF SOLIDS	36.6	47.8	15.0	40.0	50.0	–
FLAME RETARDANT	UREA PHOSPHATE	ANTIMONY TRIOXIDE	UREA PHOSPHATE	UREA PHOSPHATE	–	–
% OF SOLIDS	18.2	12.5	35.0	20.0	–	–

[1] 20% SOLIDS IN SPRAY, MAGNESIUM CHLORIDE CATALYST, DICYANDIAMIDE BUFFER.

Table 4. Chemical Composition.

SAMPLE	E	F	G	H	CONTROL 1	CONTROL 2
CHEMICAL COMPOSITION OF SPRAY [1]						
THERMOSETTING RESIN	DIMETHYLOL DIHYDROXY ETHYLENEUREA	METHYLATED METHYLOL MELAMINE	METHYLATED METHYLOL MELAMINE	METHYLATED METHYLOL MELAMINE	METHYLATED METHYLOL MELAMINE	–
% OF SOLIDS	49.5	40.0	50.0	37.5	50.0	–
THERMOPLASTIC RESIN	VINYL ACRYLATE	VINYL ACRYLATE	ACRYLIC	VINYL CHLORIDE ACRYLATE	VINYL ACRYLATE	–
% OF SOLIDS	21.7	40.0	34.5	37.5	50.0	–
THERMOPLASTIC RESIN	VINYL CHLORIDE	–	–	–	–	–
% OF SOLIDS	8.6	–	–	–	–	–
FIRE RETARDANT	UREA PHOSPHATE	UREA PHOSPHATE	DIAMMONIUM PHOSPHATE + BORAX	BORATED AMIDO POLYPHOSPHATE	–	–
% OF SOLIDS	20.2	20.0	15.5	25.0	–	–
CATALYST	$ZnCl_2$	$MgCl_2$	$MgCl_2$	$MgCl_2$	$MgCl_2$	–
BUFFER	–	DCY	DCY	DCY	DCY	–

[1] 20% SOLIDS IN SPRAY.

DCY = DICYANDIAMIDE

Table 5. Performance Characteristics.

SAMPLE		A	B	C	D	CONTROL 1	CONTROL 2
PERFORMANCE CHARACTERISTICS [2]							
RECOVERY FROM CYCLIC LOADING 65% R.H.	0'	91.5	84.0	87.0	90.7	89.5	65.9
	2'	93.6	88.1	90.8	92.7	91.6	65.9
	4'	93.6	90.2	90.8	92.7	93.7	68.9
RECOVERY FROM CYCLIC LOADING 100% R.H.	0'	72.3	62.5	66.0	67.8	62.5	40.4
	2'	74.5	66.5	67.8	71.4	64.5	43.6
	4'	76.6	68.7	69.7	71.4	66.6	45.8
COMPRESSION % (1.0 LB./SQ. IN. LOAD)		61.9	66.7	65.6	63.1	64.8	78.8

[2] 2.0 LBS./CU. FT. NOMINAL DENSITY.

FLAME RETARDANCE

Recent research on the Cotton Flote process has centered upon means of improving the ability of the products to withstand damage by pyrolysis [8, 9, 10, 11, 12, 13, 14, 15, 16] while retaining all of the desirable properties that have accrued from the previous research.

Fibrous nonwovens that are used in cushioning applications are not subjected to multiple launderings, consequently the flame retardants needed, while they have to permanently confer resistance to burning, do not have to be of the so-called durable type.

Some of the resins used in the Cotton Flote process such as the methylated methylol melamines, certain acrylics, vinyl acrylate-chloride, vinyl chloride and vinylidene chloride are known to contribute some flame retardance to cotton fibers, the amount is not considered adequate for the products to be considered flame retardant. Semidurable flame retardants that seem to offer promise for Cotton Flote products include urea-phosphates, borated urea formaldehyde, antimony trioxide, halogen containing polymeric materials, borated amino polyphosphates, and phosphoric acid derivatives of dicyandiamide.

The compatibility of the flame retardant with the other resins used in the production of Cotton Flote is of primary concern. Tables 3 and 4 show some typical resin-flame retardant formulations used to impart enhanced resilience, dimensional stability, tensile strength, integrity, and flame retardance to nominal 2 lb./ft.^{3} Cotton Flote products.

From Tables 3 and 4 it can be seen that the flame retardant concentration varied from as low as 12.5% of the total solids in the formulation to as high as 35%. Converting these figures to dry add-on based upon a wet add-on of 100% gives values of 3.5% to about 6% by weight of flame retardant in the finished product. Control 1 is Cotton Flote without the addition of a flame retardant; Control 2 is conventional untreated nonwoven cotton batting.

As can be seen from Tables 5 and 6, the addition of the flame retardant to the formulation does not significantly affect the resilience performance of

Table 6. Performance Characteristics.

SAMPLE		E	F	G	H	CONTROL 1	CONTROL 2
PERFORMANCE CHARACTERISTICS [2/]							
RECOVERY FROM	0'	88.3	88.1	88.5	89.0	89.5	68.9
CYCLIC LOADING	2'	92.0	90.8	90.5	93.7	91.6	68.9
65% R.H.	4'	92.0	92.3	90.5	95.3	93.7	70.0
RECOVERY FROM	0'	77.0	65.5	69.8	65.0	62.5	43.4
CYCLIC LOADING	2'	80.8	69.0	71.5	71.5	64.5	45.6
100% R.H.	4'	82.5	69.0	73.4	73.0	66.6	48.8
COMPRESSION % (1.0 LB./SQ. IN. LOAD)		62.1	60.1	62.9	61.6	64.8	78.9

[2/] 2.0 LBS./CU. FT. NOMINAL DENSITY.

the products. All of the Cotton Flote products including the control had recoveries from cyclic compressive loading characteristics that compare favorably with those of Cotton Flote products that do not contain a flame retardant.

Tables 7, 8, 9, and 10 show the flame retardance behavior of the samples whose chemical and physical properties were given in Tables 3 through 6. Samples were tested after exposure to four environmental conditions:

1. One hundred percent relative humidity for three days at 100°F.
2. Sixty-five percent relative humidity for 24 hours at 70°F.
3. One hundred and fifty-eight degrees Fahrenheit at ambient humidity for 16 hours.
4. In addition, the samples were incrementally wetted with water until they were completely soaked, then they were dried at 158°F and tested for flame retardancy.

A modification of the AATCC Test Method 34-1964 (vertical) was used to evaluate the effectiveness of the flame retardant treatments.

Figure 8 shows the acceptable limits used for screening research products for flame retardance propertties.

Table 7. Flame Retardance Characteristics.

SAMPLE	A	B	C	D	CONTROL 1	CONTROL 2
CONDITIONED 70°F – 65% R.H.						
OVERALL CHAR (CM.)	5.7	5.8	2.2	5.2	11.7	B.E.L.
MAIN CHAR (CM.)	3.5	3.6	1.0	3.0	7.5	"
SOAK WITH H_2O – DRIED 158°F						
OVERALL CHAR (CM.)	4.0	3.0	4.0	5.0	11.1	"
MAIN CHAR (CM.)	2.0	1.3	1.0	2.0	8.5	"
100% R.H. – 3 DAYS – 100°F						
OVERALL CHAR (CM.)	3.0	2.8	3.0	3.0	12.8	"
MAIN CHAR (CM.)	1.0	1.8	1.0	1.5	8.2	"

Table 8. Flame Retardance Characteristics.

SAMPLE	A	B	C	D	CONTROL 1	CONTROL 2
158°F – 16 HOURS						
OVERALL CHAR (CM.)	4.0	7.0	5.1	6.0	16.0	B.E.L.
MAIN CHAR (CM.)	1.0	3.0	1.0	1.0	6.0	"
AFTER 7 MONTHS STORAGE 1/						
OVERALL CHAR (CM.)	3.5	5.0	2.0	4.0	10.5	"
MAIN CHAR (CM.)	1.5	2.0	1.0	1.0	5.5	"

1/ CONDITIONED 70°F – 65% R.H. BEFORE TESTING.

In general, as the density decreases from 6.7 to 2.1 pounds per cubic foot, the amount of flame retardant needed increased from 3% by weight to 8 or 9% by weight to make the product acceptably resistant to burning.

The addition of a properly selected flame retardant to the treating formulation does not adversely affect the spray ability or the pot-life of the formulation.

MINIMUM LEVEL OF FLAME RETARDANCE	
AFTERFLAME TIME	5 SECONDS
AFTERGLOW TIME	5 SECONDS
MAIN CHAR LENGTH	10 CM.(4 IN.)
OVERALL CHAR LENGTH	15 CM.(6 IN.)

Figure 8. *Acceptable limits used in screening products for flame retardance.*

SUMMARY

The Cotton Flote nonwoven process is extremely versatile. Products having densities of 1.5 to 6.0 pounds per cubic foot are possible. The selection of the formulation permits the production of products have various degrees of resilience, tensile strength and flame retardance. The dimensional stability and integrity can be modified to meet the specifications of the end-use.

The products can be die-cut to fit any desired two-dimensional configuration, and molded to meet any three-dimensional shape.

Table 9. Flame Retardance Characteristics.

SAMPLE	E	F	G	H	CONTROL 1	CONTROL 2
CONDITIONED 70°F — 65% R.H.						
OVERALL CHAR (CM.)	5.8	5.3	5.0	2.5	11.7	B.E.L.
MAIN CHAR (CM.)	2.7	3.5	2.3	1.0	7.5	"
SOAK WITH H_2O — DRIED 158°F						
OVERALL CHAR (CM.)	5.0	4.5	8.7	2.5	11.1	'
MAIN CHAR (CM.)	2.5	3.0	6.0	1.2	8.5	"
100% R.H. — 3 DAYS — 100°F						
OVERALL CHAR (CM.)	5.1	2.7	5.3	2.0	12.8	"
MAIN CHAR (CM.)	1.8	1.3	3.8	1.1	8.2	"

Table 10. Flame Retardance Characteristics.

SAMPLE	E	F	G	H	CONTROL 1	CONTROL 2
158°F — 16 HOURS						
OVERALL CHAR (CM.)	9.0	5.5	9.0	2.5	16.0	"
MAIN CHAR (CM.)	3.0	2.0	2.0	1.0	6.0	"
AFTER 7 MONTHS STORAGE [1]						
OVERALL CHAR (CM.)	5.0	5.0	5.5	—	10.5	"
MAIN CHAR (CM.)	2.0	2.0	2.8	—	5.5	"

[1] CONDITIONED 70°F — 65% R.H. BEFORE TESTING.

REFERENCES

1. N. B. Knoepfler, H. K. Gardner, Jr., H. L. E. Vix and E. L. Patton, "Cotton Research Aims at Recapturing Batting Markets," *Bedding 84* (3): 40-44, 46, 48, 50 (1963).
2. N. B. Knoepfler and H. L. E. Vix, "Recapturing Cotton Batting Markets by Chemical Engineering Research," *The Cotton Gin and Oil Mill Press 65*(20): 28-32 (1964).
3. N. B. Knoepfler, "The Development of Cotton Flote, a New Treated Cotton Batting," Agr. Res. Serv. Pub. ARS 72-31 (1964).
4. N. B. Knoepfler and H. L. E. Vix, "Cotton Batting Improved by Resin and Latex Application," *Oil Mill Gazetteer 69*(5): 26-31 (1964).
5. N. B. Knoepfler and P. A. Koenig, "Progress in the Commercialization of Improved Cotton Batting Through Developmental Research," *The Cotton Gin and Oil Mill Press 67*(8): 7, 8, 10 (1966).
6. L. L. Holzenthal, C. H. Haydel, and N. B. Knoepfler, "Drying and Curing Chemically Treated Cotton Batting," Agr. Res. Serv. Pub. ARS 72-45 (1966).
7. N. B. Knoepfler, L. L. Holzenthal, P. A. Koenig, and H. L. E. Vix, "Engineering Aspects of the Production and Performance of Cotton Flote Products," *American Dyestuff Reporter 57*(11): 394-939 (1968).
8. N. B. Knoepfler, P. A. Koenig, and W. T. Gentry, Jr., "Study Shows How Cotton Flote Retards Fire," *Bedding* 94(1): 78, 80, 82 (1968).
9. N. B. Knoepfler, P. A. Koenig, and W. T. Gentry, Jr., "Flame Resistant Cotton Flote Products for Furniture and Mattresses," *Western Furniture Manufacturing 23*(12): 17-19 (1968).
10. N. B. Knoepfler, P. A. Koenig, and H. L. E. Vix, "Flame Retardant Cotton Flote Products for Automobile Seats, Mattresses and Furniture," *Oil Mill Gazetteer 73*(8): 9-13 (1969).
11. P. A. Koenig and N. B. Knoepfler, "Flame Retardant Cotton Batting Products," *American Dyestuff Reporter 58*(17): 30-34, 36 (1969).
12. P. A. Koenig, "Flame Retardant Cotton Flote," Proc. Flame Retardant Cotton Batting Workshop, New Orleans, La., November 15, 1968: Agr. Res. Serv. Pub. ARS 72-72 (Oct. 1969) pp 29-39.
13. N. B. Knoepfler, "Flame Retardant Cotton Batting," Proc. Eighteenth Cottonseed Processing Clinic, New Orleans, La., Feb. 3-4, 1969: Agr. Res. Serv. Pub. ARS 72-74 (Sept. 1969) pp 74-85.
14. N. B. Knoepfler, "New Flame Retardant Cotton Batting Products," Proc. Conf. of Collaborators Sou. Agr. Expt. Sta., New Orleans, La., March 24-25, 1969: Agr. Res. Serv. Pub. ARS 72-76 (Dec. 1969) pp 63-74.
15. N. B. Knoepfler, "Flame Retardant Multiple Density Cotton Flote Products for Furniture and Mattresses," Agr. Res. Serv. Pub. ARS 72-86 (Oct. 1970) 8 pp.
16. P. A. Koenig and N. B. Knoepfler, "New Flame Retardant Non-Woven Cotton Products for Cushioning Applications," *Matls. Eng. and Sci. Div. Am. Inst. Chem. Engrs. Biennial Conf. Proc. Preprint* (1970) pp 339-346.

N. B. Knoepfler

Nestor B. Knoepfler is Head of Cotton Products Investigations of the Engineering and Development Laboratory in the Southern Marketing & Nutrition Research Division of the U.S. Department of Agriculture. He received his B.E. degree in Chemical Engineering from Tulane University in 1940. He served in the U.S. Navy from 1941 to 1946 and was employed by Southern Photo Craft from 1946 to 1950. He was a National Cottonseed Products Association Fellow at the Southern Utilization Research and Development Division from 1950 to 1952 and joined the staff in 1952. His experience includes cotton and cotton linters, cotton batting products, and oilseed processing and process development.

Y. UEHARA

Faculty of Engineering, Yokohama National University, 31-1, Ohka 2, Minami-ku, Yokohama, Japan

E. YANAI

Fire Research Institute, 14-1, Nakahara 3, Mitaka, Tokyo, Japan

SMOKE, CO AND CO_2 FROM CELLULOSIC MATERIALS TREATED WITH DAP IN A FIRE ENVIRONMENT

(Received May 8, 1972)

ABSTRACT: Smoke, CO and CO_2 produced from cellulosic materials treated with DAP were examined using four types of flammability testing apparatus. As the treatment concentration became higher, the emission of smoke and CO were either reduced or increased, depending on the intensities of the heat sources. It seems that this inversion occurred at the limit of fire retardant ability of materials. Hence it is recommended to use the apparatus with high and low intensities of heat sources for a more complete flammability test.

INTRODUCTION

FIRE RETARDING OF flammable materials is very useful for preventing fire or suppressing flame propagation. The fire-retardant treatment offers a number of advantages, but it involves inherently a risk of increasing the emission of smoke and other toxic gases in the event of a fire. Thus an important problem to resolve is how to reduce such emissions [1]. The smoke and toxic gases produced in a fire environment are governed by the kinds of raw materials, and the fire retardants used, and also by such factors as the magnitude of fire source, the ambient temperature and the oxygen concentration in the environment. For an analysis of the conditions of generation of smoke and toxic gases from materials rendered fire retardant, all these factors should be taken into consideration; however they are related with one another in a complex manner so that they cannot be evaluated easily.

There are a number of measurements reported of the smoke and gases from materials treated with fire retardants as well as those untreated. These results indicate that the treatment tends to produce a greater amount of smoke and gases [2-4]. However, such measurements are for the most part concerned with commodities and those carried out systematically with variations in the kind and concentration of fire retardants are rather few. Among these works, Saito [5] and Eickner et al [6] described the smokes produced from wood or plywood materials treated with inorganic salts as they were burnt in an electric furnace or 8-ft. tunnel furnace. According to these results, the smoke level varies in a complex manner depending on the kind and concentration of fire retardant and the burning (pyrolysis) temperature; and thus it cannot be described consistently throughout its process of generation.

According to Saito, the retardant, diammonium phosphate, which we shall discuss, reduces the smoke level at low exposure temperatures but increases it at high temperatures with increasing concentration of treatment. Eickner et al [6] found that as the treatment concentration increases, the smoke level would decline at first but then increases sharply. Also, Mickelson et al [7] and DiPietro et al [8] have reported with respect to synthetic polymeric materials treated with increasing amounts of bromine-containing fire retardants that the smoke level tends to increase with increasing treatment concentration. These results were obtained under predetermined burning conditions and may show the variation in the generation of smoke depending on the kind and amount of retardant used; however they fail to show what influences the burning conditions encountered in an actual fire may have on the smoke level.

The present paper examines this problem by fixing the kinds of polymeric materials and retardants applied thereto, and the oxygen content in the atmosphere, and by measuring the emission of smoke and gases from the materials as affected by the retardant concentration and also the magnitude of the source of fire.

EXPERIMENTAL

Test Specimens

As test specimens of cellulosic materials, an undyed rayon material (Japanese Industrial Standard L-0803, 83 g/m^2) and an undyed canvas material (No. 3, 830 g/cm^2), both treated with diammonium phosphate (DAP), were used. For treatment, the test material was immersed in a DAP aqueous solution of predetermined concentration for 24 hours, then centrifuged and air dried in a room. Thereafter, it was dried in an electric dryer kept at 50 ± 2°C for 24 hours and then allowed to cool for more than 2 hours in a desiccator containing silica gel. The combination of cellulosic material and DAP was chosen since it has a long history of use; it is presently used as one of the best finishing processes; and it is readily prepared in ordinary laboratories.

The concentration of the treatment and the deposition of the ingredient are approximately in a linear relationship, but under the same treatment concentration the canvas material contains less DAP than the rayon material. The fire retardance achieved is shown in Figure 1 as a function of the treatment concentration by the Limited Oxygen Index Test (LOI). As can be seen, the rayon material yields an LOI value up to 55, while the canvas material gives a value of only 43 with a saturated solution of the reagent and generally lower values at lower concentrations of the reagent, than those of the rayon material.

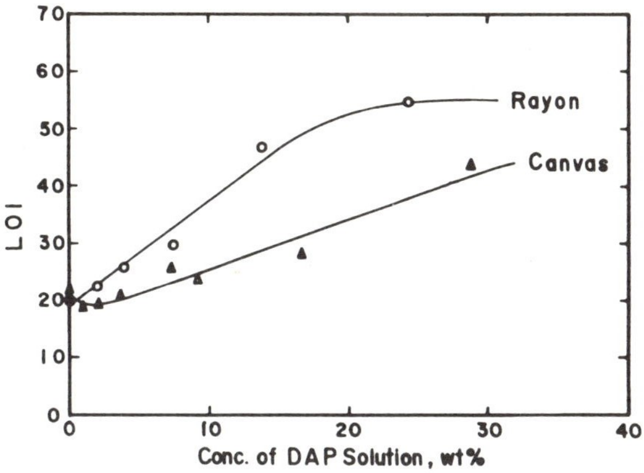

Figure 1. *Relative effectiveness of DAP in rayon and canvas.*

Flammability Test Apparatus and Procedure

The complexity of the aspects of a fire is apparent from the fact that there are a number of flammability test methods available for reproducing them. The present fire retardant treatment for textiles is designed for resisting a small fire source and delaying the time until flashover. But, where smoke and gases are to be handled, it is necessary to consider the influences on the other rooms after the flashover. Thus flammability test apparatus of the following four types were used to correspond to such various stages of fire as much as possible.

1. 45 Degrees Method.—This is a method employing an apparatus designed for measurement of the flammability of thin cellulosic materials against a small fire source, and in Japan, authorized measurements under the Fire Protection Law are made by this method. The apparatus is shown in Figure 2. A test specimen of 35 by 25 cm is supported by a holder and is held in the apparatus at an angle of 45 degrees. Upon heating for 1 minute with a micro-burner using propane as fuel, the material is acceptable if it produces a charred area less than 30 cm^2; and it is unacceptable if the charred area is 30 cm^2 or more. In the present study, for the purpose of measuring the smoke and gases, the smoke passage had a smoke level meter inserted and a gas sampling tube provided. Further, the temperature and velocity of the air stream passing through the smoke passage were also measured.

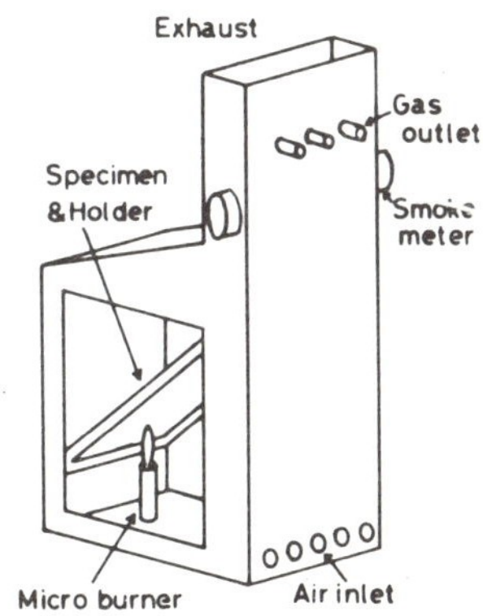

Figure 2. *Apparatus for 45 degrees method.*

2. Electric Furnace Method.—According to this method, a sample material (rayon) of about 1 gram is subjected to pyrolysis or combustion in an electric furnace (having an outer diameter of 19 cm, a height of 33 cm and a sample inlet port of 3 cm diameter) contained in a box and heated to a predetermined temperature; the thermogravimetric change of the sample is observed, while the smoke and gases collected in a smoke box (70 X 70 X 70 cm and a capacity of 0.28 m^3) are measured. The method permits observation of the burning behavior of a cellulosic product in various stages of a fire from smoldering to flashover and after by changing the furnace temperature from low to high. The smoke level meter was placed in the smoke box, but the sampling of gases was made externally as in the case of the method 1. The test apparatus is shown schematically in Figure 3.

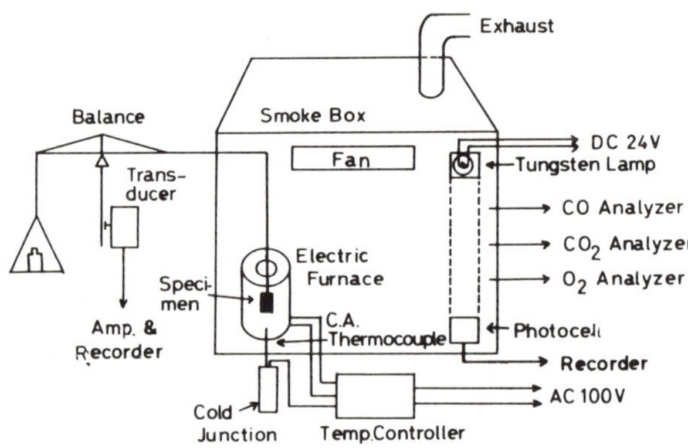

Figure 3. *Apparatus for electric furnace method.*

3. Large Vertical Furnace Method.—The foregoing method 2 is applicable to burning at high temperatures, but it destroys the sample rather quickly. Therefore, to observe the burning behavior, with particular attention to the propagation of flame upon flashover, a large vertical furnace containing a high heat source was fabricated. The apparatus is a modification of the specifications of DIN 4102 to suit to our laboratory condition, as shown in Figure 4 [8]. Four-test specimens of a size of 20 by 100 cm were used. They were spread in a holder respectively in such a manner as to face one another, kept in equilibrium by a balance and burnt from the bottom with city gas (5000 Kcal/m^3) fed to a gas burner at a rate of 7 m^3/hr. along with air fed from a blower at a rate of 30 m^3/min. The temperature within the furnace was measured at six points, and the smoke and gas concentrations were measured by sampling from the smoke passage. Here, only the canvas sample was used.

4. LOI Method.—The LOI method was initially developed by Fenimore et al

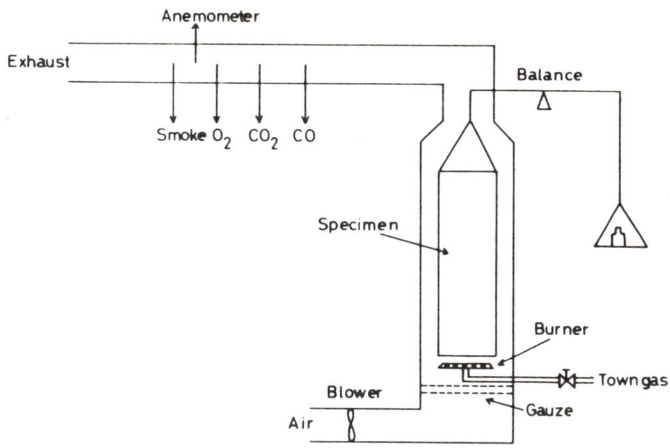

Figure 4. *Large vertical furnace.*

[10] and, through a number of research studies, has come to be used extensively as a flammability test method [11]. In the present study, the LOI apparatus was modified to incorporate a method developed by Uehara and Suga [12] to permit measurement of smoke levels. The apparatus is shown in Figure 5. The smoke level meter is mounted on an extension of the tube, and in this respect the present method differs from the method employed by DiPietro et al [8] for measurement of the smoke level under the same LOI method.

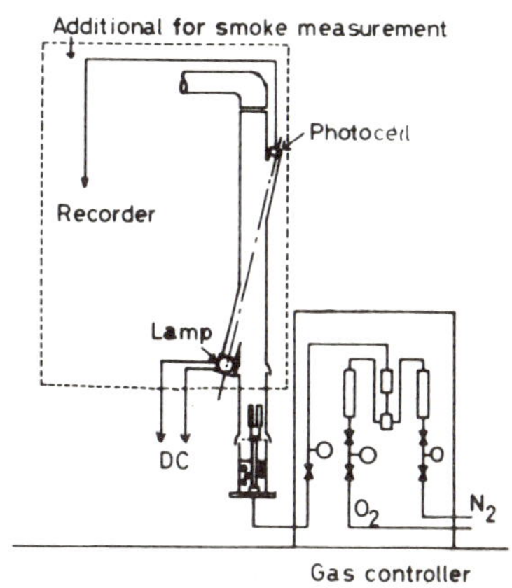

Figure 5. *Apparatus for LOI method.*

The methods described in the foregoing subparagraphs 1 through 3 are concerned with combustion under conditions where the air supply is adequate. In this respect, the LOI method may not reflect an actual fire environment and thus may present a problem as a method for comparative study. However, the results of measurements so far conducted on a variety of polymeric materials indicate clearly that the smoke condition represented is in agreement with that obtained by combustion at high temperatures [12]. The LOI method permits measurements with ease as compared with the electric or vertical furnace method so that it was employed in order to confirm the results provided by the other methods. It is also evident that the smoke level measurement by this method has desirable features in that the response is quick, that the reproducibility is excellent and that upon clearance of the smoke, the transmissivity returns instantaneously to the original value of 100 percent [12]. In the present study, this method was applied to the rayon only, and no gas analysis was carried out.

In the test procedure, a specimen of 6 by 15 cm was placed on a holder, and was burnt in a concentration of oxygen corresponding to LOI and LOI ± 0.5% for the particular sample to measure the smoke level and examine the relationship between the smoke level and the treatment concentration. Further, with the treatment concentration fixed, the oxygen concentration was increased to determine the change in the smoke level.

Measurement of Smoke and Gas Concentrations

For the determination of the smoke level in these various tests, a system utilizing the attenuation of a beam of light was employed, using a tungsten lamp as a source of light and a selenium photocell for reception. The length of the light path was changed to cope with the size of the apparatus employed, that is, it was 25 cm for the 45-degree method, 50 cm for the electric furnace and LOI method, and 100 cm for the vertical furnace. For representing the present experiments, the maximum value of the attenuation coefficient or the minimum value of transmittance and the total smoke concentration were regarded to be approximately in parallel relationship.

The concentrations of carbon monoxide and carbon dioxide gases were determined by a non-dispersion type infrared analyzer, made by Horiba Seisakusho, model LIA-2. For monitoring the oxygen content in the environment, an oxygen analyzer of the type of oxygen concentration cell, a product of Nippon Insulators Mfg. Co., model KS-1202, was used.

The temperature was measured by a 0.3 mm diameter copper-constantan or chromel-alumel thermocouple, and the air flow through the smoke channel was measured by a hot wire anemometer, product of Kokusai Dengyo Co., with automatic temperature compensation up to 300°C.

RESULTS AND DISCUSSIONS

45 Degree Method

Figure 6 indicates curves of the concentrations of smoke and gases changing with time as produced from burning of the rayon material having a DAP deposition of 0.9 percent. As can be seen, the smoke and gases increase or decrease simultaneously, demonstrating the features as the differential values.

The values of charred area, total smoke concentration and total gas concentration obtained from these experiments are shown in Figures 7 and 8. It will be noted from Figure 7 that in burning of the rayon, the charred area clearly decreases with increasing treatment concentration, accompanied by a decrease of CO as well as CO_2. It is natural that a smaller amount of combustion produces smaller amounts

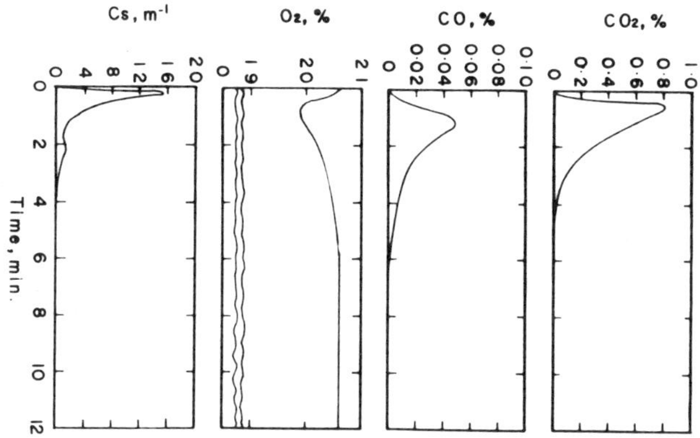

Figure 6. *Smoke and gases emitted from treated rayon vs time curve in 45 degrees method (DAP deposit; 0.9%).*

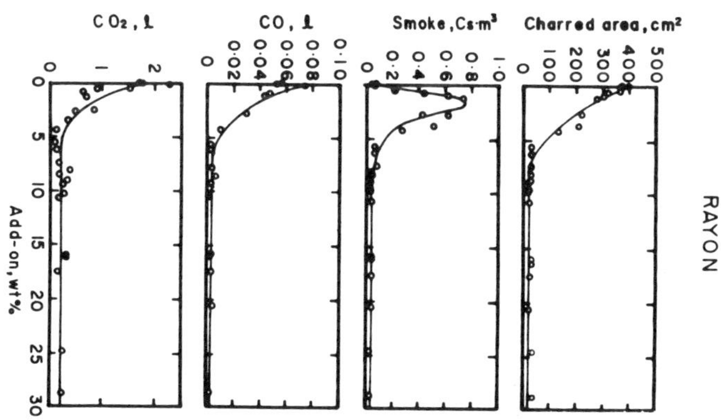

Figure 7. *Dependence of total smoke and gases on DAP deposit, (rayon, 45 degrees method).*

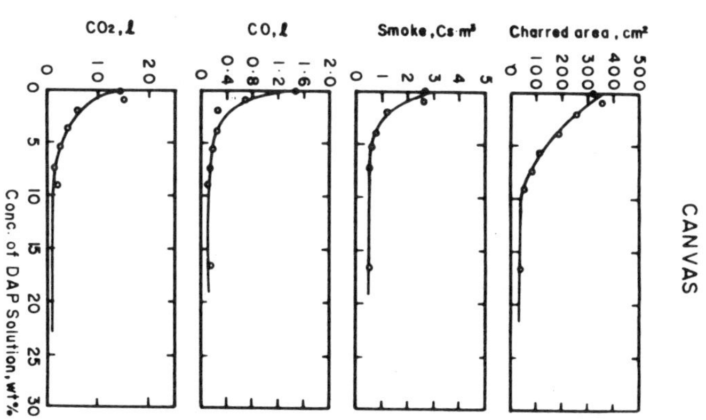

Figure 8. *Dependence of total smoke and gases on DAP deposit, (canvas, 45 degrees method).*

of gases. However, the smoke level increases then declines with increasing treatment concentration. The treatment concentration giving a peak value of smoke level is substantially noneffective in making the rayon fire retardant. It is considered that at this concentration, the retardant is scarcely able to suppress the burning so that the material burns fast, producing pyrolytic products of the retardant and resulting in an increase of the smoke. This is proved by the fact that upon achievement of the fire retardation, the smoke emission is small.

The results with the canvas are shown in Figure 8. Here, the emissions of smoke and gases decrease with increasing retardant concentration, the maximum emissions being given by the untreated material. From such different results provided by the materials of the same cellulosic nature, it seems that this method is not suitable for materials of thick texture like the canvas in the present case particularly with respect to the size or intensity of the source of heat and the inflow of air. In fact, this test apparatus is hardly able to burn the untreated canvas material.

Electric Furnace Method

The results by this method are given in Figures 9 through 13. Figure 9 shows the curves of concentrations of smoke and gases as a function of time upon pyrolysis at 600°C of the rayon treated with a saturated solution of DAP. With a sharp decrease of sample weight, the smoke increased rapidly, but the gases increased slowly. Here,

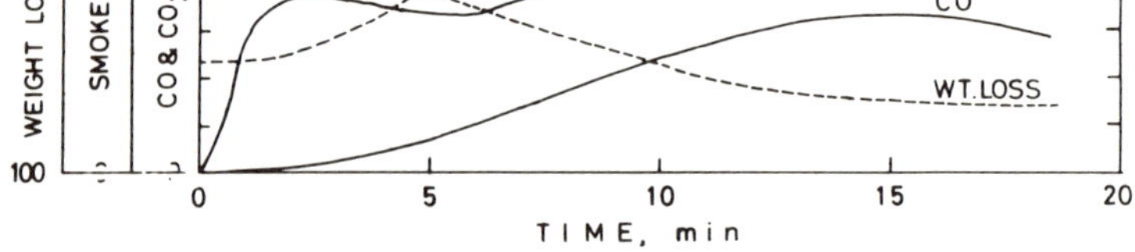

Figure 9. *Pyrolysis curve of rayon treated with saturated DAP solution (electric furnace method).*

a system of collecting the smoke in a smoke box or the so-called integral system was employed so that the response is by no means fast. Further, the smoke once collected in the smoke box should never decrease, but in fact it decreased as the smoke concentration curve indicates. This may be explained by condensation following agitation of the smoke. It must be noted that the treatment with a concentrated solution of DAP is characteristic of producing smoke in two stages.

This seems to correspond to the first sharp weight loss by flaming or rapid pyrolysis, followed by the gradual weight loss by glowing of carboneous materials. With respect to the gases, such a phenomenon was not observed.

Figure 10 shows the relationship between the smoke level (maximum value) and the treatment concentration plotted with the temperatures as parameters. As can be seen, the curves show, with a few exceptions, that higher treatment concentrations

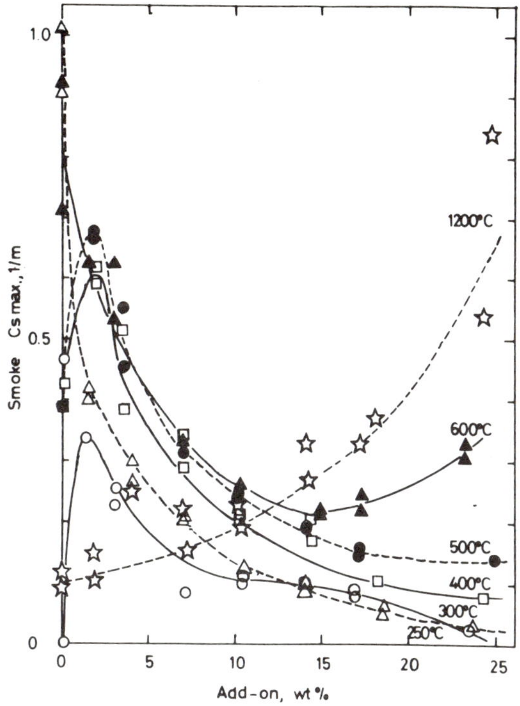

Figure 10. ***Dependence of smoke concentration on DAP deposit (electric furnace method).***

result in smaller smoke concentrations. However some interesting characteristics are apparent.

First, at 250°C, the untreated rayon is substantially entirely free of smoke. But, with the treated rayon, the smoke increases with increasing treatment concentration, having a peak at 2.5 to 5.0 percent retardant. The reason is, as is well known, that while the untreated material undergoes little pyrolysis at this temperature, the treated material starts to decompose at a lower temperature. Further, the rayon with an insufficient concentration of treatment decomposes to 80 percent finally at such temperature, and the rate of pyrolysis is fast, whereas the material with a sufficient treatment undergoes less pyrolysis and at a smaller rate, thus producing the foregoing result.

The temperature at which the untreated rayon shows a maximum smoke level is 300°C. At this temperature, the cellulose begins to decompose [13] and smolder but is still short of flaming. At 300°C, less smoke is produced with higher treatment concentrations. This trend of curves is in agreement with the result reported by Saito [5]. At 400 to 500°C, the curve of smoke concentration is similar to that at 250°C, but the absolute value is higher. At 600°C, the curve again has a trend similar to that of the curve at 300°C. In the temperature range of 250°C to 600°C, the amount of smoke increases generally as the temperature becomes higher in that the rate of pyrolysis increases with temperature. Within this temperature range, the relationship between the smoke level and the treatment concentration is not similar to that reported by Saito [5].

At 1200°C, the aspect is entirely different, the smoke increasing greatly with increasing treatment concentration. This occurs as the rate of pyrolysis becomes very high at such a temperature on the one hand and the residue of about 15 percent of the material effected by a high treatment concentration at lower temperature is no longer preserved on the other. That is, the material rendered fire retardant is forced to burn, thus resulting in a great amount of smoke. It is of interest to speculate at what temperature the curve given at 600°C tends to convert to a curve such as given at 1200°C. Unfortunately, the furnace had a failure so that it was not possible to determine this with the same furnace. However, according to the results of tests carried out with an electric furnace of the same type having a slightly greater sample inlet port (5.5 cm diameter) as shown in Figure 11, such conversion of the trend occurs at a temperature between 400°C and 600°C. This indicates that there is certainly a transition from the state in which the increasing treatment concentration effects a reduction of smoke to the state in which the increasing treatment concentration effects not a reduction but an increase of smoke. This fact is significant in considering the smoke emitted from cellulosic materials.

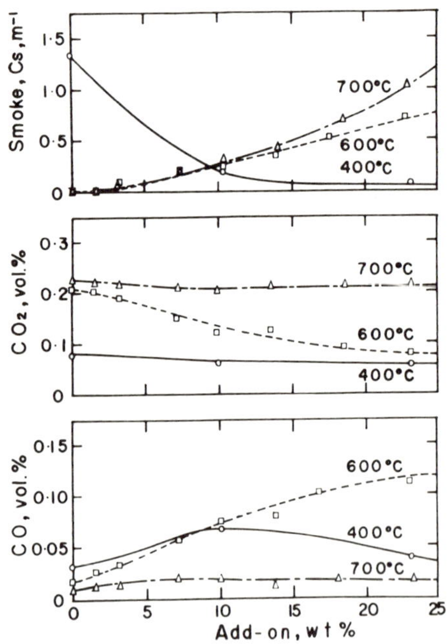

Figure 11. *Dependence of smoke and gases on DAP deposit (electric furnace method with sample inlet port of 5.5 cm diameter).*

Now referring to the curves of emission of carbon monoxide, at temperatures within the range of 250°C to 600°C, (Figure 12) the emission is small with the

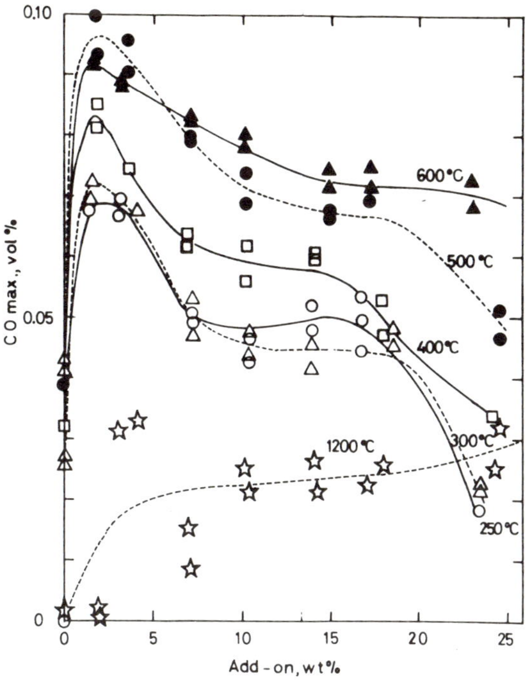

Figure 12. *Dependence of CO concentreation on DAP deposit (electric furnace method).*

untreated material, rises sharply with the treatment to give a peak at a deposition of 2.5 to 5 percent of the retardant, and then declines as the treatment concentration increases. These curves indicate that the materials having insufficient fire retardant treatment undergo incomplete burning and produce a greater amount of gas than those untreated or the treated sufficiently. Within this temperature range, the emission of carbon monoxide increases apparently with elevation of the temperature. In other words, it may be said that the treated materials have some resistance to burning.

With the furnace temperature at 1200°C, the trend is reversed as in the case of the smoke, the emission of carbon monoxide increasing with increasing treatment concentration. However, the emission is much smaller than in the case of the temperature range of 250°C to 600°C.

Lastly, the results obtained with respect to carbon dioxide are shown in Figure 13. At 250°C, the untreated material undergoes no pyrolysis so that there is no emission of carbon dioxide as in the cases of smoke and carbon monoxide, but with the treated materials, the gas is emitted approximately to the same degree regardless of the extent of treatment. At 300°C, the untreated material produces the largest amount of carbon dioxide, while with the treated material, the emission has a

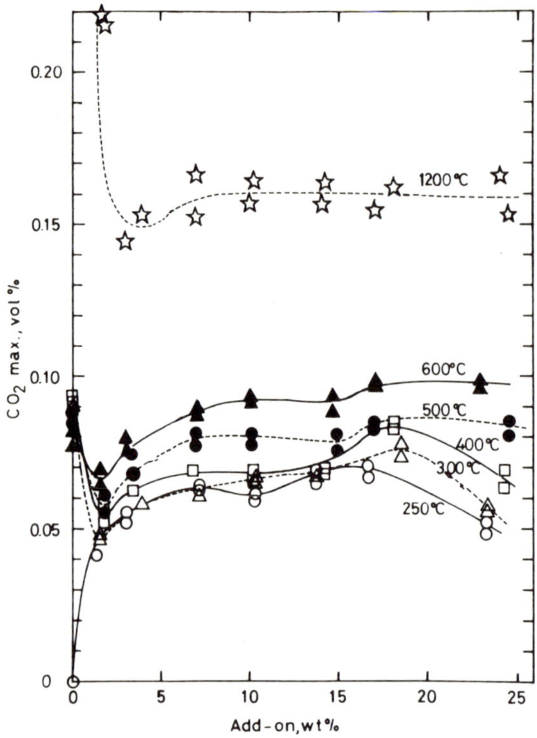

Figure 13. *Dependence of CO_2 concentration on DAP deposit (electric furnace method).*

minimum value at a retardant concentration of 2.5 percent, and then increases to the same level regardless of the treatment concentration. This trend prevails up to about 600°C, with higher temperatures inducing greater emission of carbon dioxide generally. At 1200°C, the heat exceeds the limit of whatever fire retarding treatment, and the materials burn at an accelerated rate so that the maximum value of carbon dioxide becomes exceedingly great. At this temperature, the greatest emission is from the untreated material, and there is no substantial difference observed between the emissions from the materials treated with a critical concentration of retardant and above, this being approximately 60 percent of the emission from the untreated material. The same trend is observed in Figure 11, the emission of carbon dioxide increasing with increasing temperature.

Large Vertical Furnace Method

Measurements by this method are shown in Figure 14. This represents the results of burning the canvas material treated with a 28.6 percent aqueous solution of DAP. The smoke presents a complex curve, but the gases have only one peak or valley, as the case may be.

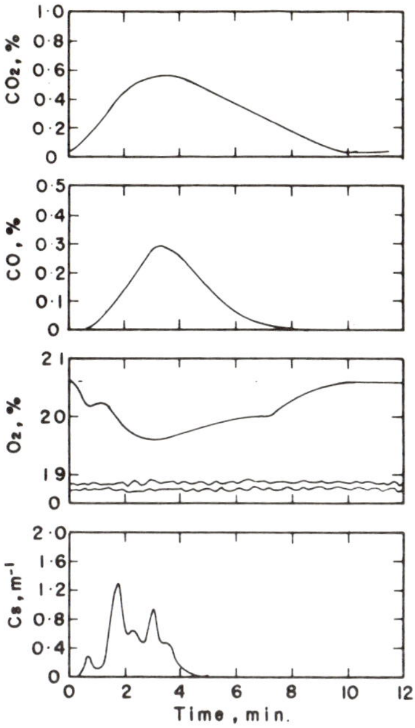

Figure 14. *Smoke and gases emitted from canvas treated with 28.6% DAP solution vs time curve (vertical furnace method).*

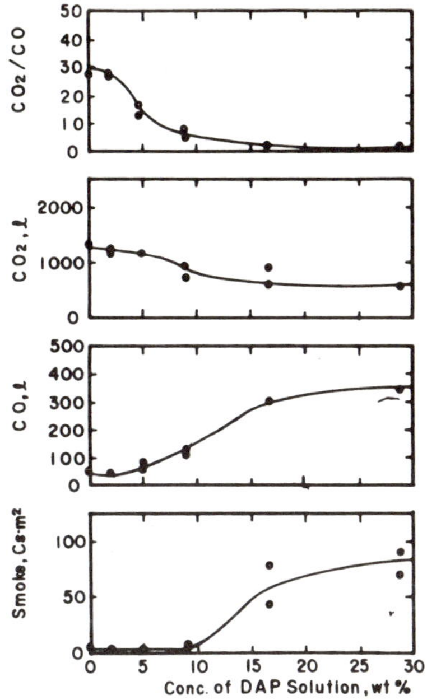

Figure 15. *Dependence of total smoke and gases on concentration of DAP solution (vertical furnace method).*

The results are summarized in Figure 15 in which the smoke and gases are indicated in the total amounts respectively. The emission of carbon dioxide decreases, although gradually, with increasing treatment concentration. This trend is the same with the result described in the foregoing. The emission of carbon monoxide increases definitely with increasing treatment concentration. This trend is reverse to the results obtained by the 45-degree method and in the temperature range of 250°C to 600°C under the electric furnace method. Rather, it is similar to the trend provided at 1200°C under the electric furnace procedure. Such a trend is also true with the smoke. These results indicate that when a treated material is exposed to a strong fire, it may exhibit an effect of suppressing the combustion as seen from the ratio of emissions of carbon dioxide and carbon monoxide, but it produces an increasing amount of smoke and carbon monoxide. In fact, when a treated material is burnt in this furnace, it burns out or chars completely in spite of the effort towards nonflammability. This result is inherently expectable of cellulosic materials treated with fire retardants and presents problems involving the limit for nonflammability treatment and emission of smoke and carbon monoxide when such limit is exceeded. Thus it seems to imply a fate of the treated materials when exposed to a big fire.

LOI Method

The tests by this method were carried out in order to confirm the results obtained at high temperatures in the electric furnace tests and those with the large vertical furnace. Figure 16 shows the relationship between the time and the smoke level from the data obtained when the test specimens are burnt in environments

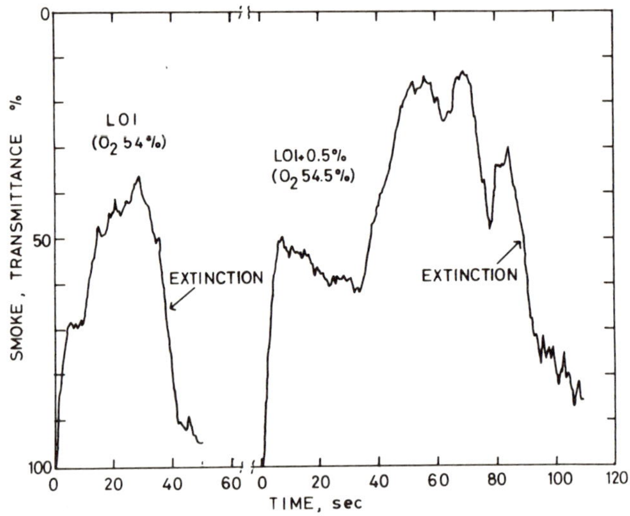

Figure 16. ***Smoke concentration vs time curve in LOI method.***

containing oxygen at concentrations corresponding to LOI AND LOI + 0.5% for the respective specimens. These two concentrations were chosen in that in the former the combustion occurs in a form close to self-extinguishing, while in the latter the materials burn completely to the last. In fact, in an oxygen content corresponding to LOI, the materials continue to burn for only about 40 seconds, but with additional 0.5 percent oxygen, they burn completely and produce much smoke. The results here indicated are for the rayon having a retardant deposition of 28.6 percent. As can be seen, the way in which smoke is produced is very complex similar to the cases of the 600°C tests in the electric furnace and the vertical furnace. According to Figure 17 which shows the results summarized, the materials burnt in the LOI oxygen concentration scarcely emit smoke up to nearly 10 percent of the treatment concentration, but where this treatment concentration is exceeded, the smoke level increases sharply. At the LOI + 0.5% oxygen concentration, the smoke emission follows a similar curve, the difference being a little faster rise and generally to a higher level. It appears that the greater the treatment concentration the higher the smoke level due probably to the treated materials being forced to burn.

In Figure 18 are shown the maximum smoke levels and burning rates of the cellulosic materials given the highest concentrations of treatment in Figure 17 when they are burnt in an atmosphere of oxygen concentration greater than LOI. As the

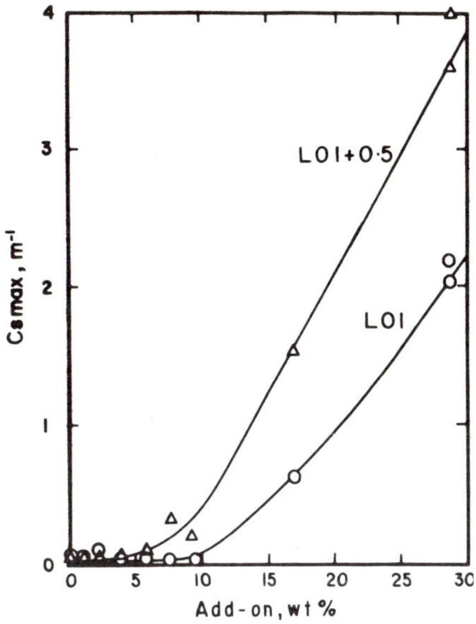

Figure 17. *Dependence of smoke concentration on DAP deposit (LOI method).*

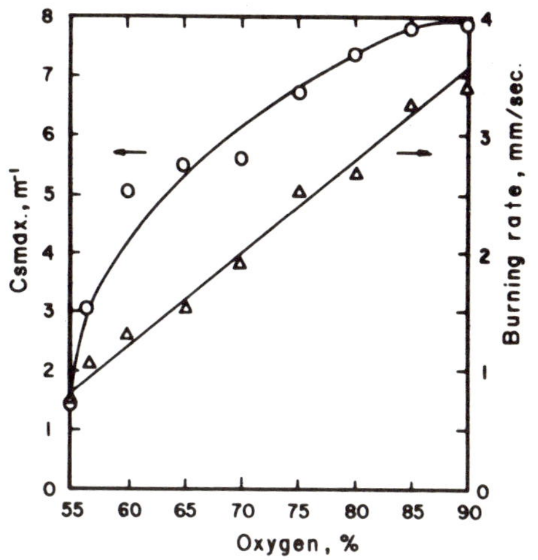

Figure 18. *Dependence of smoke concentration and burning rate on oxygen concentration (LOI method).*

oxygen concentration increases, both increase proportionately. When an organic material is burnt in an atmosphere containing a high concentration of oxygen, it is generally considered that the emission of smoke is low. The result of the present experiment disproves this prediction. With the treated materials, it shows that the emission of smoke increases. This may be explained by the increasing rate of combustion with increasing oxygen concentration. The rate of emission of smoke is given generally by the product of the rate of heat decomposition and the amount of smoke per unit weight. Therefore, as it is expected that the amount of smoke will increase by the treatment for fire proofing, while the rate of combustion increases as shown in Figure 18, the foregoing result may be said to be a reasonable consequence.

Comparison of Results Obtained by Four Methods

Comparing the results of the foregoing four methods, they may be divided into two categories: one being the case in which the smoke and carbon monoxide decrease with increasing treatment concentration, and the other being the case in which they increase directly with the treatment. The former result is provided by the 45-degree method and in the low temperature region of the electric furnace method; and the latter result is provided in the high temperature region of the electric furnace method, by the large vertical furnace and the LOI methods. These results show that the fire retardant treatment acts advantageously against the combustibility including smoke and carbon monoxide when the flame applied to the treated materials is small or the temperature of heat is low but plays a negative role against the smoke or gases if the flame is large enough or the heat is of high temperature. In other words, the materials treated for fire retardance resistive to combustion or suppress the emission of smoke and gases up to the limit of the effect of fire retardant treatment, but when they are exposed to the situations in which the limit is exceeded, desirable results are not provided. Rather the fire retardance acts in a negative way after the flashover point.

This poses a serious problem in fire retarding cellulosic materials. The fire retardance of cellulosic materials has so far been designed to cope with the initial stages of a fire, and the tests have been concerned with the resistance to a small flame accordingly. So far as the object of fire retarding the cellulosic materials remains in such scope, the diammonium phosphate treated cellulose provides a satisfactory performance even if the emission of smoke and gases is considered inclusively. However, if the fire retardance is to be defined, to include prevention of the smoke and gases in the stage after flashover, such treatment of cellulose with diammonium phosphate may not be sufficient. The object of fire prevention should be clearly established before determining which position is to be taken. But it will be difficult to derive a conclusion immediately. Tentatively, therefore, it will be necessary to have a knowledge of the combustibility, smoke producibility and toxic gas emission of the treated materials in either instance where they exhibit the effect of fire retardance against a small fire source or where they are exposed to a big fire source exceeding the limit of their fire retardance. For such a purpose, if the flame propagation test is to be conducted, a combination test of the 45-degree method and large vertical furnace or LOI method should be employed. However, the large vertical furnace method involves a large apparatus and a troublesome procedure so that a combination of the 45-degree method and LOI method is recommended. The electric furnace method is designed for testing the combustibility of materials. In such a case, it is necessary to carry out the test in many stages of temperature from low to high temperature. As currently employed in Japan, at least three stages of 350°C, 500°C and 750°C should be adopted for measurement.

Measurement of Smoke and Gas Concentrations

For measurement of the smoke and gas concentrations, there are two methods available: the differential method for direct measurement of the air flow through the smoke channel and the accumulation or integral method for measurement of the accumulation in a smoke box. The former method has an advantage in that it responds quickly to the rate of generation of smoke but involves a problem with respect to the cross sectional dimension of the light path passing across the smoke channel in that should the smoke or gases be unevenly distributed in the smoke channel, correct measurement may not be achieved. In the present experiments, however, such a phenomenon was not observed, and at least the smoke passed, evenly distributed, through the smoke channel. However, this method has another shortcoming in that when it is desired to know the total amount of smoke or gases emitted from a material, the rate of the air stream through the smoke channel must be measured. For calculating the total amount, a troublesome procedure of multiplying the concentration and the instantaneous value of the wind velocity and integrating the product must be followed.

The other integral method is free of such difficulties because the smoke and gases generated are wholly stored in a smoke box. However, since the smoke and gases emitted from a small sample are to be accumulated, the response is not fast, and for identifying the rate of generation, a process of integrating curves is required. Further, the smoke tends to agglomerate in the smoke box, and the agitation by means of a stirrer in order to maintain the represented as the difference between the actual amount and the loss. This is a serious defect.

Then, if it is possible to know the concentration of the smoke in the smoke channel exactly, and by taking a method of interlinking to a simple computer when it is desired to know the total amount of smoke, the product of the concentration and the rate of air flow, and its integration with time, will be readily obtained. Thus, the differential method is considered to be desirable, since it permits one to place the instrument close to the source of generation of smoke. Also there is quick response and good reproducibility and these factors are the basic requirements for good measurement.

SUMMARY

Cellulosic materials treated with diammonium phosphate were subjected to tests of burning with various types of the source of heat under conditions where the air supply was adequate and the following results were obtained.

1. As the treatment concentration becomes higher, the emissions of smoke and carbon monoxide are either reduced or increased, the former situation occurring when the sample is exposed to a small fire or a heat of relatively low temperature and the latter being observed in the case of a large source of fire or a high temperature.

2. So long as the intensity of a source of fire remains within the range of control of the retardant, the emissions of smoke and carbon monoxide are small, but they increase when the limit is exceeded. This is a natural consequence from the nature of fire retarding treatment.

3. Accordingly, it is necessary for measurement of the smoke and gases to burn the sample with a source of fire corresponding to a heat or temperature within the controllable range and that outside such range. For this purpose, a combination of the 45-degree method and the LOI method or measurements at 350°C, 500°C and 750°C under the electric furnace method are recommended. Any treated materials producing a large amount of smoke and gases should be disapproved however they may suppress burning.

4. For determination of the smoke and gas concentrations, the differential method is recommended from its fast response, correctness and good reproducibility.

5. In considering fire retardant treatments including prevention of smoke and gases, the conventional concept of fire retardation should be re-examined, and no tests should be designed before determination of what is aimed at by the particular fire retardant treatment. Otherwise, there is a danger of inducing a reverse conclusion, as referred to in subparagraph 1.

ACKNOWLEDGMENT

The authors wish to thank Mr. T. Miyata and T. Takishima, Fire Research Institute, for the design and construction of the large vertical furnace, Mr. S. Nago and S. Shinozaki, Yokohama National University, for their experimental supports in the electric furnace method, and Mr. S. Suga, Toyo Rika Instruments Co., for smoke level measurement in the LOI method.

REFERENCES

1. Chem. and Eng. News, Mar. 30, p. 34, 1970.
2. H. Imazu, Industrial Materials (Japanese), *16* 77 (1967).
3. D. Gross, L. Loftus, T. G. Lee and V. E. Gray, Smoke and Gases Produced by Burning Aircraft Interior Materials, Nat'l. Bur. Stds. Building Science Series 18, (1969).
4. C. H. Hilado, Flammability Handbook for Plastics, Technomic Pub., Westport, Conn., 1969.
5. F. Saito, Bull. Fire Prevention Soc. of Japan, *19* 189 (1969).
6. H. W. Eickner and E. L. Schaffer, Fire Technol., *3* 90 (1967).
7. J. W. Lyons, The Chemistry and Uses of Fire Retardants, p. 370, Wiley-Interscience, New York, 1970.
8. J. DiPietro and H. Stepmiczka, SPE Journal, *27* 23 (1971).
9. H. Dorn, Untersuchung des Brandverhaltens von Kunstoffolien untershiedlicher Dicke, AGF Bericht Nr. 2, Dec. 1966.
10. C. P. Fenimore and F. J. Martin, Combustion and Flame, *10* 135 (1966).
11. J. L. Isaac, J. of Fire and Flammability, *1* 3 (1970).
12. S. Suga, Plastics (Japanese), *21* (9) 19 (1971).
13. A. E. Lipska, The Pyrolysis of Cellulosic and Synthetic Materials in a Fire Environment, AD-645858, Dec. 1966.

Y. Uehara

Yoichi Uehara received his BSc in Engineering degree from Kyoto University in 1957. After working in Osaka Municipal Fire Bureau as an inspector for nine years, he moved to the Fire Research Institute. He is now an associate professor in the Department of Safety Engineering, Faculty of Engineering, Yokohama National University and, in addition, he is also the chief of the Combustion Section of the Fire Research Institute, Ministry of Home Affairs, Japanese Government. He is studying the inhibition mechanisms of halogenated hydrocarbons, spreading of flammable gases, fire of liquefied flammable gases and their extinguishment, pyrolysis of treated polymers and smoke and gases produced from fire-proofing materials in a fire environment.

E. Yanai

Eiji Yanai is a research fellow of the Fire Research Institute. He received his BSc degree from Science University of Tokyo in 1971. He works with Mr. Uehara, concerning pyrolysis of treated organic polymers and smoke and gases emitted from fire-proofed materials in a fire environment.

Kikuo Sumi and Yoshio Tsuchiya

National Research Council of Canada
Division of Building Research

COMBUSTION PRODUCTS OF POLYMERIC MATERIALS CONTAINING NITROGEN IN THEIR CHEMICAL STRUCTURE

(Received January 31, 1972)

ABSTRACT: Five materials that contain nitrogen were burned in a flask at 800°C, and the amounts of hydrogen cyanide, carbon monoxide and carbon dioxide produced were quantitatively analyzed by gas chromatography. The harmful effects of these products to the resulting atmosphere were then evaluated. The results vividly illustrate the potential danger of HCN to a fire environment when materials containing nitrogen are involved in fire.

INTRODUCTION

A large number of fire deaths are caused by inhalation of toxic gases and vapors produced by thermal decomposition and oxidation reactions. Information on toxic compounds produced by the combustion of various materials that are found in buildings is of importance in determining the products that are responsible for fire deaths and in assessing the potential danger due to the formation of these harmful substances. At present, experimental data on the quantities of toxic gases and vapors produced by combustion are still lacking for many organic materials. For example, little information is available on the amount of hydrogen cyanide (HCN) produced from the decomposition of materials containing nitrogen in their chemical structure. Hydrogen cyanide is a highly toxic compound. The concentration of HCN that is fatal to man in 30 minutes is reported to be 135 ppm [1], while that of CO that is fatal in exposures of less than one hour is reported to be 4000 ppm [2]. Although the lethal concentration of gases and vapors estimated by various toxicologists are not in complete agreement, these examples show that HCN is much more toxic than CO. The above data suggest that HCN is thirty times as toxic as CO on a volume basis.

In 1933, Olsen et al [3] analyzed the combustion products of a number of materials including wool and silk. A weighed sample in a porcelain boat was inserted into a heated silica tube through which sufficient air was drawn to burn the sample with a flame. The combustion of wool yielded 0.5 to 5.0% CO, 1.26 to 2.52% HCN, 1.25 to 3.70% ammonia (NH_3), 0.02 to 0.41% hydrogen sulfide (H_2S) and 0.05 to 0.12% sulfur dioxide (SO_2); and the combustion of silk yielded 3.0 to 4.4% CO, 2.22 to 6.80% HCN, and 3.09 to 3.58% NH_3. These quantitative data on

HCN and CO indicate that the greatest toxic hazard from the combustion of wool or silk is HCN poisoning.

In 1963, Dufour [4] made a comprehensive literature survey of data concerning the combustion products of building materials, and included experimental data on polymers containing nitrogen from unpublished reports by Hobbs and Patten [5] and McDermott and Critchfield [6]. Hobbs and Patten analyzed the combustion products of several materials. The combustion apparatus consisted of a furnace and a "Vycor" tube maintained at 800°C. Either compressed air or a mixture containing 11.7% O_2 and 88.3% N_2 was passed through the tube containing the sample. The combustion of one gram of wool yielded 0.446 g of CO and 0.007 g of HCN; one gram of silk yielded 0.634 g of CO, 0.036 g of HCN and 0.053 g of NH_3; and one gram of nylon yielded 0.304 g of CO, 0.0076 g of HCN and 0.032 g of NH_3. The results are in conflict with those obtained in the early work of Olsen [3], and suggest that the hazard due to CO is far greater than that due to HCN when wool, silk or nylon are involved in fire.

McDermott and Critchfield [6] investigated the amount of HCN produced by the combustion of wool, nylon and flexible urethane foam, but did not determine the amount of CO produced. Nagao et al [7] investigated the amount of HCN produced by the combustion of wool, silk and polyacrylonitrile. They, too, did not determine CO produced. Thus, two sets of conflicting data are available on the amounts of HCN and CO produced by materials containing nitrogen.

The purpose of the present investigation was to quantitatively analyze HCN, CO and CO_2 produced by the combustion of five organic materials containing nitrogen, and to assess the relative importance of these products in creating a dangerous atmosphere.

EXPERIMENTAL

1. Materials

- (a) Acrylic fiber
- (b) Nylon-6
- (c) Wool
- (d) Urea-formaldehyde foam
- (e) Rigid urethane foam

2. Combustion

The combustion experiments were conducted in a 3-neck flask, having a capacity of 5 liters, shown in Figure 1. The air in the flask (open to the atmosphere) was heated by a nichrome wire wound around a half section of a quartz tube. The air was heated for 15 minutes to raise the temperature and maintain the hot zone at about 800°C. A weighed specimen was then pushed into the hot zone and a stopper was placed in the appropriate neck of the flask. Heating was continued for 2.5

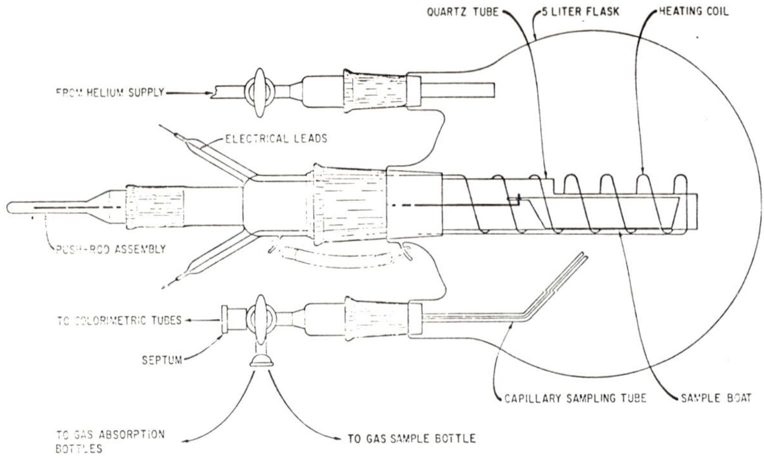

Figure 1. ***Combustion apparatus.***

minutes, and the combustion flask was allowed to cool for 30 minutes. During the cooling period the pressure in the flask was maintained at atmospheric by allowing helium to flow into the flask.

3. Analysis

At the end of the cooling period, a sample of gas was taken in an evacuated bottle, and quantitatively analyzed for CO, CO_2 and HCN. Gas chromatography was the main method used in this analysis. Carbon monoxide and carbon dioxide were analyzed using a thermal conductivity detector, with a molecular sieve 5A packed column for CO and a Porapak N packed column for CO_2. Hydrogen cyanide was analyzed using a thermal conductivity detector and a Porapak S packed column. For low concentrations of HCN ($<$0.3% by volume), colorimetric detector tubes were used by pumping the products directly from the combustion flask. The detector tubes were also used to explore the presence of nitrogen dioxide.

RESULTS AND DISCUSSIONS

The experimental conditions that were adopted for the quantitative determination of toxic products require some explanation. The procedure adopted was that favorable for the formation of toxic combustion products in order to find out under extreme conditions if HCN is a potential hazard at fires. A combustion temperature of 800°C was selected to study the products formed from flaming combustion. At this temperature, flaming started soon after the specimen was inserted in the combustion flask and continued for 30 to 60 seconds, for all five samples. Preliminary experiments conducted at various temperatures indicated that the quantities of toxic substances formed were influenced by the relative thermal stability of the samples, and the highest amounts of toxic combustion products from given weights of materials were produced at the maximum working temperature of the apparatus of 800°C. Preliminary experiments also indicated that HCN that is produced by combustion reactions diminishes if the heating time is extended for an appreciable period after flaming ceases. A series of experiments was therefore conducted with materials containing nitrogen to find the amount of HCN produced as a function of heating time ranging from 0.5 to 5.0 minutes. The maximum amount of HCN was found when the heating period was 2 to 3 minutes. A heating time of 2.5 minutes was therefore adopted.

The experimental results showing the amounts of HCN, CO and CO_2 produced at various sample weights ranging from 400 to 2400 mg are presented in Table 1. They are reported in grams per gram of sample, and are based on means of two or more repeat tests. The standard deviation of the determination of each component was about 20% of the values reported in this table. The weight and hence the sample/air ratio was varied to find the relative contribution of HCN and CO to the toxicity of the resulting atmosphere over a wide range of conditions that may exist at fires. The possible yield of NO_2 was investigated by using colorimetric tubes having a range of 1 to 1000 ppm. Nitrogen dioxide was not detected with any of the five materials following combustion of 800 mg of specimen.

The toxicity of a combustion product, based on both the nature of the compound and the quantity evolved, was evaluated by a relationship suggested in earlier papers [8, 9]. According to this relationship,

$$T = \frac{c_e}{c_f}$$

where

- T is a toxicity index
- c_e is the experimentally determined concentration of a toxic product expressed as volume of product when one gram of material is decomposed (or burned) and the products are diffused in a volume of one cubic meter.
- c_f is the concentration of the same product that is fatal (or dangerous) to man in a 30-minute exposure.

The experimental data were translated into concentration for one gram burned in one cu.m. to provide a convenient method for evaluating not only the present results but also those obtained by other investigators under different experimental conditions. The toxicity index of a material with two or more toxic combustion products is determined by assuming that the effect of two or more harmful products is additive; that is

$$T = \Sigma\, c_e/c_f$$

Table 1. Combustion Products

Material	Sample Weight (mg)	Weight of product per weight of sample (g/g)			
		HCN	CO	CO_2	Residue
Acrylic fiber	400	0.095	0.170	1.46	0.19
	800	0.128	0.168	0.57	0.19
	1200	0.130	0.148	0.44	0.19
	1600	0.142	0.089	0.28	0.19
	2000	0.193	0.109	0.17	0.15
	2400	0.170	0.076	0.23	0.15
Nylon	400	0.070	0.38	1.42	0
	800	0.055	0.28	0.75	0
	1200	0.022	0.28	0.44	0
	1600	0.027	0.150	0.34	0
	2000	0.020	0.095	0.29	0
	2400	0.022	0.123	0.27	0
Wool	400	0.054	0.175	1.26	0.15
	800	0.037	0.22	0.85	0.14
	1200	0.022	0.19	0.54	0.16
	1600	0.027	0.114	0.30	0.17
	2000	0.026	0.131	0.33	0.16
	2400	0.018	0.117	0.25	0.15
Urea-formaldehyde foam	400	0.015	0.041	1.35	0.10
	800	0.025	0.034	0.70	0.12
	1200	0.023	0.032	0.500	0.15
	1600	0.023	0.060	0.46	0.14
	2000	0.042	0.068	0.40	0.18
	2400	0.033	0.077	0.34	0.17
Rigid urethane foam	400	0.008	0.21	1.40	0.12
	800	0.008	0.23	0.70	0.13
	1200	0.008	0.157	0.47	0.12
	1600	0.008	0.142	0.43	0.12
	2000	0.008	0.197	0.32	0.12
	2400	0.008	0.092	0.28	0.10

The additive assumption may be conservative because of possible synergistic effects. Synergism may have to be considered in the future for certain combinations of toxic products, if it is found to be significant.

The toxicity indexes of combustion products were determined from the present experimental data and c_f values given in Table 2. This information, presented in Table 3, vividly illustrates the importance of HCN as a potential contributor to the toxicity of the fire atmosphere. This Table shows that the toxicity due to HCN could be 55 times as great as that due to CO for acrylic fiber, 5 times as great for nylon, 8 times as great for wool, 26 times as great for urea-formaldehyde and twice as great for rigid urethane foam.

Table 2. Dangerous Concentrations of Combustion Products

Combustion product	Dangerous concentration for a 30-minute exposure (ppm)	Reference
HCN	135	1
CO	4000	2
CO_2	70,000 – 100,000	1

Table 3. Toxicity Index of Combustion Products

Material	Sample Weight (mg)	Toxicity index due to:			
		HCN	CO	CO_2	Total
Acrylic fiber	400	0.58	0.03	0.01	0.62
	800	0.79	0.03	<0.01	0.82
	1200	0.82	0.03	<0.01	0.85
	1600	0.88	0.02	<0.01	0.90
	2000	1.19	0.02	<0.01	1.21
	2400	1.05	0.02	<0.01	1.07
Nylon	400	0.43	0.08	0.01	0.52
	800	0.34	0.06	<0.01	0.40
	1200	0.14	0.04	<0.01	0.18
	1600	0.18	0.03	<0.01	0.21
	2000	0.12	0.02	<0.01	0.14
	2400	0.14	0.02	<0.01	0.16
Wool	400	0.33	0.04	0.01	0.38
	800	0.23	0.04	<0.01	0.27
	1200	0.14	0.04	<0.01	0.18
	1600	0.16	0.02	<0.01	0.18
	2000	0.16	0.03	<0.01	0.19
	2400	0.11	0.02	<0.01	0.13
Urea-formaldehyde foam	400	0.10	0.01	0.01	0.12
	800	0.15	0.01	<0.01	0.16
	1200	0.14	0.01	<0.01	0.15
	1600	0.15	0.01	<0.01	0.16
	2000	0.26	0.01	<0.01	0.27
	2400	0.21	0.02	<0.01	0.23
Rigid urethane foam	400	0.05	0.04	0.01	0.10
	800	0.05	0.05	<0.01	0.10
	1200	0.05	0.03	<0.01	0.08
	1600	0.05	0.03	<0.01	0.08
	2000	0.05	0.02	<0.01	0.07
	2400	0.05	0.02	<0.01	0.07

The maximum toxicity index obtained for the five samples by varying the sample weights is useful in assessing the toxic gas producing potential of materials. The data obtained at 800°C are presented in Table 4 for the five samples used in the present study as well as earlier results on white pine and polystyrene [10]. The maximum toxicity indexes of some of the nitrogen-containing materials were much greater than the values for white pine and polystyrene.

Table 4. Maximum Toxicity Index

Material	Toxicity index due to:			
	HCN	CO	CO_2	Total
Acrylic fiber	1.19	0.02	<0.01	1.21
Nylon	0.43	0.08	0.01	0.52
Wool	0.33	0.04	0.01	0.38
Urea-formaldehyde foam	0.26	0.01	<0.01	0.27
Rigid urethane foam	0.05	0.05	<0.01	0.10
White Pine	–	0.09	0.003	0.09
Polystyrene	–	0.09	0.01	0.10

Kishitani [11] conducted animal tests in which mice were exposed to the gaseous combustion products of a number of materials including urethane. Pathological examination of the mice that died from exposure to the combustion products of urethane showed that the carboxyhemoglobin content in the blood was much lower than that required to kill these animals. Thus, deaths were probably due to the combined effect of CO and other toxic combustion products.

CONCLUSIONS

Hydrogen cyanide, carbon monoxide and carbon dioxide formed by the combustion of five materials that contain nitrogen were quantitatively analyzed, and the harmful effects of these products to the resulting atmosphere were evaluated. The present results vividly illustrate the potential hazard of HCN produced when combustible materials containing nitrogen are involved in fire.

ACKNOWLEDGMENT

The authors wish to thank J. Boulanger and D. W. Morwick for assistance in conducting the experiments.

This paper is a contribution of the Division of Building Research, National

Research Council of Canada, and is published with the approval of the Director of the Division.

REFERENCES

1. Documentation of Threshold Limit Values. American Conference of Governmental Industrial Hygienists. 223 pp. (1966).
2. Y. Henderson and H. W. Haggard. Noxious Gases and the Principles of Respiration Influencing Their Action. 212 pp. (1927). The Chemical Catalog Co., Inc., New York.
3. J. C. Olsen, G. E. Ferguson and L. Scheflan. Gases from Thermal Decomposition of Common Combustible Materials. Ind. and Eng. Chem. *25*, 599 (1933).
4. R. E. Dufour. Survey of Available Information on the Toxicity of the Combustion and Thermal Decomposition Products of Certain Building Materials under Fire Conditions. Underwriters' Laboratories, Inc. Bulletin of Research No. 53, 52 pp. (July 1963).
5. A. P. Hobbs and G. A. Patten. Products of Combustion of Plastics and Other Common Solids. Dow Chemical Co. (March 1, 1962). Unpublished.
6. W. H. McDermott and F. E. Critchfield. Thermal Degradation of Polyurethane to Hydrogen Cyanide. Union Carbide Plastics Co. (Feb. 24, 1961). Unpublished.
7. H. Nagao, M. Uchida and A. Yamaguchi. (In Japanese) Thermal Decomposition of Polyacrylonitrile. J. Chem. Soc. Japan, *59,* 940 (1956).
8. Y. Tsuchiya and K. Sumi. Thermal Decomposition Products of Polyvinyl Chloride. J. Appl. Chem. *17,* 364 (1967).
9. Y. Tsuchiya and K. Sumi. Evaluation of the Toxicity of Combustion Products., *J. Fire and Flammability 3,* 46 (1972).
10. K. Sumi and Y. Tsuchiya. Toxic Combustion Products of Wood and Polystyrene. Building Research Note No. 76, Division of Building Research, National Research Council of Canada, 3 p. (Sept. 1971).
11. K. Kishitani. Study on Injurious Properties of Combustion Products of Building Materials at Initial Stage of Fire. *J. Faculty of Engineering,* Univ. of Tokyo (B), *31,* No. 1, 1 (1971).

Kikuo Sumi

Kikuo Sumi is a Research Officer of the Division of Building Research, National Research Council of Canada. He received his B.A.Sc. degree in Applied Science from the University of Toronto and his Ph.D. degree in Chemical Engineering from the University of London. He joined the National Research Council in 1951 and has experience in various aspects of fire research such as thermal decomposition of polymers, fire extinguishment and fire prevention regulations.

Yoshio Tsuchiya

Yoshio Tsuchiya is a Research Officer of the Division of Building Research, National Research Council of Canada. He received his Bachelor of Engineering and Doctor of Engineering degrees in Applied Chemistry from the University of Tokyo in 1953 and 1962 respectively. He joined the National Research Council in 1965. He has experience in the fields of industrial explosives, organic peroxides and fire research.

Flame Resistant Non-Cellulosic Fibers and Fiber Blends

S. B. Sello

Research & Development Dept.
J. P. Stevens & Co., Inc.
Garfield, N. J.

(Received August 17, 1971)

ABSTRACT

Techniques to attain flame resistant non-cellulosic textiles from high temperature resistant organic polymers, from inorganic fibers or by modification of existing synthetic fibers, are reviewed. The latter modifications are attained by copolymerization with flame resistant monomers, by incorporating flame retarding chemicals to the dope prior to spinning, or by the surface modification of synthetic fibers. The flammability behavior of cellulosic/thermoplastic fiber blends cannot be predicted from the flammability of the fiber components. Polyester/cellulosic blends can be rendered self-extinguishing by selectively treating one of the fiber components, preferably the cellulosics, with flame retarding chemicals effective on both types of fibers.

INTRODUCTION

Natural, regenerated cellulosic, inorganic, man-made fibers, and blends of these fibers vary widely in their ease of combustion and present different problems of flammability.

While many approaches have been documented in the literature and several commercial procedures are available to overcome the flammability of cellulose with durable flame retardant systems, much less progress has been attained in reducing the flammability of synthetic fibers, and especially of blends of these with cellulosic fiber. Perhaps the reason for the lesser effort in flame retarding synthetic fibers such as nylon and polyester is that these fibers were generally not considered very hazardous since they are manufactured from thermoplastic polymers which, upon ignition, melt and carry away the flame. Blending of these thermoplastic fibers with non-melting fibers considerably altered this concept.

Flame resistant fibers can be produced by various approaches:

1. Use of high temperature resistant organic polymers
2. Use of inorganic fibers
3. Modification of existing synthetic fibers

THERMALLY STABLE ORGANIC FIBERS

Several thermally stable polymers have been designed to retain physical properties for extended periods at elevated temperature, and several fire resistant polymers were formulated not to burn. These two requirements are not identical. Attempts to retard burning significantly increase the smoke and gas evolution, which might be as hazardous as the fire itself.

Aromatic repeat units are dominant features in heat resistant polymers [1]. The thermal stabilities of aromatic amide, imide and ether linkages far exceed those of their aliphatic counterparts (Figure 1).

Examples of the aromatic polyamides, polyimides and polyimidazoles are as follows [2]:

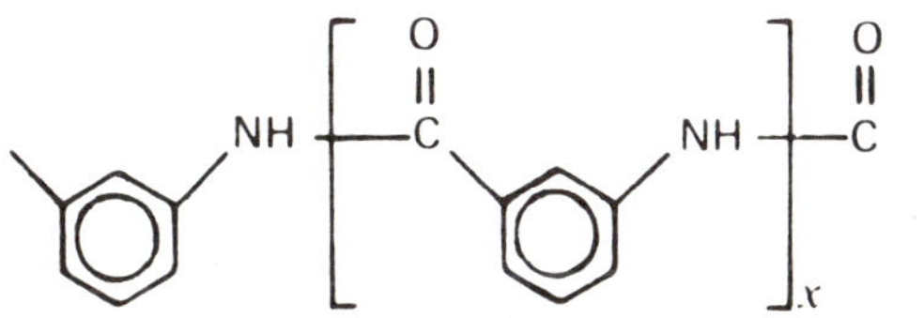

mp 424 C

Poly-m-phenylene isophthalamide

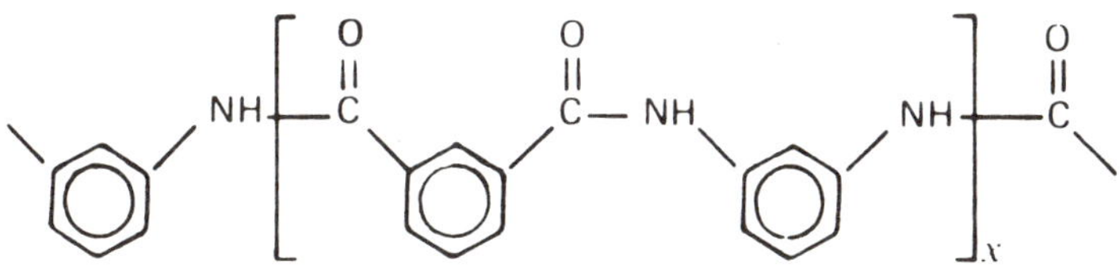

Nomex

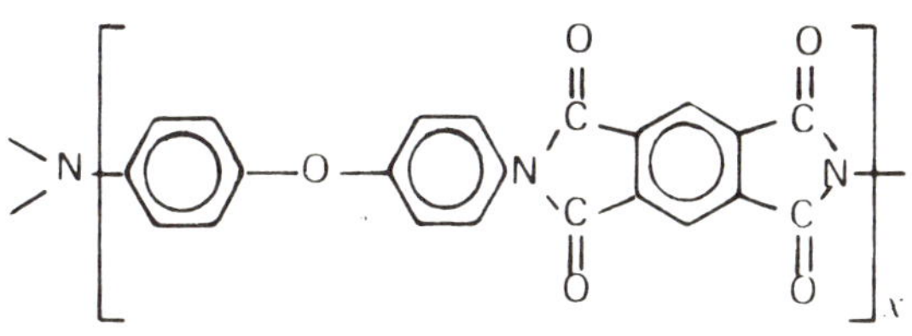

dec. temp ~ 600 C

Poly-N,N'-(p,p'-oxydiphenylene) pyromellitimide (Kapton)

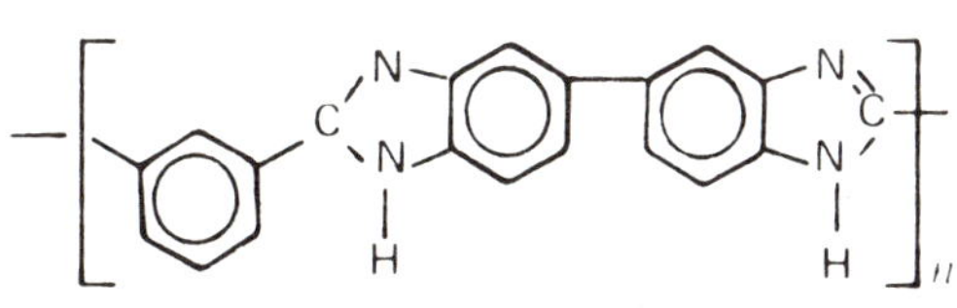

Polybenzimidazole (PBI)

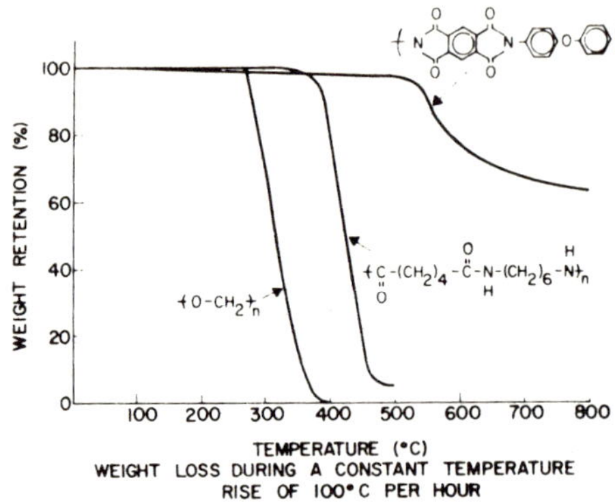

Figure 1. *Comparison of thermal stabilities of aromatic and non aromatic polymers with ether and amide linkages in the main chain.*

These fibers are not only non-burning in the usual flammability tests, but retain good properties even after prolonged exposure to elevated temperatures.

Elimination of all hydrogen atoms and other pendent groups from an aromatic polymer gives excellent high temperature oxidative resistance by preventing most bond fractures and crosslinking reactions [3].

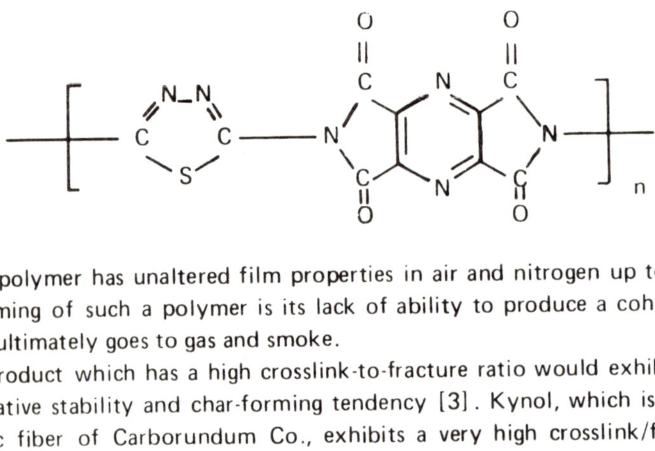

Such a polymer has unaltered film properties in air and nitrogen up to 600°C. The shortcoming of such a polymer is its lack of ability to produce a coherent residue; thus, it ultimately goes to gas and smoke.

A product which has a high crosslink-to-fracture ratio would exhibit both thermooxidative stability and char-forming tendency [3]. Kynol, which is a crosslinked phenolic fiber of Carborundum Co., exhibits a very high crosslink/fracture ratio, which renders it entirely non-burning with very little smoke-producing tendency.

Excellent high performance fibers with long term resistance to oxidation at elevated temperatures, minimal gas and smoke generation and high char-forming tendency can be attained by controlled heating of aromatic polyamides in the presence of various gases (halides and/or oxyhalides) through an oxidizing stage to pyrolysis to 2500 C to form graphite fibers [4]. The Durette procedure of Monsanto and the Fypro process of Travis Mills are based on these effects. These fibers are finding use in steel mills and for military and aerospace purposes.

DuPont recently announced [5] the development of an organic fiber (PRD-49) for high performance, high priced advanced composites in which the use of graphite and boron fibers now dominates.

INORGANIC FIBERS

The inorganic fibers are inherently flameproof, although organic finishes present on them may ignite. Glass fibers are the most extensively used textile fibers which are highly resistant to heat and flame. The structure represents a fully oxidized silicon dioxide

$$-[\overset{\overset{\displaystyle O}{\|}}{Si} - O - \overset{\overset{\displaystyle O}{\|}}{Si} - O]_n-$$

which does not produce flammable gases even when heated above 2000°C.

The beta-glass fiber [6], a novel fine denier glass fiber, is suitable for apparel and household goods. The Teklan Heat-Shield fabrics [7] of Courtaulds are made from Teklan (Courtaulds' modacrylic fiber) and Beta-Fiberglas, core spun and woven in such a way that the hand is soft and the glass fiber does not come into contact with the skin. This fabric gives protection from direct flame and high resistance to heat radiation penetration.

MODIFICATION OF EXISTING SYNTHETIC FIBERS

Several types of approaches are documented in the literature to impart flame resistance to man-made fibers. These approaches can be categorized as follows: copolymerization with flame resistant monomers, addition of flame retarding chemicals to the dope prior to spinning, modification of the fibers by chemical reaction or by deposition or insolubilization of flame retarding finishes [8]. Let us examine these approaches for the most widely used man-made fibers.

A. Acrylic Fibers

Copolymerization of acrylonitrile with vinyl chloride or vinylidene chloride results in the production of a flame resistant fiber. Such fibers which contain less than 85% but more than 35% acrylonitrile are called modacrylic fibers [8]. (Dynel, the modacrylic fiber of Union Carbide, gives excellent flame- and acid-resistance.)

Vinyl comonomers lower the resistance of the fibers to heat; therefore, their content is limited in the fiber. Copolymers of acrylonitrile with vinylidene chloride exhibit increased glass transition temperature, which results in better resistance to deforming stress at high temperature. α-chloromethyl- or α-bromomethyl acrylonitrile as a comonomer is also a good approach to reduce the flammability [9]. Bromine consistently contributes more flame retarding effectiveness to polymers than chlorine. Thus, it is not surprising to see references in the recent literature to the use of vinyl bromide as a comonomer for acrylic fiber [10]. In addition, fibers containing vinyl bromide have less shrinkage than those consisting of copolymers of acrylonitrile and vinylidene chloride.

The condensate of vinyl acetate and acrylic ester obtained by the transesterification of hexa-halopentadiene was copolymerized with acrylonitrile, and it was found that 10% Br or 25% Cl content was required to obtain a non-flammable film [11].

Copolymerization of acrylonitrile with the allyl- or vinyl-esters of phosphoric or phosphonic acids such as dialkyl- or bis-(haloalkyl)-vinyl phosphonate, monoallyl-dialkyl phosphites, etc.) are also suggested [12] approaches to attaining flame resistant acrylic fibers.

Aldehyde groups have been introduced into macromolecules by copolymerizing acrolein with acrylonitrile. Flame resistance is then imparted by reacting the aldehyde groups with dialkyl phosphite and dialkylamine to obtain a hydrolytically stable phosphonate structure [13]. Fiber containing 3.5% P was reported to be non-combustible.

Acrylic fibers are spun from solution (dimethyl acetamide and aqueous $ZnCl_2$ or sodium thiocyanate solution) at moderate temperature. Incorporation of flame retarding additives into the spinning solution is a feasible method for reducing the flammability of acrylic fibers. Phosphoric- or phosphonic acid esters, which usually contain halogen in the alkyl side chain, are considered as additives. In addition, tris-(2,3-dibromo-propyl) phosphate, trialkyl phosphates, halogenated telomers [14] obtained by reacting triallyl phosphate with tetrahalomethanes and halogen, bromine-containing phosphonates in combination with insoluble calcium phosphate and many other additives, such as mixtures of Sb_4O_6 and halogens, are suggested as effective flame retardant additives.

The surface treatment of acrylic fiber with phytic-acid [16] in conjunction with formaldehyde and urea, or with ammonium phosphates and urea derivatives, etc., are also suggested for imparting flame resistance.

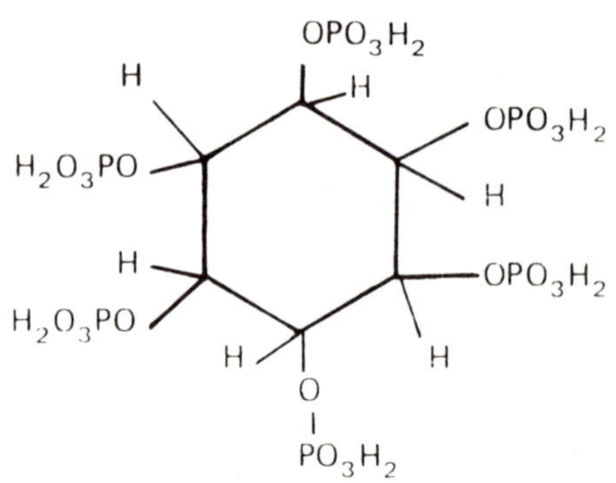

Phytic Acid

We may conclude that acrylic fibers can be rendered flame resistant by any of the following [2]:

5% P
10-12% Br
10-15% Cl
1- 2% P + 10-12% Cl or + 5-10% Br
2% Sb_4O_6 + 8% Cl or + 6% Br

These requirements are not too different from those established for the flame retarding treatment of cellulosic substrates.

B. Polyvinyl Alcohol (PVA) Fibers

It has long been known that vinyl chloride polymerizes readily and the polymer can be drawn into long filaments, which are entirely fire resistant. The copolymerization of vinyl chloride with vinyl acetate gives a polymer in which the vinyl chloride is internally plasticized. This is the basis for the Vinyon fiber of Union Carbide and American Viscose.

The Toyo Chemical Co. in Japan developed a flame resistant wool-like fiber consisting of 50% polyvinyl alcohol and 50% polyvinyl chloride. It is marketed under the trade name Cordela.

PVA fibers have been flameproofed by acetylating a portion of hydroxyl groups with bromine-containing aldehydes [17], by phosphorylation [18] to at least 3% P content, or by treatment with P- and N-containing chemicals designed for the flame retarding treatment of cellulosic substrates. The requirements to render PVA flame resistant are similar to those established for cellulosic substrates.

C. Polyamides

The two principal commercial polyamides, nylon 6 and 66, have similar glass transition temperatures (75 to 80°C) and melting points (220 to 265°C). They are spun from a melt at temperatures near 300°C. The melt spinning imposes a limitation on the type of additives or comonomers one can add, since the flame retardant additive is required to be thermally stable at the temperature of melt spinning and to decompose at the temperature of ignition.

However, there are approaches documented in the literature to use flame retarding additives. Reportedly low flammability in nylon 66 has been attained by incorporating a phosphorus monomer of the following structure

$$HOOCCH_2CH_2-\overset{\overset{\large O}{\|}}{\underset{\underset{\large \phi}{|}}{P}}-CH_2CH_2COOH$$

to yield 3.5% P in the fiber [19] or by adding derivatives of hexachloro cyclopentadiene, or chlorinated diphenyl alone and with Sb_4O_6 [20]. These additives apparently survive the melt spinning process without decomposition.

The copolymerization of halogenated dicarboxylic acids with diamines has also been suggested as an approach to attain polyamides of reduced flammability.

It is apparent from the literature that methods proposed to attain flame resistant nylon are primarily fabric finishing or aftertreatment techniques. Most of the flame retardants which are effective on cellulosic substrates are not capable of rendering nylon flame resistant. For instance, ammonium phosphate and ammonium sulfate are not sufficiently effective in imparting flame resistance to nylon. Ammonium thiocyanate, ammonium bromide and especially thiourea were found more effective. The insolubilization of at least 3% thiourea is required to attain durable flame resistance [21]. The application of thiourea can be carried out as a part of an aminoplast resin coating.

D. Polyester

Polyester fibers are melt-spun from the reaction product produced by the transesterification of terephthalic acid and ethylene glycol at nearly 300°C. Thus, the flame retarding additives must withstand this temperature without decomposition.

Dimethyl phosphite [22] is suggested as a flame retardant additive at such a level as to give 1.7% P in the fiber. Presumably transesterification occurs and some of the terephthalate is replaced with phosphite. In analogous case, ethylene-glycol-phenyl-phosphonate [23] has been proposed as an additive.

The copolymerization of halogenated dicarboxylic acids (e.g., tetrabromophthalic anhydride) with polyols, the surface chlorination of polyester fibers and films [24], and the use of tris-(2,3-dibromo-propyl)phosphate as an additive, are also suggested approaches to obtain polyester with improved flame resistance.

We may conclude that polyester fibers can be rendered flame resistant by one of the following [2]:

5% P
25% Cl
12-15% Br
1% P + 15-20% Cl
2% P + 6% Br
2% Sb_4O_6 + 16-18% Cl or + 8-9% Br

POLYESTER/CELLULOSIC BLENDS

Since the polyester/cellulosic blends became so prominent in the U.S. market, research efforts have been focused recently on developing polyester/cellulosic blends with reduced flammability. It has become apparent that problems exist with blend fabrics which are not present with 100% cellulosic systems. For example, the flammability behavior of blend fabrics cannot be predicted from the flammability of the fiber components [25, 26]. It has been shown that the treatment of one of the fiber components with a flame retarding agent specifically effective for that component does not necessarily render the two-component blend flame resistant [27, 28] unless the treated component comprises at least 85% of the blend [25].

It has also been established that certain phosphorus- and nitrogen-containing flame retardants designed for cellulosics are also capable of rendering polyester fiber flame resistant [29].

Combining unmodified cotton or polyester with cotton or polyester rendered flame resistant with chemicals effective on both types of fibers resulted in reduced levels of flame resistance, which were practically independent of whether both the treated and unmodified components of the blends were the same type of fiber or whether one was cotton and the other polyester. It suffices to flame-retard only one component, while the second one can be completely free of finish.

The flame retardants applied from aqueous medium to polyester/cotton blends are preferentially located on the cellulosic component. The adhesion of the polymer is generally much poorer to polyester than to cellulose

By selectively treating one of the fiber components, preferably the cellulosics, with a flame retardant (preferably one effective on both fibers), polyester/cotton blends can be rendered self-extinguishing. The total chemical add-on required to render polyester/cotton blend self-extinguishing is at least equal to that needed for 100% cotton fabric. If the cotton is selectively treated undesirable side effects may be anticipated, especially on blends of low cotton content.

REFERENCES

1. F. H. Winslow, L. D. Loan and W. Matreyek, "Polymer Structure and Pyrolytic Behavior," paper presented at the 161st National Meeting of A.C.S. in Los Angeles, Cal., April 1971.
2. J. W. Lyons, "The Chemistry and Uses of Fire Retardants," Wiley Interscience Publ. 1970.
3. S. S. Hirsch, "Char Smoke, Flammability and Heat Resistance," paper presented at the Polymer Conference Series, Detroit University, June 1970.
4. S. S. Hirsch, USP 3,549,307, to Monsanto (1970).
5. Chemical Engineering News March 29, 1971, p. 10.
6. W. A. Reeves: The Chemist *47*, 300 (1970).
7. The Textile Manufacturer *96* (3) 107 (1970).
8. R. C. Nametz: Ind. Eng. Chem. *62* (3) 41 (1970).
9. E. F. Stroh, USP 3,379,699 (1968).
10. G. Palethorpe, USP 3,487,058 to Monsanto (1969).
11. J. Kovacs and C. S. Marvel, J. Polymer Science Part A-1, *5*, 1279 (1967).
12. T. W. Tarkington and N. T. Anderson, USP 2,949,432 to Chemstrand (1960).
13. Z. A. Rogovin, Krimrcheskie Volokna *1966* (3) 27.
14. G. Palethorpe, USP 3,318,978 (1967).
15. F. J. Lowes, USP 3,242,124 (1966).
16. J. J. Hirschfeld, Belg. P. 631,087 to Monsanto (1963).
17. K. Matsubayaski, Sen-i Gàkkaishi *19* (1) 27 (1963).
18. J. H. Johnson, USP 3,210,147 (1965).
19. B. A. Arbuzov, et al, Brit. P. 1,012,009 (1965).
20. R. R. Lunt, Jr., USP 3,440,211 to duPont (1969).
21. B.C.M. Dorset, Text. Manuf. *87*, 243 (1961).
22. H. Coates, Brit. P. 862,539 to Albright & Wilson, Ltd. (1961).
23. Y. Iwakura, et al, Jap. P. 20,771 to Dainippon Print. Ink Manuf. Co. (1961).
24. W. J. Jackson, Jr. and J. R. Caldwell, USP 3,454,672 to Eastman Kodak Co. (1969).
25. W. Kruse and K. Filipp, Mell. Tex. Ber. *49*, 203 (1968).
26. G. C. Tesoro and J. Rivlin, Text. Chem. and Col. *3* 156 (1971).
27. W. Kruse, Mell. Tex. Ber. *50*, 460 (1969).
28. G. C. Tesoro and C. H. Meiser, Jr., Text. Res. J. *40*, 430 (1970).
29. P. Linden, S. B. Sello and H. S. Skovronek, Proceedings of the Fourth Annual Meeting of the Information Council on Fabric Flammability, pp. 173-188 (December 1970).